Ecological Studies
Analysis and Synthesis

Edited by
W.D. Billings, Durham (USA) F. Golley, Athens (USA)
O.L. Lange, Würzburg (FRG) J.S. Olson, Oak Ridge (USA)
H. Remmert, Marburg (FRG)

Volume 58

Ecological Studies

Ecology of Biological Invasions of North America and Hawaii

Edited by
Harold A. Mooney
James A. Drake

Contributors
H.G. Baker, F.A. Bazzaz, D.L. Dahlsten, A.P. Dobson,
J.A. Drake, P.R. Ehrlich, J.J. Ewel, S.P. Hamburg, R.N. Mack,
R.M. May, H.A. Mooney, M.P. Moulton, P.B. Moyle, G.H. Orians,
D. Pimentel, S.L. Pimm, P.J. Regal, J. Roughgarden,
D. Simberloff, P.M. Vitousek

With 25 Figures

Springer-Verlag
New York Berlin Heidelberg London Paris Tokyo

Harold A. Mooney
Department of Biological Sciences
Stanford University
Stanford, California 94305 USA

James A. Drake
Department of Biological Sciences
Stanford Univeristy
Stanford, California 94305 USA

Cover illustration from Myers RL (1984) In: Ewel KC, Odum HT (eds), Cypress Swamps. Reproduced with permission of University of Florida Press.

Library of Congress Cataloging in Publication Data
Ecology of biological invasions of North America
  and Hawaii.
  (Ecological studies ; v. 58)
  Based on a symposium held in Asilomar, Calif.,
Oct. 21–25, 1984.
  Includes bibliographies and index.
  1. Animal introduction—North America—Congresses.
2. Plant introduction—North America—Congresses.
3. Pest introduction—North America—Congresses.
4. Ecology—North America—Congresses.   I. Mooney,
Harold A.   II. Drake, James A., 1954—
III. Series.
QH102.E284   1986   574.5'24   86-6630

Media conversion by David E. Seham Associates Inc., Metuchen, New Jersey.
Printed and bound by Arcata Graphics/Halliday, West Hanover, Massachusetts.
Printed in the United States of America.

9 8 7 6 5 4 3 2 1

ISBN 0-387-96289-1 Springer-Verlag New York Berlin Heidelberg
ISBN 3-540-96289-1 Springer-Verlag Berlin Heidelberg New York

Preface

The diversity of the earth's climates superimposed upon a complex configuration of physical features has provided the conditions for the evolution of a remarkable array of living things which are linked together into complex ecosystems. The kinds of organisms comprising the ecosystems of the world, and the nature of their interactions, have constantly changed through time due to coevolutionary interactions along with the effects of a continually changing physical environment.

In recent evolutionary time there has been a dramatic and ever-accelerating rate of change in the configuration of these ecosystems because of the increasing influence of human beings. These changes range from subtle modifications caused by anthropogenically induced alterations in atmospheric properties to the total destruction of ecosystems. Many of these modifications have provided the fuel, food, and fiber which have allowed the expansion of human populations. Unfortunately, there have been many unanticipated changes which accompanied these modifications which have had effects detrimental to human welfare including substantial changes in water and air quality. For example, the use of high-sulfur coal to produce energy in parts of North America is altering the properties of freshwater lakes and forests because of acidification.

The massive restructuring of the biotic components of these systems has been to a large extent purposeful with, for example, the substitution of a select few plants and animals of direct human use, for an array of those which are not. The rate of substitution has increased as human populations have grown

and as the means of transport of humans and their favored domesticated organisms has improved. Through evolutionary times there have been organisms,
such as migratory birds and animals, which in their yearly cycles transcended
the climatic boundaries of a given ecosystem unintentionally carrying with them
individuals and propagules of other species. People have increased this interchange many-fold, often, and importantly, bringing individuals from one ecosystem type into a comparable one between which there has been essentially
no interchange in the past. Often such interchanges occurred when these natural
systems were most vulnerable, having been disrupted by human activity. Unfortunately, it has now become clear that in many cases these purposeful interchanges have not had the intended effect and further they have been accompanied by many accidental interchanges, some of which have had disastrous
economic and environmental impacts.

Because of the ability of invading species to alter the structure and function
of communities and ecosystems an understanding of invasion processes is essential. SCOPE (Scientific Committee on Problems of the Environment) of the
International Council of Scientific Unions (ICSU) has established an international program to investigate the ecology of biological invasions. This volume
represents a component of this program and is based on a symposium which
was held in Asilomar, California, on October 21–25, 1984, to explore biological
invasions into North America and Hawaii.

The composition of the biota of North America and Hawaii, almost undisturbed until two centuries ago, has been changing at an unprecedented rate due
primarily to the disruptions of natural systems by human activities coupled
with both the deliberate and accidental introduction of foreign species. In this
volume, some indications of these changes and their causes are given for selected
North American geographic regions. For example, Florida is a landscape where
introduced species are extremely common, comprising up to 25% of certain
taxonomic groups. Some of these introduced species have had a significant
effect on native plant and animal populations, in several cases eliminating them
completely. The lowlands of Hawaii provide another example of habitat which
have been modified extensively and are now composed almost entirely of introduced species. Similarly, in California coastal regions introduced species
make up almost 50 percent of the floras. In the Intermountain West, ecosystem
structure has been totally altered through the influence of a few particularly
successful introduced plant species.

Disruptions of the natural systems are often responsible for the increased
success of these introductions, as for example in Florida, where there has been
clearing, altered burning cycles, and large drainage projects. An altered grazing
regime by the introduction of domesticated animals and an infusion of new
plant introductions triggered the massive change in the intermountain steppe
vegetation. Similarly, grazing and agriculture have drastically altered the California coastal ecosystems. No doubt these disruptions will continue and become
ever more pervavsive, resulting in large-scale rearrangements of the biota. As
discussed in this book, methods of control of invaders are costly and not always
successful.

What are the impact of these invaders? Such a question is comparatively easy to answer for economically important crop ecosystems where impacts of "pests" can be measured in terms of reductions in yield and hence given a market value. For example, estimates for yearly economic losses in the United States are in the order of billions of dollars as calculated by Pimentel in this volume. Information on impacts on natural ecosystems is not readily available nor is it so easily quantified. Vitousek shows, however, how such central eco-system properties as productivity, soil structure, and nutrient and water cycling can be strongly altered by invading animals and plants. His analysis provides an important beginning for assessing ecosystem-level impact of invaders.

A question addressed by a number of the contributors to this volume is whether predictions can be made as to which particular species will be successful invaders. Certainly, a central question because most natural systems are in some state of disruption and new taxa are continually being introduced through one mode or another. It becomes obvious from reading the diverse contributions in this book that we as yet do not have the capacity to make precise predictions of the potential establishment and success of a given invading species. Some authors maintain that there is little likelihood in the near future of gaining this capacity. Others are more optimistic and maintain that species with certain suites of physiological and population attributes are more likely to invade than others and that from such information a risk analysis can be developed. However, it is now becoming apparent that the information required for predicting invasion success will include species properties, interactions between species, and properties of the system being invaded. Hence, the numbers of variables to consider in predicting the success of any potential invader in a given eco-system is great and the task of analysis is not trivial. Yet the effort is surely worthwhile in view of the increased rate of introductions of foreign genetic material into native ecosystems including perhaps organisms which have been purposefully modified through genetic engineering. Not all successful invaders will have a substantial ecosystem impact but we must be forewarned of those that might. It is hoped that this volume will provide some indication of what is known about invasion processes in general, as well as giving directions for future study.

H.A. Mooney
J.A. Drake

Contents

Contributors

BAKER, H.G.   Department of Botany, University of
                              California, Berkeley, California 94720
                              USA

BAZZAZ, F.A.   Biological Laboratories, Harvard
                              University, Cambridge, Massachusetts
                              02138 USA

DAHLSTEN, D.L.   Division of Biological Control, University
                              of California, Berkeley, California 94720
                              USA

DOBSON, A.P.   Department of Biology, Princeton
                              University, Princeton, New Jersey 08544
                              USA

DRAKE, J.A.   Department of Biological Sciences,
                              Stanford University, Stanford, California
                              94305 USA

EHRLICH, P.R.   Department of Biological Sciences,
                              Stanford University, Stanford, California
                              94305 USA

EWEL, J.J.   Department of Botany, University of
                                    Florida, Gainesville, Florida 32611 USA

HAMBURG, S.P.   Department of Systematics and Ecology,
                                    University of Kansas, Lawrence, Kansas
                                    66045 USA

MACK, R.N.   Department of Botany, Washington State
                                    University, Pullman, Washington 99164
                                    USA

MAY, R.M.   Department of Biology, Princeton
                                    University, Princeton, New Jersey 08544
                                    USA

MOONEY, H.A.   Department of Biological Sciences,
                                    Stanford University, Stanford, California
                                    94305 USA

MOULTON, M.P.   Department of Biological Sciences, Texas
                                    Tech University, Lubbock, Texas 79409
                                    USA

MOYLE, P.B.   Department of Wildlife and Fisheries
                                    Biology, University of California, Davis,
                                    California 95616 USA

ORIANS, G.H.   Institute for Environmental Studies and
                                    Department of Zoology, University of
                                    Washington, Seattle, Washington 98195
                                    USA

PIMENTEL, D.   Department of Entomology, Cornell
                                    University, Ithaca, New York 14853 USA

PIMM, S.L.   Department of Zoology, University of
                                    Tennessee, Knoxville, Tennessee 37996
                                    USA

REGAL, P.J.   Department of Ecology and Behavioral
                                    Biology, University of Minnesota,
                                    Minneapolis, Minnesota 55455 USA

ROUGHGARDEN, J.   Department of Biological Sciences,
                                    Stanford University, Stanford, California
                                    94305 USA

SIMBERLOFF, D.   Department of Biological Sciences,
                                  Florida State University, Tallahassee,
                                  Florida 32306 USA

VITOUSEK, P.M.   Department of Biological Sciences,
                                  Stanford University, Stanford, California
                                  94305 USA

1. The Patterns of Invasions: Systematic Perspective

1. Introduced Insects: A Biogeographic and Systematic Perspective

D. Simberloff

1.1. Introduction

Introduced insects attract public attention primarily when they become "pests." The impact of fire ants, killer bees, gypsy moths, Japanese beetles, and Colorado potato beetles on our economy and environment is well known. The last have even played roles in international politics. The Ministry of Agriculture under the Third Reich generated antagonism against the British before the start of World War II with the rumor that English planes had dropped larvae of the beetle on areas where the German farmers had massive stores of potatoes. And, during one of the most frigid moments of the Cold War, the Soviet Union claimed that the United States had bombed eastern Europe with the Colorado potato beetle, which the Russian press called "the six-legged ambassador of Wall Street" (Kahn 1984).

The above are examples of "successful" (if not purposeful) introductions, and there are thousands of others—at least 1554 in the continental United States alone (Sailer 1983). Yet propagules of enormous numbers of insect species have landed on new shores and have quickly been extinguished. Either they did not reproduce at all, or recruitment was so low that the population dwindled to zero. Such "attempted" introductions are not nearly so well documented as the successful ones are (Simberloff 1981), and this bias in the literature makes it difficult to approach the matter of why some introductions succeed and others fail. Yet this question, in one form or another, is the one that most interests

ecologists confronting the phenomenon of introduced species, and was raised in the earliest general examinations of introductions (e.g., Elton 1958; Baker and Stebbins 1965).

The purpose of this chapter is to describe the biogeography of insect introductions, including origins, target sites, and rates of dispersal. What follows will be a description of some available data, a number of patterns that these data seem to suggest, and hypotheses on success and failure that one can begin to test with such data. In particular, I will discuss several versions of the "biotic resistance" hypothesis, the contention that the key to success or failure of an invader is the extent of resistance by the resident biota and that this degree of resistance is predictable to some extent.

About 1,111,225 species of insects have been described worldwide (Arnett 1983), yet estimates by trained entomologists of the number of species in existence rarely fall below 2,500,000 and range up to 30,000,000 (Erwin 1982). In a group so poorly known, it is not surprising that few data sets are sufficiently complete to allow testing of the sorts of hypotheses that spring to mind. Most literature reports on introduced insects consist of more or less anecdotal case histories of particular introductions, usually successful ones. Two works transcend this form, however. One is the ongoing record of insects introduced within the 48 contiguous states compiled by Sailer (1978, 1983); the other comprises information on all attempts at biological control of pests through introduction of their insect enemies (Clausen 1978). Only for the biological control data is there sufficient information to begin to estimate success and failure rates.

Sailer's data present a number of provocative patterns. For example, of 1554 successful introductions in the continental United States, approximately 66% are of western palearctic origin. On the other hand, only 14% are from South and Central America and the West Indies. This may seem surprising, since the tropical American entomofauna, especially that of South America, is very rich. I have found no exact estimate of neotropical insect diversity, but it is very high (e.g., Erwin 1982; Illies 1983) and at a minimum would be not much less than that of the palearctic entomofauna. Furthermore, Antillean, Central American, and South American insects appear on geographical grounds alone to have as ready access to the United States as do European and Asian insects.

Sailer believes that patterns of human commerce and travel have a significant impact on this distribution of origins. Up to 1800 there were only about 36 introduced insect species in the United States. It is probable that because voyages to America took so long only species that could survive and reproduce on humans, domestic animals, stored products, and ballast could reach our shores (Lindroth 1957; Sailer 1978, 1983). The few species that invaded at this time must have originated mostly from Europe, as the vast majority of ships that reached America through the late 18th century were European. Even as trade and travel from other parts of the world increased, and the volume and speed of intercontinental movement turned the trickle of immigrants into a flood, the majority of successful insect immigrants were still palearctic. The main reason, Sailer (1978) suggests, is that, until very recently, most commerce has continued to be with Europe. Second, the climate and crops of Europe are

similar to ours, and this may be why, for example, 40 of the 68 new insect and arachnid species introduced since 1970 have been European. Lattin and Oman (1983) argue that climate is no more important than broad vegetational similarities between potential source and target regions. For example, the landscape of the southeastern and southern United States is typified by grasses and other herbs, broadleaf deciduous trees, and needleleaf evergreens, a vegetation broadly similar to that of western Europe and much of central Asia.

Aspects of species' biology that might have contributed to this disparity of origin include two that are particularly intriguing. First, reversal of seasons must present a barrier to many potential invaders from the southern hemisphere (Lattin and Oman 1983). A species arriving in New York in summer from Argentina or New Zealand could well be in its winter diapause, and its eclosure could occur during our winter. On the other hand, the freezing temperatures that would confront an active form, such as an adult moth, brought in on a plane to Kennedy Airport in January would be just as forbidding as the hostile agricultural inspectors and cab drivers. Second, a large part of southern hemisphere land is tropical, and little of the continental United States is tropical. Not only will there thus be relatively little suitable habitat for tropical insects, but these species probably have narrower temperature tolerances than do temperate ones, so that conditions during transport and on landing are more likely to cause mortality.

The taxonomic distributions of known insect colonists and of insects of the world in general are heavily influenced by the numbers of specialists working on different groups. However, if we neglect these biases, or assume they are equal for the two cases, we can compare the distribution of insect orders that have successfully colonized the United States to the distribution of a world estimate of 1,111,225 species (Arnett 1983). There is much too wide a disparity (Fig. 1.1) to entertain the hypothesis that colonists are just a random sample of the species pool. In particular, there are among the introduced species far too many of certain phytophages, especially Homoptera and Thysanoptera, too many Coleoptera and Hymenoptera, and far too few Diptera and Lepidoptera. Leston (1957) suggests that high-level insect taxa have inherent properties that predispose some of them to be better colonists than others. There is a tradition in colonization studies of seeking to enumerate on either deductive or inductive grounds those traits that make some species good colonists and others poor ones (Baker and Stebbins 1965). Since such traits are often distributed partly taxonomically, Leston's suggestion is reasonable.

Surely, however, factors other than inherent colonizing ability are important, especially because of the human role in many insect introductions. A large fraction of the systematic distribution of known successful colonists can be attributed to the timing of commerce, plant introduction, and biological control (though the known distribution is partly an artifact of differential intensity of study for different orders, as noted above). The proportional representation among introduced species at different times (Fig. 1.2) shows, for example, that beetles predominated early, doubtless because they came in ship ballast (Sailer 1978, 1983) and stored grain. Homopteran dominance was established in the

D. Simberloff

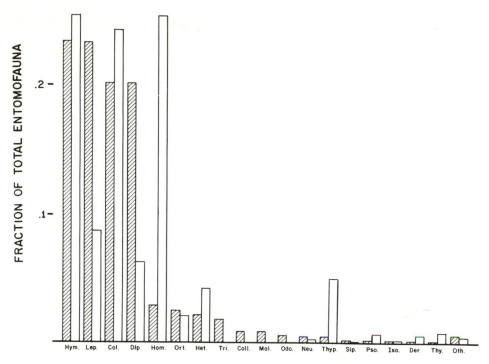

Figure 1.1. Proportional representation of the various orders among the insects of the world (cross-hatched; data from Arnett RH 1983) and successful introduced insects of the United States (clear; data from Sailer RI. In: Graham C, Wilson C (eds), Exotic Plant Pests and North American Agriculture, pp 15–38, 1983. Reproduced with permission of Academic Press.)

middle of the 19th century, a consequence of both the faster oceanic crossings afforded by steamships and the explosion of nursery stock imported into the United States, primarily from Europe (Sailer 1978). The decreasing rate of addition of introduced species in general, and Homoptera and Thysanoptera in particular, beginning about 1920, probably results from enforcement of plant quarantine laws (Sailer 1978). The tremendous burst of new Hymenoptera at the same time surely reflects increasing biological control efforts, especially the use of parasitic wasps. The resurgence of Coleoptera at this time may be due to predatory beetles introduced for the same purpose.

   Such explanations, though plausible and useful for placing insect introductions in a geographical, temporal, and taxonomic framework, do not provide the sort of hypotheses that might lead to the predictions we desire. If they are correct, they are too general to tell us much about attempted introductions on a case-by-case basis. They tell us, for example, that if we attempt to introduce many more wasps than other families, we will end up with proportionally more wasps. If we provide more habitats and food plants for scale insects, as well

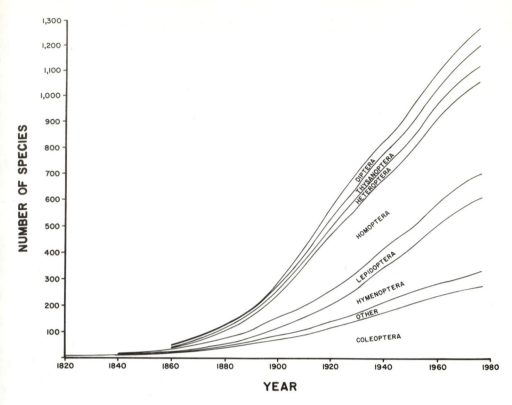

Figure 1.2. Proportional representation of the various orders among successful introduced insects of the United States at times. (Data from Sailer RI. Bull Entomol Soc Am 24:3–11, 1978. Reproduced with permission.)

as transportation, we will end up with more scales. However, in addition to wasps and scales that have successfully invaded the continental United States and other regions, there are many that reached the same locations but failed to produce ongoing populations; we would like to know why. Even the broad considerations and explanations outlined above that would seem to narrow the scope of predictive hypotheses do not narrow them too far. For example, southern hemisphere species may be at an inherent phenological disadvantage when they invade the United States, but some such invaders *have* been successful—for example, about 60 species from southern South America, Australia, and South Africa have established themselves here. Some of them have been important biological control agents. Control of the black scale *(Saissetia olea)*, a devastating citrus pest in California, was achieved by two wasp parasitoids from South Africa, the encyrtids *Metaphycus helvolus* and *M. lounsburyi* (Clausen 1978). Thus it would have been unwise to restrict our search for beneficial insects to the northern hemisphere.

1.2. Biotic Resistance of Island Faunas

Several narrow hypotheses revolve around the notion that different locations or different times present differing obstacles to potential invaders and that invaders from different sources have characteristically different abilities for invasion. For example, island biotae are often thought to have been shaped by much less stringent selective pressures, so that islands are more easily invaded than mainland, and island species are less likely to be successful invaders than are mainland species. Carlquist (1965) articulates this view:

> Continental species, steeled by competition, not only stand their ground, they are often preadapted to places far beyond the ranges they occupy. The depredations of introduced animals such as the Colorado potato beetle in Europe or the European starling in America bear adequate testimony to this. Island creatures, reared in the hothouse-like conditions of isolation, have no such reserve of rampant capabilities. Rather, they often fall victim—not only of man and the plants and animals he introduces onto islands, but often victims of the shortsighted qualities of their own evolution.

Wilson (1965) was first to generate precise and testable hypotheses from this reasoning. Observing that islands have suffered higher extinction rates than continents, and have received more of their species from continents than vice versa, he asked what might be the reason for this "faunal dominance." Continental species might be genetically stronger than island species, in proportion to the relative areas of the particular island and continent, or the dominance might result simply from the greater numbers of species on continents, or the greater numbers of individuals. Wilson then refined this conjecture into four alternative hypotheses, depending on the exact reasons why mainland species are numerically dominant to island ones: (1) If it is simply a question of a random flux of species between mainland and island, the ratio of exchange rates (mainland-to-island colonists/island-to-mainland colonists) should be proportional to the relative numbers of species. (2) If dominance rests on the greater numbers of individuals on the mainland, the ratio of exchange rates should rest on the proportion of numbers of individuals in the two sites, which in turn would be proportional to the ratio of areas. (3) If genetic superiority of the mainland species to the island ones causes the faunal dominance, one might hypothesize that the ratio of exchange rates would be greater than in either of the first two hypotheses. In particular, if the competitive ability of species increases linearly with the number of individuals, the ratio of exchange rates increases linearly with the area ratio. (4) Finally, if competitive ability increases disproportionately with number of individuals, there would be a geometric increase in the exchange ratio with increase in the area ratio. In fact, hypothesis 3 is not stated quite correctly, since linear increase in competitive ability with increasing number of individuals should produce more than linear increase in the ratio of exchange rates with increase in the area ratio, but not so great an increase as in hypothesis 4.

Wilson found few data with which to test these hypotheses. His taxon cycle

for Melanesian ponerine ants (Wilson 1961) was essentially an earlier version of this concept of "faunal dominance." Competitive pressure in the interior of larger sites such as New Guinea or tropical Asia drives species into marginal shore habitats. These "expanding species" are preadapted for overwater colonization of smaller islands in Melanesia, and when they reach these islands they are competitively superior to resident species (Brown and Gibson 1983), so much so that they occasionally cause extinction. No ponerines of Fiji or the Solomons have successfully invaded New Guinea or Asia, while the exchange ratio of New Guinea to tropical Asia is 0.125. The area ratio for these two sites is 0.123, while the ratio of numbers of resident species is 0.543. Wilson believes that these data are too scarce for a hypothesis test, but that they tentatively suggest hypothesis 3. I believe these data seem more in accord with hypothesis 2. In any event, Wilson seems to conclude by interpreting the biology of the situation as in hypothesis 2: "The results seem to suggest that species from large source regions do not have a higher average intrinsic competitive ability over those from small source regions."

The sites of many early successes of biological control were islands, and Imms (1931) suggested that biological control could be effective only in simplified insular settings. However, DeBach (1965) and Huffaker et al. (1976) found that subsequent data disprove this conjecture, with the most recent tabulation (Laing and Hamai 1976) showing about 1.5 times as many successful biological control projects for both insect pests and weeds on continents as on islands. These data do not record attempted control projects that failed, nor do they always state the native region of the introduced insect (as opposed to the source from which the introduction occurred). Also, they may be biased by proportionally more attention to mainland projects. Consequently it is difficult to frame or test the sorts of hypotheses that Wilson initiated with data available at present.

Sailer (1978) suggests a provocative comparison. The contiguous United States have 1554 successful introduced insects, and there are approximately 89,358 named native species in North America north of Mexico (Arnett 1983), so the introduced component is about 1.7%. On the other hand, in Hawaii 3638 native insects and about 1476 introduced ones were known by 1948 (Zimmerman 1948). There are thus approximately as many introduced species in Hawaii as in the entire contiguous United States, and they constitute 29% of the total Hawaiian entomofauna. Sailer attributes this difference to a kind of "environmental resistance" that continents present to invaders and that islands lack. By this term he means not only greater likelihood of competitive exclusion from the larger resident fauna, but also the host of predatory insects, birds, and spiders, as well as other enemies awaiting an insect as soon as it leaves a major continental airport as opposed to, for example, Honolulu International. Tristan da Cunha has suffered a fate similar to that of Hawaii; it has only 84 native insects and 32 introduced ones, or 28% of the entire entomofauna (Holdgate 1960). For neither Hawaii nor Tristan is there any inkling of how many insect species have landed propagules but failed to become established, so these data must be viewed impressionistically: it *seems* as if it is easier for species to invade islands, and one hypothesis for why this should be so is that biotic

resistance (competitive, predatory, pathogenic, etc.) is lower on islands. A critical need is information on success and failure rates.

Examination of the data for Hawaii reveals grounds for skepticism about biotic resistance. In Table 1.1. are presented the fractions of the total entomofauna belonging to the different orders for the world, North America north of Mexico, and Hawaii, as well as the fractions for introduced insects for North America and Hawaii. Many orders are absent from the native Hawaiian entomofauna, and the remainder of the Hawaiian entomofauna is somewhat disharmonious relative to those of North America and the world. For example, Diptera constitute 0.198 of the world's insect species, and 0.189 of those of North America, but only 0.068 of Hawaiian species. However, one does not see order-by-order patterns suggesting greater resistance to invasion in North America. For example, introduced Thysanoptera are very overrepresented in both areas relative to native ones, and introduced Lepidoptera are greatly underrepresented in both areas relative to native ones. Hawaii has a greater proportion of introduced Lepidoptera than does North America even though it has

Table 1.1. Proportions of native and introduced entomofaunas of North America and Hawaii in the different orders, plus proportions of world entomofauna

Order	World	North America Native	North America Introduced	Hawaii Native	Hawaii Introduced
Protura	0.000	0.001			0.001
Thysanura	0.001	0.001	0.007	0.001	0.003
Collembola	0.009	0.008			0.022
Plecoptera	0.002	0.006			
Ephemoptera	0.002	0.008			0.001
Odonata	0.006	0.005		0.008	
Embioptera	0.001	0.000	0.001		0.001
Orthoptera	0.025	0.013	0.021	0.012	0.014
Zoraptera	0.000	0.000			0.001
Isoptera	0.002	0.001	0.001		0.003
Dermaptera	0.001	0.000	0.005		0.008
Coleoptera	0.198	0.265	0.239	0.335	0.257
Strepsiptera	?	0.001			0.001
Thysanoptera	0.004	0.008	0.050	0.007	0.043
Psocoptera	0.002	0.003	0.007	0.007	
Mallophaga	0.009	0.028		0.001	0.030
Anoplura	0.000	0.001	0.001		0.003
Hemiptera	0.021	0.040	0.043	0.049	0.030
Homoptera	0.029	0.072	0.222	0.094	0.108
Trichoptera	0.018	0.016			0.001
Lepidoptera	0.230	0.126	0.086	0.234	0.102
Mecoptera	0.001	0.001			
Diptera	0.198	0.189	0.061	0.068	0.104
Siphonaptera	0.002	0.004	0.001		0.005
Hymenoptera	0.231	0.194	0.251	0.170	0.259
Neuroptera	0.005	0.004	0.003	0.015	0.004

Data from Arnett (1983), Sailer (1983), and Zimmerman (1948).

a greater proportion of native Lepidoptera. This is not consistent with the competitive aspect of the resistance hypothesis. Similarly, there are relatively more native beetles in Hawaii than in North America, but the same holds true for introduced beetles there. On the other hand, there are fewer native Diptera in Hawaii than there are in North America, and the fraction of the introduced insect species that are flies is greater for Hawaii than for North America; this observation is consistent with competitive resistance. On the whole, though, one sees for Hawaiian introductions the same common and uncommon orders that one sees for North America.

1.3. A Null Hypothesis

I would like to present a null hypothesis (not necessarily mutually exclusive with Sailer's) for why there are more successful introduced insect species on islands, as a fraction of total fauna, than on mainland, followed by some general points that may bear on the matter and some data gleaned from the literature. The hypothesis is this: For any species of insect not found at some site (either island or mainland), there is an intrinsic probability that a propagule will initiate establishment, and this probability does not depend on what other species (competitors, predators, etc.) are found at the site. Rather, it depends on the biology of the species and the availability of suitable habitat on the site (in the term "habitat," I include plant species for insects that require particular plants for food or shelter and prey species for predatory and parasitic insects). Since native insular entomofaunas tend to be small, the introduced species that establish themselves constitute a large fraction of the total entomofauna. However, for the most part the same species would have established themselves even if the native insular fauna had been larger.

I am motivated to test this null hypothesis against the alternative of biotic resistance by several observations. First, in the compilation on biological control (Clausen 1978) that I will use below are a very few instances of failed introductions that are believed to have failed because of competition between the species and a native species or previously or subsequently introduced species. For example, the Australian vedalia beetle *(Rodolia cardinalis)* that controlled cottony-cushion scale *(Icerya purchasi)* in California is said to have eliminated an Australian congener *(R. koebelei)* that had established itself about the same time (Clausen 1978). It is similarly rare for parasitoids to have an effect on the success or failure of an introduction. For species introduced to control weeds, Goeden and Louda (1976) could find no case where a parasitoid was crucial. Nor were there introductions that failed because of disease, with one possible exception. There are a few more cases in which introductions are thought to have failed because of predation. One is the introduction of the English geometrid defoliators *Anaitis plagiata* and *A. efformata* into Australia to help control Klamath weed *(Hypericum perforatum)*. Predation by native ants apparently prevented establishment (Clausen 1978). Goeden and Louda (1976) give several other examples in which predation by native species contributed to failure of

an introduction. But there do not appear to be more of these examples, proportionally to number of attempted introductions, on mainland than on islands, though I have not made an exhaustive compilation. Furthermore, there are many more examples, on both islands and mainland, where an introduction seems to have failed because the habitat (especially with regard to temperature and humidity) turns out to be inappropriate. Of course, the greatest number by far of failed introductions are cited with no apparent cause of failure, and even in those cases where a cause is suggested, the suggestion is almost always a tentative one, based on a naturalist's feel rather than direct experiment.

Second, it is probably true that, on average, there are relatively fewer species at higher trophic levels on islands than on mainland (Williamson 1981). This need not mean that predatory pressure is less on islands. For example, Allan et al. (1973) found that Puerto Rican second growth vegetation has a far higher proportion of arthropod individuals that are predators (especially spiders) than does comparable vegetation on the Central American mainland. I do not mean to say that this example is typical. However, direct estimates of the relative amount of predatory pressure on islands and mainland are rare, and, in the absence of such estimates, it is far from obvious that a new arrival in the Honolulu Airport is less likely to fall victim to a predator than is a new arrival at Kennedy.

Third, part of the apparent invasion ability noted by Sailer is an artifact of the size of the resident pool. Since the continental United States has so many more insect species than does Hawaii (Table 1.2.), the same number of introduced species in the two places would show Hawaii to have a much greater percentage of introduced component. Thus, if (1) the null hypothesis were correct, and (2) exactly the same insect species were introduced into the two areas, and (3) Hawaii presented no less biotic or other resistance to invaders than did the continental United States, one would see exactly the pattern that is observed: a much higher proportion of introduced species in the Hawaiian fauna.

This is not all there is to the matter. Sailer (1978) suggests that there probably have been many more attempted introductions in the continental United States than in Hawaii, largely because of the greater amount of commerce in the former. He is likely correct. On the other side of the coin, Hawaii has a very small and

Table 1.2. Areas and numbers of recorded insect species for selected regions

Region	Area (km²)	Number of Insects[a]
North America north of Mexico	19,299,161	89,358
Australia	7,704,172	54,071
Great Britain excluding Northern Ireland	230,616	21,833
Hawaiian Islands	16,708	3,638

[a] Sources: North America, Arnett 1983; Australia, C.S.I.R.O. 1970; Great Britain, Williamson 1981; Hawaiian Islands, Zimmerman 1984.

disharmonious native flora (Carlquist 1974) with many more genera and families absent than in the continental United States. There are only 1729 native flowering plant species, compared to 16,673 for North America north of Mexico (Williams 1964). As these lacunae were redressed by plant introduction from all over the world (Elton 1958), the proportional change in the vegetation was probably relatively greater for Hawaii than for the continental United States. But these new plants constitute habitats for insects, and some of the new insects that were thus permitted to survive themselves constituted habitats for yet other insects. In short, the relatively greater habitat change in Hawaii may have increased intrinsic probabilities of success for many more species, an increase that was independent of the amount of predatory or competitive resistance presented by the resident insect community.

A similar picture can be painted for vertebrate parasites. One notices (Table 1.1.) no introduced Mallophaga in North America, but a native component of 0.028 (ca. 2500 species). On the other hand, Hawaii has only five native species (0.001 of the native fauna), but at least 45 introduced species (0.030 of all introductions). One could argue that the reason for the success of introduced Mallophaga in Hawaii and apparent failure (we do not know how many propagules were actually introduced) in North America lies in the different degrees of competitive resistance in the two regions. However, Hawaii has undergone 162 introductions of bird species (of which 70 are successfully established), more than have been introduced in all of North America (Long 1981). A similar situation probably holds for mammals. These introductions provided not only transport for Mallophaga, but also habitats that had not previously existed.

1.4. Preliminary Tests

To test the view that mainland species are somehow "stronger" and/or island communities present less biotic resistance, one would need information on success and failure rates. One would want to know for a large number of related species, some from islands and some from mainland, what happened when they were introduced into both island and mainland target areas. One would want the species to be related because, for example, one would not want the introduced island species to be characteristically different from the introduced mainland species in some way that would predispose the island species to survive and the mainland species to become extinct. For example, one would not want the mainland test species to be mayflies and island test species to be carabid beetles.

Consequently, I scanned records in Clausen's compilation (1978) to find the six genera with the largest numbers of species that had been moved, in various directions, among islands and continents (with Australia considered a continent). These comprised a coccinellid beetle *(Scymnus)* and five parasitic wasps *(Opius, Coccophagus, Aphytis, Bracon,* and *Apanteles)*. For each genus about 20 species were listed, with patterns of introduction typified by *Aphytis* (Fig. 1.3). One sees that often a species was moved from one site to a second site, and

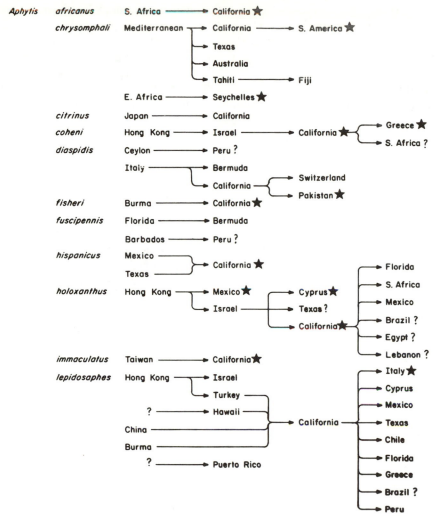

Figure 1.3. Attempted introductions of wasps of the genus *Aphytis;* star indicates un-successful introduction. (Data from Clausen CP. In: Agriculture Handbook 480, USDA, 1978.)

from there to other sites. In such instances, the source was considered to be the original site. If the idea of strong mainland species and invasible island communities is correct, then among the four possible source-to-target intro-ductions (source and target can each be either island or continent), probability of success should be as indicated in Table 1.3.. There is no clear prediction about whether the probability of success should be greater in mainland-to-mainland transfers or in island-to-island ones, but the other five comparisons allow unequivocal predictions. For example, mainland-to-island transfers should have a higher probability of success than mainland-to-mainland ones because the target in the former instance presents less resistance. On the other hand,

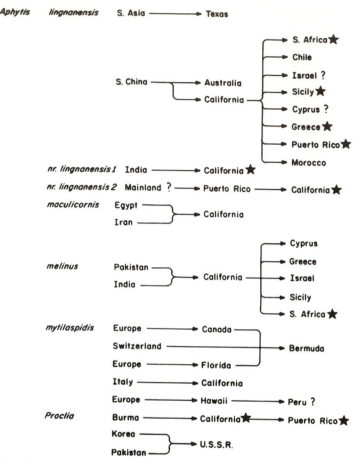

Figure 1.3. *Continued*

mainland-to-mainland transfers should have a higher probability of success than
island-to-mainland ones because the source species are stronger.

A word of caution: Most biological control efforts on mainland or island are
in agricultural communities, and these may be more similar to one another than
would be pristine communities. Thus, they may be less likely to manifest "resistance" differences. However, these are the only insect introduction data
that allow an assessment of failed as well as successful introductions. So, for
better or worse, I have used them.

Table 1.3. Predicted relative success rates for introductions, based on
hypothesis of biotic resistance

Rate	Route
Highest	Mainland to island
Intermediate	Mainland to mainland, island to island
Lowest	Island to mainland

Table 1.4. Success rates for introductions of insect genera widely deployed in biological control

Genus	Cases	M → I		M → M		I → I		I → M	
Opius	73	0.46	(24)	0.14	(37)	0.67	(6)	0.50	(6)
Aphytis	66	0.69	(16)	0.67	(48)	—	(0)	0.50	(2)
Bracon	47	0.36	(11)	0.14	(28)	0.33	(6)	0.00	(2)
Apanteles	44	0.73	(11)	0.59	(27)	0.40	(5)	1.00	(1)
Six genera	281	0.49	(71)	0.43	(176)	0.47	(19)	0.33	(15)

Data from Clausen CP. In: Agriculture Handbook 480, USDA, 1978.
Numbers of attempts in parentheses. M, mainland; I, island.

Scymnus and *Coccophagus* had too many cells with zero or small numbers, but the other four genera had success probabilities as depicted in Table 1.4.. One could have pointed to *Aphytis* and *Bracon* as generally supporting the hypothesis, though one would want to test for statistical significance. On the other hand, *Opius* and *Apanteles* do not fit the predictions (though the latter result rests on a single island-to-mainland introduction that was successful). For the six genera totalled, there were 281 trials, with results as indicated in Table 1.4. The probabilities are ordered as the hypothesis predicts, though the values are very close. I performed a multiway contingency test (Zar 1984), with three dimensions of two levels each: source (mainland or island), target (mainland or island), and outcome (success or failure). H_0 represented a situation where target, source, and outcome are mutually independent, whereas H_A referred to one where they are not all mutually independent. H_0 was rejected at $0.025 > Pr > 0.01$ ($\chi^2 = 11.435$, df = 4). This result led to a new H_o: success or failure is independent of target and source. H_A was likewise revised to mean that success or failure is not independent of target and source. The contingency test yielded $\chi^2 = 1.606$, df = 3; $0.9 > Pr > 0.75$. Consequently, I cannot reject the hypothesis that the success or failure of an introduction does not depend on whether source or target areas are island or mainland.

This result contradicts conventional wisdom, but it is not unassailable. It could be, for example, that the parasitic wasps that dominate these data are not typical, in relative colonization ability of island and mainland species, of other taxa. Or perhaps for these particular data the island sources tend to be those that have particularly strong competitors, and the island targets tend to be those that have particularly resistant entomofaunas. It did not appear to me that this was so, but any study that lumps together all islands and all continents cannot lead to strong quantitative conclusions.

I have also attempted to exploit the Clausen (1978) data set in the spirit of Wilson's (1965) attempt, by examining success rates of colonizations between targets and sources of different size, whether or not they are islands. It seems to me that, as stated, Wilson's hypotheses about species flux between two sites deal only with successful colonizations, whereas the underlying reasoning, particularly for hypotheses 3 and 4, is really relevant to relative success rates. For example, if species at a larger, more species-rich site are all more competitive than those at a smaller site, one might hypothesize that no species from the

small site will successfully invade the large site (which, in fact, seems to be true for the Solomons and Fiji ponerines). The available ponerine data do not address unsuccessful introductions. They do not allow one to reject the hypothesis that the reason there are no Solomons and Fiji ponerines in New Guinea and Asia is that no propagules dispersed there. They also allow the possibility that there has never been a failed New Guinean or Asian propagule on Fiji or the Solomons.

From Clausen (1978), one can begin to assess success rate as well as numbers of successes. These data are probably not as complete for failures as for successes (a point to which I will return below), and source areas for some attempted introductions are not known. Nevertheless, these seem to be the best data available for this purpose and may provide the outlines of a pattern. With this goal, I tabulated for each of four sites (North America north of Mexico, Australia, Great Britain excluding Northern Ireland, and the Hawaiian Islands) all successful and failed introductions from each of the other three sites. There were no attempts into Great Britain from any site or from Hawaii to Australia, but the record of attempts and success rates for the other routes is presented in Table 1.5., while areas and total numbers of species for the sites are given in Table 1.2.

For only two routes (North America to Australia and North America to Hawaii) were introductions attempted in both directions. In both instances the relative exchange rates were as one would have predicted from the relative areas or numbers of species, though the Hawaii-to-North America rate, which rested on only three attempts and one success, was almost as high as the North America-to-Hawaii rate. Because of this fact, and the presence of only two points, I will not attempt to extrapolate a curve of the sort in the above equation. Suffice it to say that, as in the mainland–island examination above, there are plenty of successes and failures in any direction. Comparisons with a single source or a single target generally neither support conventional wisdom nor suggest quantitative hypotheses. For example, the success rate for North American insects introduced into Australia is 0.500, whereas North American insects introduced into Hawaii were successful only 0.455 of the time—although Australia is 461 times as large as Hawaii and has 15 times as many native insect

Table 1.5. Numbers of attempted insect introductions, and success rates, over selected routes

Route	Attempts	Success Rate
North America to Australia	26	0.500
Australia to North America	32	ca. 0.375
North America to Hawaii	55	0.455
Hawaii to North America	3	0.333
Australia to Hawaii	21	0.571
Great Britain to Hawaii	3	1.000
Great Britain to Australia	13	0.308
Great Britain to North America	26	0.538

Data from Clausen CP. In: Agriculture Handbook 480, USDA, 1978.

species, and North America is in the same hemisphere as Hawaii but in a different one from Australia. It would certainly be preferable to restrict comparisons of this sort to inhabitants of particular habitats (e.g., tropical forest), but the insect data are insufficient for accurate habitat assignment.

1.5. Biotic Resistance in Disturbed and Undisturbed Habitats

Elton (1958), drawing heavily on insect examples, argued that continental invasions tend to succeed most frequently in disturbed ecosystems. He suggested that an "ecological resistance," an array of competitors, predators, parasites, and diseases, opposes a newly arrived species, and that this resistance is lowered in the simplified setting of disturbed areas as on islands. Sharples (1983) reiterates the view that disturbed areas, like islands, are particularly vulnerable to invasion because their simplified biotas are less resistant. Neither author presents other than anecdotal evidence to support this contention, however. A test of the hypothesis would require tabulating successes and failures in both disturbed and undisturbed sites, and I can find no insect data that would allow such a test.

For now, I propose a hypothesis analogous to the one above for islands: Patterns of successful insect introduction on disturbed and undisturbed sites do not depend primarily on different amounts of biotic resistance. Rather, each potential invader has a probability of successfully colonizing each site, and this probability rests largely on the nature of its habitat requirements and habitat availability at the site, and only secondarily on what other species are present.

The fact that agricultural and other systems managed by humans have suffered many well-publicized and studied invasions is at least partly due to the economic importance of these systems to us—this is why we manage them. Consequently these systems are studied more intensively than are natural systems, and harmful effects are publicized more widely. In addition, agricultural systems often consist of many species of introduced plants, and these constitute new habitats for very many phytophagous insects, which, in turn, constitute new habitats for entomophages. Finally, there have probably been far more deliberate attempts to introduce insects into agricultural systems, as biological control agents, than into pristine ones. A valid test of the hypothesis that disturbed habitats are more easily invaded than are undisturbed ones would require natural disturbed habitats, not one that presents a battery of new microhabitats. One might, for example, examine relative success of invaders into fire disclimax forests and climax forests not subjected to continual disturbance.

1.6. Biotic Resistance and the Build-up of Entomofaunas

Hall and Ehler (1979) and Ehler and Hall (1982) have used the Clausen data base to attempt to answer a related question about environmental resistance. They asked whether the success rate of introductions of enemies of Homoptera,

Coleoptera, and Lepidoptera implicated the existence of competitive exclusion of new propagules by previously introduced ones. This study is an entomological analog to that of Moulton and Pimm (1983) on the Hawaiian introduced avifauna. Ehler and Hall made two comparisons: (1) sequential releases of single species in the presence or absence of species previously released for the same purpose, and (2) sequential releases of one or more than one species in the presence of previously released species. Their results were generally consistent with the hypothesis of competitive exclusion. For example, mean establishment rate for sequential releases of two or more species given five to nine incumbent introduced species is only 0.10, which is much less than the mean establishment rate for single species released in the absence of potentially competing incumbent introduced species.

Their results have aroused controversy. Several authors (e.g., Keller 1984) have cited Ehler and Hall as saying that competition has probably occurred frequently, while they actually said only that the data are consistent with the hypothesis that competitive exclusion has operated sometimes. To clarify their view, Ehler and Hall (1984) have stated that they do *not* believe that competitive exclusion has been a frequent occurrence in past biological control efforts; rather, they make no claims about whether or how frequently it has happened. They also point out that there is no firm evidence in any biological control case that a potential control agent has been excluded by a previously introduced agent. Complete exclusion is, of course, a stringent demand, and experimental examples in the control literature strongly indicate competitive limitation of population size or geographical range, even when complete exclusion does not occur. A particularly compelling example is control of California red scale *(Aonidiella aurantii)* by three aphelinid wasps, *Aphytis chrysomphali* from the Mediterranean, *A. lingnanensis* from the Far East, and *A. melinus* from northwest India and Pakistan (Clausen 1978). *A. chrysomphali* was first introduced around 1900, and was quite successful in mild, coastal areas. In 1947 *A. lingnanensis* was released, and within 3 years had colonized all citrus areas of California. It gradually replaced *A. chrysomphali* in interior and intermediate climate areas, and by 1958 *A. chrysomphali* had disappeared from all of southern California except for a few small coastal pockets. However, sufficient economic control was still lacking in interior regions. In 1957 *A. melinus,* naturally adapted to more extreme climate, was introduced and quickly displaced *A. lingnanensis* from interior areas and reduced its density in intermediate areas. Part of the basis for the competitive superiority of *A. melinus* over *A. lingnanensis* in areas where both can survive the climate resides in the ability of the former to use smaller host individuals (Luck et al. 1982).

The controversy that Ehler and Hall have engendered, in spite of their rather tentative conclusions, probably stems from the association of their study with a long and heated debate in the biological control literature. The argument has revolved around whether a parasitoid or predator that exerts the best control (from an economic standpoint) of a pest can be replaced by competition from another introduced species that might not be as economically beneficial. Another aspect of this debate has involved the notion that control by more than one

natural enemy is an inherently unstable situation. The debate was initiated by Pemberton and Willard (1918), who proposed that only the single, best enemy be introduced, and has recrudesced right up to the present with essentially the same suggestion for the same reason—competition among the enemies will lessen control. In spite of abundant field evidence that this has never happened (reviewed by Doutt and DeBach 1964; Huffaker et al. 1971, 1976; DeBach 1974) and examples such as control of *Operophtera brumata,* the winter moth (Clausen 1978), in which two or more parasitoids have persisted and exerted consistent control for extended periods, theorists (e.g., Watt 1965; Turnbull 1967; Levins 1969) continue to raise the specter of disaster in biological control from multiple introductions.

Keller (1984) warns against drawing strong conclusions from survey rather than experimental data, echoing a plea by Goeden and Louda (1976), who found adequate experimental support for hypotheses of biotic resistance in only two of 23 cases in which predation, parasitism, and disease were claimed to have interfered with weed control by introduced insects. Keller particularly asks that one consider whether alternative hypotheses might not explain the same data equally well. Two possible deficiencies with the Ehler-Hall data set might introduce biases. Successful introductions were probably reported more assiduously than failed ones, and it is possible that failures in multiple species introductions were reported at a higher rate than single species failures. There are also aspects of typical biological control procedures that could give a misleading picture from the sorts of statistics that Ehler and Hall use. First, when only one or a few species are introduced, there is likely to have been a preintroduction study that suggests that these are likely to succeed, and early candidates that the study found a priori unpromising were discarded. This contrasts with a shotgun approach in which many species are introduced at once, with at most a cursory consideration of their probabilities of survival and successful control, in the hopes that at least one will do the job. Second, successful insects are likely to be shipped elsewhere for introduction alone, while unsuccessful ones are less likely to be tried again. Thus, multiple releases tend to act as a filter, leading to release programs of one or a few species that are a priori more likely to be successful.

Finally, biological control programs often keep releasing species until successful control is established, then stop. The order of species releases can thus bias observed establishment rates. To show this, Keller proposes as a null hypothesis a version of the hypothesis I have proposed above, stating that each species has intrinsic probabilities of success at a site no matter what potential competitors are present when it is released. He imagines three species, of which one would be destined to succeed and two would be destined to fail on a particular target no matter which of the other species were there. If all three are introduced, the overall success rate will always be $1/3$, whether they are released together or in any order. On the other hand, if they are released singly, but in random order, until success is achieved, the expected average success rate is $(1 + 1/2 + 1/3)/3 = 11/18$. Furthermore, species that are rare or difficult to culture

or have restricted geographical ranges or ecological requirements are likely to be the last ones in a sequential release, and are also intrinsically less likely to survive no matter what other species are present.

Keller notes that these sources of bias would produce a high probability of type II error (accepting a false hypothesis) for a test of the competition hypothesis with the Clausen data. He also observes that, when insects are introduced for biological control, the pest populations are usually high, and therefore not a limiting resource. Though high levels do not by themselves guarantee an adequate supply of hosts for their enemies, they at least suggest the hypothesis that, if there is competition at all, it is for something else. As Keller points out, there is usually no obvious limiting resource, so much more detailed field study is required. The main response by Ehler and Hall (1984) is that the null model lacks realism. Although this may be true, and it would usually be good to have more realistic models, this is not a damning criticism, for three reasons (Simberloff 1983, 1984). First, any model is an abstraction, and does not mirror nature in many ways. So long as the aspects that are unrealistic do not affect the model's performance with respect to issues of interest, the lack of realism need not be critical. Second, as emphasized by Keller (1984), we should not examine just one model and seek evidence to confirm it. Instead, we should consider several alternatives and seek evidence that would distinguish among them. When this is done, all the models should be subjected to the same scrutiny. There seems to me no prior reason why Keller's model is less realistic than Ehler and Hall's. Finally, even if a model is unrealistic, it may aid us to see just which sorts of data *can* help to winnow the available hypotheses, and which sorts, by virtue of being compatible with many hypotheses, are not likely to be very useful.

Tallamy (1983) constructed a scenario similar to Ehler and Hall's but for hymenopteran and dipteran parasites of just one introduced species, the gypsy moth *(Lymantria dispar)* in the United States. His conclusion is that these parasites had reached an equilibrium of about 10 species (both native and introduced) by ca. 1930. Very few introductions succeeded after 1914 (by then, six species had persisted), though over 60 species were introduced. Tallamy believes that a kind of competitive resistance built up as the community increased, until subsequent introductions could succeed only at the expense of extinctions of resident species. Washburn (1984) points out that parasites introduced early were more likely to succeed independently of which other species were present. He also finds at most one unequivocal extinction for gypsy moth parasites that had actually been established, and no evidence that an equilibrium has been achieved.

1.7. Further Considerations

We see that broadscale biogeographical patterns exist for introduced insects, but they do not seem to lead to the sorts of narrow, quantitative predictions we desire.

Nor do I think that the biogeographical literature on rates of spread of successful introductions will take us much further toward this goal. Many examples exist where a species' spread approximates concentric circles that become progressively more warped for one reason or another, usually, I believe, because as the circles enlarge they inevitably abut against heterogeneities in the physical environment (cf. Rapoport 1982). The Colorado potato beetle *(Leptinotarsa decemlineata)* in Europe is a good example (Nowak 1971, Fig. 28), as is the Japanese beetle *(Popillia japonica)* in the United States (Elton 1958, Figs. 14 and 15). The rate at which such ranges expand is, of course, related to dispersal capabilities of the individual species, and sometimes to behavioral characteristics. Though exact predictions do not seem possible, one can often guess in advance which of two successfully introduced species will expand its range fastest if one knows enough about the biology of the species.

On the other hand, some introduced insects spread irregularly from the outset, or after a brief period of circular range expansion. It is not uncommon for several foci to arise more or less simultaneously, each to serve as a base for slower circular growth or for subsequent further long-distance leaps. The aphid *Hydaphis tatarica* was restricted to a small area of southern Russia, limited by the range of its host, Tatarian honeysuckle *(Lonicera tatarica)*. It was apparently spreading gradually westward when it was scientifically described in 1935, having reached the Moscow region. However, as the honeysuckle was planted as an ornamental throughout much of central and eastern Europe, the aphid's range increased dramatically and irregularly (Nowak 1971). Parts of the range were not even originally contiguous.

Pielou (1979) has drawn a useful distinction between dispersal of the first type, which she terms "diffusion," and of the second type, called "jump dispersal." She also points out that, in practice, range expansion often is a combination of the two; certainly this is true for *Hydaphis tatarica*. Aside from calling attention to these two modes, I would ony add that success rate of introduced species does not seem to be correlated with dispersal mode. In particular, there are many examples in the literature in which a slow diffuser is nevertheless a successful colonist, as evidenced by success in different sites, few failed attempts, and characteristically large population size once a region is occupied.

I found it striking that many very successful introductions started with an inoculum of remarkably few individuals, often from one site. For example, the braconid wasp *Aphidius smithi,* a highly successful and widespread colonist of the continental United States (Clausen 1978), was introduced in 1958 as 17 females shipped from New Delhi by G.W. Angalet to the Introduced Beneficial Insects Investigations Laboratory at Moorestown, New Jersey (Angalet and Coles 1966). By 1959 74,000 individuals had been reared, and these were released in much of the eastern United States plus Arizona, Texas, and Washington in an attempt to control the pea aphid *(Acyrthosiphon pisum)*. Several specimens were also sent from Moorestown to the Department of Biological Control at the University of California, Riverside, where there was subsequent rearing

and release. By 1977 *A. smithi* was established from central Mexico through Ontario, Canada (Angalet and Fuester 1977), and even displaced a native species during its spread (Hagen et al. 1976), despite the initial genetic depauperation.

One immediately wonders about inbreeding depression in these cases. If it is such a universal and dangerous phenomenon (Schonewald-Cox et al. 1983), why has it not operated more effectively here? Sailer (1978) believes that it has occasionally been effective. He found that at least 10 established immigrant insects subsequently became extinct; possibly inbreeding depression contributed. He notes, however, that for Hymenoptera (such as *Aphidius*) the problem may not be nearly as severe as for other groups, since they are haplo-diploid (Sailer, personal communication). Thus it is possible that genes that would be deleterious in homozygous condition may already have been selected out. Brückner (1978) has found inbreeding depression in honeybees, but does not rule out a possible reduction in disadvantageous recessive alleles in haplo-diploid systems. I think such a reduction is very likely. Inbreeding depression really has two components. The major one (Ralls and Ballou 1983) is a decrease in general fitness with decreasing heterozygosity; this is a sum of small effects, and homozygosity at any one locus contributes only a small amount. The minor component is contributed by alleles that are lethal or very deleterious in homozygous condition, and this component is probably reduced in haplo-diploids. Also, Hamilton (1967) lists a thrips and many parasitic wasps that typically are spanandrous and inbreed very highly by sibmating. This breeding system would also quickly eliminate highly deleterious alleles (or else it would quickly eliminate the entire population!).

It is worth noting again (Fig. 1.1.) that the two insect orders vastly overrepresented among successful introductions in the United States are Homoptera and Thysanoptera. It happens that haplo-diploidy occurs not only in all Hymenoptera, but also in white flies (Aleyrodidae) and some scale insects (Coccoidea), two homopteran groups that are very heavily represented among introductions, and in thrips (Thysanoptera), as well as in scolytid and micromalthid beetles (Hartl and Brown 1970). Above I have argued that the phytophagous habit plus transportation history have been key to the apparent success of introduced homopterans and thrips. Perhaps their sex-determination mechanism has also played a role.

1.8. Conclusions

I would still look to ecology for the major explanations of why some insect species are successful invaders and others are not. However, as is clear from my arguments and examples, I doubt that the biogeographical and taxonomic aspects of ecology will help to produce good predictive hypotheses. I am also skeptical that population and community ecology will give us the easy answers that we want, though it could be that refinement of the approach of many of the papers in *The Genetics of Colonizing Species*—formulating lists of traits

that good colonizers should have—will improve on those early attempts. Many ecologists have argued that the outcome of an introduction is unpredictable (references in Sharples 1983), and I see little reason to think that advances in ecological theory will change this situation in the next decade. I remain convinced that the reasons for success or failure of any attempt *can* be determined by extensive field study, but that these reasons will reside in aspects of the particular species and system that are so idiosyncratic that they will defeat any attempt at concise generalization. In fact, the variety of insect introductions is so great that I think any general theory with relatively few parameters will make incorrect predictions about success and failure many times.

1.9. Acknowledgments

I thank Reece I. Sailer and John D. Lattin for numerous insights on introduced insects and leads on sources of data, and Louise E. Robbins, Sharon Y. Strauss, James Murray, and two anonymous referees for suggestions on this manuscript.

1.10. References

Allan JD, Barnthouse LW, Prestbye RA, Strong DR (1973) On foliage arthropod communities of Puerto Rican second growth vegetation. Ecology 54:628–632

Angalet GW, Coles LW (1966) The establishment of *Aphidius smithi* in the eastern United States. J Econ Entomol 59:769–770

Angalet GW, Fuester R (1977) The *Aphidius* parasites of the pea aphid *Acyrthosiphon pisum* in the eastern half of the United States. Ann Entomol Soc Am 70:87–96

Arnett RH (1983) Status of the taxonomy of the insects of America north of Mexico: a preliminary report prepared for the subcommittee for the insect fauna of North America project. Privately printed

Baker HG, Stebbins GL (eds) (1965) The Genetics of Colonizing Species. Academic Press, New York, 588 p

Brown JH, Gibson AC (1983) Biogeography. CV Mosby, St. Louis, 643 p

Brückner D (1978) Why are there inbreeding effects in haplodiploid systems? Evolution 32:456–458

Carlquist S (1965) Island Life. Natural History Press, Garden City NY, 447 p

Carlquist S (1974) Island Biology. Columbia University Press, New York, 660 p

Clausen CP (1978) Introduced parasites and predators of arthropod pests and weeds: a world review. Agriculture Handbook 480, US Department of Agriculture, Washington DC, 545 p

C.S.I.R.O. (Division of Entomology) (1970) The Insects of Australia: A Textbook for Students and Research Workers. Melbourne University Press, Melbourne, 1029 p

DeBach P (1965) Some biological and ecological phenomena associated with colonizing entomophagous insects. In: Baker HG, Stebbins GL (eds), The Genetics of Colonizing Species. Academic Press, New York, pp 287–303

DeBach P (1974) Biological Control by Natural Enemies. Cambridge University Press, Cambridge, 323 p

Doutt RL, DeBach P (1964) Some biological control concepts and questions. In: DeBach P (ed), Biological Control of Insect Pests and Weeds. Reinhold, New York, pp 118–142

Ehler LE, Hall RW (1982) Evidence for competitive exclusion of introduced natural enemies in biological control. Environ Entomol 11:1–4

Ehler LE, Hall RW (1984) Evidence for competitive exclusion of introduced natural enemies in biological control: an addendum. Environ Entomol 13:v–vii

Elton CS (1958) The Ecology of Invasions by Animals and Plants. Methuen, London, 181 p

Erwin TL (1982) Tropical forests: their richness in Coleoptera and other arthropod species. Coleopterists Bull 36:74–75

Goeden RD, Louda SM (1976) Biotic interference with insects imported for weed control. Annu Rev Entomol 21:325–342

Hagen KS, Viktorov GA, Yasumatsu K, Schuster MF (1976) Biological control of pests of range, forage, and grain crops. In: Huffaker CB, Messenger PS (eds), Theory and Practice of Biological Control. Academic Press, New York, pp 397–442

Hall RW, Ehler LE (1979) Rate of establishment of natural enemies in classical biological control. Bull Entomol Soc Am 25:280–282

Hamilton WD (1967) Extraordinary sex ratios. Science 156:477–488

Hartl DL, Brown SW (1970) The origin of male haploid genetic systems and their expected sex ratio. Theor Pop Biol 1:165–190

Holdgate MW (1960) The fauna of the mid-Atlantic islands. Proc R Soc B152:550–567

Huffaker CB, Messenger PS, DeBach P (1971) The natural enemy component in natural control and the theory of biological control. In: Huffaker CB (ed), Biological Control. Plenum, New York, pp 16–67

Huffaker CB, Simmonds FJ, Laing JE (1976) The theoretical and empirical basis of biological control. In: Huffaker CB, Messenger PS (eds), Theory and Practice of Biological Control. Academic Press, New York, pp 41–78

Illies J (1983) Changing concepts in biogeography. Annu Rev Entomol 28:391–406

Imms AD (1931) Recent advances in entomology. Churchill, London, 374 p

Kahn EJ (1984) The staffs of life. II. Man is what he eats. New Yorker, Nov. 12, pp 56–106

Keller MA (1984) Reassessing evidence for competitive exclusion of introduced natural enemies. Environ Entomol 13:192–195

Laing JE, Hamai J (1976) Biological control of insect pests and weeds by imported parasites, predators, and pathogens. In: Huffaker CB, Messenger PS (eds), Theory and Practice of Biological Control. Academic Press, New York, pp 685–743

Lattin JD, Oman P (1983) Where are the exotic insect threats? In: Graham C, Wilson C (eds), Exotic Plant Pests and North American Agriculture. Academic Press, New York, pp 93–137

Leston D (1957) Spread potential and the colonisation of islands. Syst Zool 6:41–46

Levins R (1969) Some demographic and genetic consequences of environmental heterogeneity for biological control. Bull Entomol Soc Am 15:237–240

Lindroth CH (1957) The Faunal Connections Between Europe and North America. John Wiley, New York, 344 p

Long JL (1981) Introduced Birds of the World. Universe Books, New York, 528 p

Luck RF, Podoler H, Kfir R (1982) Host selection and egg allocation behaviour by *Aphytis melinus* and *A. lingnanensis:* comparison of two facultatively gregarious parasitoids. Ecol Entomol 7:397–408

Moulton MP, Pimm SL (1983) The introduced Hawaiian avifauna: biogeographic evidence for competition. Am Natur 121:669–690

Nowak E (1971) The range expansion of animals and its causes. Zeszyty Naukowe 3:1–255 (translated from Polish, U.S. Department of Commerce, Springfield VA, 164 p)

Pemberton CE, Willard HF (1918) A contribution to the biology of fruit-fly parasites in Hawaii. J Agric Res 15:419–465

Pielou EC (1979) Biogeography. John Wiley, New York, 351 p

Ralls K, Ballou J (1983) Extinction: lessons from zoos. In: Schonewald-Cox CM, Chambers SM, MacBryde B, Thomas L (eds), Genetics and Conservation. Benjamin/Cummings, Menlo Park, California, pp 164–184

Rapoport EH (1982) Areography: Geographical Strategies of Species. Pergamon Press, Oxford, 269 p

Sailer RI (1978) Our immigrant insect fauna. Bull Entomol Soc Am 24:3–11

Sailer RI (1983) History of insect introductions. In: Graham C, Wilson C (eds), Exotic Plant Pests and North American Agriculture. Academic Press, New York, pp 15–38

Schonewald-Cox CM, Chambers SM, MacBryde B, Thomas L (eds) (1983) Genetics and Conservation. Benjamin/Cummings, Menlo Park, California, 722 p

Sharples FE (1983) Spread of organisms with novel genotypes: thoughts from an ecological perspective. Recombinant DNA Tech Bull 6:43–56

Simberloff D (1981) Community effects of introduced species. In: Nitecki MH (ed), Biotic Crises in Ecological and Evolutionary Time. Academic Press, New York, pp 53–81

Simberloff D (1983) Biogeography: the unification and maturation of a science. In: Brush AH, Clark GA (eds), Perspectives in Ornithology, Cambridge University Press, Cambridge, pp 411–455

Simberloff D (1984) Properties of coexisting bird species in two archipelagoes. In: Strong DR, Simberloff D, Abele LG, Thistle AB (eds), Ecological Communities: Conceptual Issues and the Evidence. Princeton University Press, Princeton, New Jersey, pp 234–253

Tallamy DW (1983) Equilibrium biogeography and its application to insect host-parasite systems. Am Natur 121:244–254

Turnbull AL (1967) Population dynamics of exotic insects. Bull Entomol Soc Am 13:333–337

Washburn JO (1984) The gypsy moth and its parasites in North America: a community in equilibrium? Am Natur 124:288–292

Watt KEF (1965) Community stability and the strategy of biological control. Can Entomol 97:887–895

Williams CB (1964) Patterns in the Balance of Nature. Academic Press, New York, 324 p

Williamson M (1981) Island Populations. Oxford University Press, Oxford, 286 p

Wilson EO (1961) The nature of the taxon cycle in the Melanesian ant fauna. Am Natur 95:169–193

Wilson EO (1965) The challenge from related species. In: Baker HG, Stebbins GL (eds), The Genetics of Colonizing Species. Academic Press, New York, pp 7–27

Zar JH (1984) Biostatistical Analysis. Prentice-Hall, Englewood Cliffs, New Jersey, 718 p

Zimmerman EC (1948) Insects of Hawaii, Vol. I, Introduction. University of Hawaii Press, Honolulu, 206 p

2. Fish Introductions into North America: Patterns and Ecological Impact

P.B. Moyle

2.1. Introduction

The transplantation of fishes has a long history in Western culture, beginning with the Romans, who brought carp *(Cyprinus carpio)* from the Danube River to Italy (Balon 1975). As Christianity spread throughout Europe, carp entered local fish communities, initially as escapees from monastery ponds. More deliberate colonizations were made in Scandinavia, where the introduction of indigenous salmonids into alpine lakes was apparently a regular practice by the 12th century A.D. (Nilsson 1972). Small-scale introductions continued until the 1840s, when the discovery, in France, of artificial propagation techniques for fish made introductions possible on a much larger scale (Regier and Applegate 1972). The idea of enhancing wild fish populations through the introduction of hatchery-reared fish quickly spread to North America and by the 1870s private, state, and federal fish hatcheries were common. The most important species reared at this time was the carp, which by 1890 was found throughout North America, thanks to the development of railroads and fish transport cars. Other species were raised as well, however, so that by 1873 a railroad car containing 300,000 fish of 10 species was on its way to California (Sheeley 1917). Such cars usually carried Pacific coast salmonids back to the eastern seaboard on the return trip. This practice of taxon redistribution (Regier and Applegate 1972) continues to the present time, not only with fish but with other aquatic organisms as well (e.g., Carlton 1974). However, efforts to understand the effects of fish

introductions from an ecological perspective are few (Christie et al. 1972; Zaret and Paine 1973; Magnuson 1976; Li and Moyle 1981; Schoenherr 1981).

The purposes of this chapter are (1) to review the regional, habitat, and taxonomic patterns of successful fish introductions in freshwater, saltwater, and estuarine environments in North America, and (2) to examine the nature of some fish assemblages into which introductions have been successful.

2.2. Freshwater Introductions

2.2.1. Regional Patterns

When state and regional fish lists are examined (Table 2.1.), it becomes evident that introduced species typically make up fewer than 10% of the fish species found in areas east of the Rocky Mountains but 30 to 60% of the fish species in areas west of the Rockies. In western North America, the importance of introduced species decreases in Canada and Alaska.

In eastern North America, introduced species generally do not dominate fish communities in terms of numbers or biomass of individuals. There are, however, some interesting and spectacular exceptions to this "rule" in the east, such as the fish communities of the Great Lakes, many eastern trout streams, and the canals and lakes of Florida. Many eastern fish communities have been expanded through the addition of fishes inhabiting nearby waters but unable to surmount natural barriers without the help of humans. (This explains the comparatively high percentage of introduced species found in Maryland and Delaware, which contain mainly small depauperate coastal streams but have major drainages nearby.) Most such additions have been sunfishes (Centrarchidae), catfishes (Ictaluridae), pikes (Esocidae), and perches (Percidae). The transfer of these game fishes over the past 100 or more years has been so common that the limits of the original ranges of many species are obscure (Lee et al. 1980). The widespread use of "minnows" for bait has also led to the localized spread of fish species. For example, during an intensive 2-year study of the fish assemblage of a small Minnesota lake that was popular with bait fishermen, I collected a few individuals of four species of fish that clearly did not "belong" there, but had presumably arrived in bait buckets (Moyle 1973). Such subtle faunal enrichment has no doubt been taking place for over a hundred years or more in the east, but its effects are poorly known. For the most part, the local fish faunas appear to have adjusted to enrichment without much loss of native species, a result that would be predicted given the stochastic nature of local fish assemblages (Grossman et al. 1982). Thus, when a species of shiner (Cyprinidae: *Notropis*) was introduced into an Alabama creek with "explosive success," it failed to displace a closely related species despite a high degree of ecological similarity (Mathur 1977).

In contrast to the east, wholesale replacement of native fish faunas by introduced species is common in the west. Throughout the Colorado River drainage, for example, introduced species are dominant and most of the native species are listed as threatened or endangered. The replacement species are largely

Table 2.1. Total number of species and percent introduced species in selected states and areas of North America

State	No. Species	Percent Introduced	Source
New York	186	6	Werner 1980; Daniels, personal communication
Pennsylvania	159	4	Cooper 1983
Maryland and Delaware	90	22	Lee et al. 1976
North Carolina	197	4	Menhinick et al. 1974
Louisiana	148	3	Douglas 1974
Florida	143	20[a]	Gilbert, personal communication
Alabama	170	5	Smith-Vaniz 1968
Arkansas	193	6	Buchanan 1973
Missouri	198	4	Pflieger 1975
Oklahoma	165	2	Miller and Robison 1973
Texas	163	7	Hubbs 1976
Kansas	130	5	Cross 1967
Wisconsin	157	7	Becker 1983
Illinois	199	7	Smith 1979
Ohio	166	6	Trautman 1981
Minnesota	199	9	Phillips et al. 1982
Wyoming	78	25	Baxter and Simon 1970
Idaho	67	42	Simpson and Wallace 1982
Utah	49	47	Sigler and Miller 1963
Montana	80	35	Brown 1971
Washington	76	39	Wydoski and Whitney 1979
Oregon	92	32	Bond 1973
California	124	36	Shapovalov et al. 1981
Nevada	80	51	Deacon and Williams 1984
Arizona	80	59	Minckley 1973
Alaska	56	2	Morrow 1980
Canada			
Atlantic Drainage	142	6	Scott and Crossman 1973
Hudson Bay Drainage	94	9	Scott and Crossman 1973
Arctic Drainage	55	4	Scott and Crossman 1973
Pacific Drainage	67	22	Scott and Crossman 1973

Species were counted only if they had extant, reproducing populations. Anadromous species were included, but euryhaline marine species were not.
[a] If three species whose introduced status in the Escambia drainage is questionable are omitted the percentage drops to 17.

from eastern North America, plus carp (from Europe), tilapia (from Africa), rainbow trout (from the Pacific coast), and a few others. A similar situation exists in the San Joaquin River drainage of California, except that substantial populations of native fishes persist in tributary streams (Moyle and Nichols 1974). In Clear Lake, California, the original fauna of 12 species has shown a steady decline over the past 100 years to the point where only four of the native fishes are still abundant. At least 14 introduced species are thriving in the lake (Moyle 1976b). The increase in total species observed in Clear Lake seems to be characteristic of western waterways in general over the past 100 years.

Why do introduced species dominate the fish assemblages so much more in

the west than the east? One reason is simply that the speciose eastern fish fauna already contained many species (e.g., centrarchid basses, sunfishes, catfishes) with characteristics prized by Americans, whereas the western fauna did not. The east is dominated by the huge Mississippi—Missouri drainage, an ancient and diverse center for fish speciation, so species richness even in small streams tends to be high (20 to 40 or more species). This has not precluded a few introduced species from outside the drainage, such as the common carp, from becoming extremely widespread and abundant, but most such introductions seem to have resulted in only small, localized populations, such as exist for the European cyprinids tench *(Tinca tinca)*, bitterling *(Rhodeus sericeus)*, ide *(Leuciscus idus)*, and rudd *(Scardinius erythropthalmus)* (Courtenay et al. 1984). Similarly, the recent expansion of the ranges of many native species presumably has resulted from repeated introductions into peripheral areas, rather than natural spreading from a few transfers. Overall, there have been many shifts or increases in composition of local fish faunas, but community structure and dynamics do not seem to have changed dramatically, presumably because the native fish species are well-adapted for living in complex and shifting assemblages (Grossman et al. 1982). Exceptions to this generalization occur where unusually effective predators or competitors have been introduced into systems that have been stressed by human activities (Great Lakes; Florida). Eastern fish assemblages in which introduced species are most abundant occur in areas peripheral to the main Mississippi drainage, places where the native fish faunas are less rich.

In contrast, the west is made up of a number of distinct drainages containing highly endemic fish faunas with relatively few species; a typical stream will have only five to 10 species (Moyle and Cech 1982). These species are often trophic or habitat specialists adapted for surviving in streams that show extreme seasonal and year-to-year variability in flows. Where these conditions persist, the native fish assemblages thrive. However, most western streams have been dammed, diverted, and otherwise modified, creating permanent standing water (reservoirs) and/or more constant flow regimens than previously existed. These conditions favor introduced species, many of which are most abundant in lakes and river backwaters in their native range. The introduced fishes seem to eliminate the native species through predation and competition, coupled with higher reproductive success (e.g., Schoenherr 1981). Few studies exist, however, that experimentally demonstrate the mechanism of species replacement in these situations (Taylor et al. 1984). New reservoirs in the west are often dominated by native nongame fishes for the first few years of their existence but these species gradually decline as the reservoir ages and introduced species become abundant. A notable exception to this generality exists in the Pit River of northeastern California, where the native fishes continue to dominate a series of reservoirs built in the 1920s through the 1950s. The reason for this seems to be that these reservoirs are used strictly for power production and so their water has a low residence time, much like a giant riverine pool. Upstream from these reservoirs, the river is relatively unchanged and meanders sluggishly through a wide flat valley. Here, however, the fauna is dominated by introduced

centrarchids, catfishes, and cyprinids, which displaced a native fauna better adapted for living in faster-flowing water (Moyle and Daniels 1982). The native fauna persists in canyon reaches above and below the valley reaches.

2.2.2. Habitat Patterns

Certain types of habitats in North America seem to favor introduced fishes: reservoirs and other artificial nonflowing waters, coldwater lakes, coldwater streams, desert springs, and other isolated habitats, subtropical waters, and large rivers. Reservoirs, especially in the west, are often dominated by introduced species because enormous effort is usually expended to make sure preferred species do become established when a reservoir is built. This includes such activities as massive and repeated planting of desired species and poisoning of the river being dammed to eliminate undesired native species. Thus in 1962, over 700 km of the Green River in Wyoming and Utah were poisoned to make sure the fishes introduced into Flaming Gorge Reservoir would not be subject to predation and competition from native nongame fishes; ironically, several of these nongame fishes are now on the federal list of endangered species (Ono et al. 1983). One species that does not need any help in becoming established in reservoirs is the common carp because its omnivority, high fecundity, long life, and spawning habits make it superbly adapted for reservoir conditions (Moyle 1984).

Coldwater lakes and coldwater streams often contain several species of introduced trout and/or salmon, placed there to provide more variety for anglers. In the west, most lakes and streams above 1,500 to 2,000 m harbored no fish at all until trout were planted. Even where native trout were abundant, other species were usually added, typically to the detriment of the native populations (e.g., Fausch and White 1981). However, it is common to find coldwater streams containing brook trout (*Salvelinus fontinalis,* native to the east), rainbow trout (*Salmo gairdneri,* native to the west), and brown trout (*S. trutta,* native to Europe) which manage to coexist through a combination of differing temperature preferences and spawning times (e.g., Seegrist and Gard 1972). Similar non-coevolved salmonid assemblages are characteristic of many coldwater lakes (numerous papers in Stevenson 1972). Whether or not such assemblages can continue to exist indefinitely is not known, as most have existed less than 50 to 100 years and are subject to continuous heavy exploitation. In addition, many of the bodies of water they occupy are subject to constant anthropomorphic environmental change.

Desert springs often contain a curious variety of introduced species, mainly gamefishes such as largemouth bass *(Micropterus salmoides)* and various "tropical" aquarium fishes. The tropical fishes are apparently placed in these springs by people who realize that warm temperatures and other benign conditions make them ideal places to release unwanted aquarium fishes. Thus, Courtenay and Deacon (1983) recorded 17 species of fish from one Nevada spring over a 26-year period. While this spring apparently had no native fishes, elsewhere in the southwest native spring fishes have declined or become extinct

following introductions (Ono et al. 1983). Similar replacements have occurred in warm desert streams (Schoenherr 1981).

Waters that might be classified as subtropical are found in North America in Florida, southern California, Arizona, and Mexico. These waters seem particularly susceptible to the successful introduction of exotic "tropical" freshwater fishes, especially of the families Cichlidae and Poeciliidae. Although the most famous example of one of these invasions is that of the walking catfish *(Clarias batrachus)* in Florida (Courteney et al. 1974), much more widespread (and ecologically significant) has been the explosive spread of various *Tilapia* (Cichlidae) species in many of the subtropical areas. In the past 10 years these cichlids have become by far the most abundant species in many waters while native fishes, especially centrarchids, have declined severely (Taylor et al. 1984). For example, Noble et al. (1975) found that dense populations of blue tilapia *(T. aurea)* inhibited reproduction of native largemouth bass. Taylor et al. (1984) argue that the success of "tropical" introduced fishes in subtropical areas of North America is the result in good part of their ability to breed throughout the year, while the native fishes, which evolved in more temperate areas, have short, distinct breeding sessions.

Large rivers are vulnerable to the establishment of introduced species both because of their diversity of habitat and because they are often much altered by human activities. Even the Mississippi River, the most speciose of temperate rivers, has been invaded successfully by such species as common carp, goldfish *(Carassius auratus)*, and, most recently, grass carp *(Ctenopharyngodon idella)*. Many major rivers in the west are dominated by introduced species.

2.2.3. Taxonomic Patterns

Fishes successfully introduced in North America represent at least 24 families (Table 2.2.). The total number of species that have become established in North America outside their historical range is probably in excess of 150. A majority of these species come from the families Salmonidae, Cyprinidae, Ictaluridae, Poeciliidae, Cichlidae, Centrarchidae, and Percidae, but most of the other families have at least one species that has achieved spectacular success in a new environment. As might be expected from such a taxonomically diverse group, the successful species differ widely from one another in spawning habits, degree of parental care, adult size, feeding habits, and native habitat (Table 2.2.).

There does not seem to be any one biological feature that the successful introduced species have in common, although a majority have characteristics highly prized by U.S. and Canadian citizens such as edibility, sporting qualities, aesthetic characteristics, and supposed ability to control mosquitoes or aquatic weeds (Moyle 1976a; Welcomme 1984). Biologically, the species either (1) are very hardy, so can survive transport and thrive in disturbed environments, (2) are very aggressive, so can eliminate native fishes through a combination of predation and competition, (3) are ecologically or behaviorally distinct from the native fishes, so the native fishes either do not interact with them or are

Table 2.2. Biological characteristics of families of fishes from which successful introductions have been made in North America. Characteristics are those of successfully introduced species, not of entire family

Family[a]	Spawning Location	Type of Parental Care[b]	Typical Adult Standard Length (cm)	Adult Feeding Mode[c]	Native Habitat (adults)[d]
Petromyzontidae (1)	Benthic	Brood hider (A.2.1)	30–50	A	A
Clupeidae (4)	Pelagic	None (A.1.1)	6–45	C	A, B
Osmeridae (2)	Benthic	None (A.1.5)	6–20	C	C
Salmonidae (15)	Benthic	Brood hider (A.2.1)	20–90	A, B, C	A, C
Characidae (1)	Vegetation	None (A.1.5)	5–10	D	D
Catostomidae (6)	Benthic	None (A.1.3)	15–50	D	B, C
Cyprinidae (25)	Benthic	Various (A.1, A.2, B.1, B.2)	5–50	B, C, D	B
Ictaluridae (8)	Benthic	Guarder (B.2.1)	5–120	A, B, D	B
Clariidae (1)	Benthic	Guarder (B.2.2)	22–30	D	D
Loricariidae (1)	Benthic	Guarder (B.2.5)	17–42	E	D
Cobitidae (1)	Benthic	None (A.1.3)	10–20	D	B
Atherinidae (1)	Vegetation	None (A.1.5)	5–10	C	B
Esocidae (4)	Vegetation	None (A.1.5)	25–120	A	B
Umbridae (1)	Vegetation	None (A.1.5)	10–20	B	C
Poeciliidae (10)	Internal fertilization	Livebearer (C.2.1)	3–8	B, D	B, D
Cyprinodontidae (8)	Vegetation	None (A.1.5)	3–8	B, D	B, D
Gasterosteidae (2)	Vegetation	Guarder (B.2.7)	3–8	B	C
Cichlidae (13)	Benthic	Guarder (B.2.1) Mouth brooders (C.1.3)	5–15	D, E	D
Gobiidae (3)	Benthic	Guarder (B.2.1)	8–12	B	B, D
Percichthyidae (4)	Pelagic	None (A.1.2)	25–150	A	A, B
Centrarchidae (18)	Benthic	Guarder (B.2.1)	25–150	A, B	B
Percidae (6)	Benthic/vegetation	None (A.1.2) t Brood hider (A.2.1)	5–35	A, B	B
Anabantidae (1)	Pelagic	Guarder (B.1.4)	4–6	B	D
Cottidae (1)	Benthic	Guarder (B.2.1)	6–8	B	C

a Numbers refer to number of species with introduced populations, a minimum estimate based on information in Lee et al.1980

b Numbers refer to reproductive guilds of Balon (1975)

c A, piscivorous; B, insectivorous (includes benthic invertebrates), C, planktivorous, D, omnivorous, E, herbivorous

d A, anadromous, B, warm water lakes and rivers, C, coldwater lakes and streams, D, tropical and subtropical waters

unable to deal with a new style of predation or competition, (4) have reproductive strategies that seem to confer on them an unusual degree of "fitness," (5) are preadapted to distinctive local environmental conditions, (6) are able to disperse and colonize new areas rapidly, or, (7), possess a combination of several of these characteristics. For example, the mosquitofish (Poeciliidae: *Gambusia affinis*) is a small hardy fish that has been planted worldwide to control mosquitoes in rice fields, ditches, ponds, and other manmade waterways. Because it is a live bearer, its young have fairly high survival rates and the adults do not depend on specialized substrates for spawning. It frequently escapes to natural waters, where it can eliminate similar native species through a combination of predation on their young and (perhaps) competition for food and space (Schoenherr 1981). Similarly, the success of the sea lamprey *(Petromyzon marinus)* in the Great Lakes seems to have been due to "inexperience" of the native fishes in dealing with the lamprey's particular style of predation, which resulted in the devastation of the native fish populations. Conceivably, the recent success of the introductions of Pacific salmon (*Oncorhynchus* spp.) into the Great Lakes may be related in part to their having evolved in an environment where large lampreys are common.

The ability to disperse rapidly from the point of introduction may be one of the most important characteristics of successful introduced species. Individual common carp and grass carp, for example, have been known to move hundreds of kilometers in river systems. Grass carp were first stocked in the wild in the Mississippi drainage in 1970 and 1971; within 10 years individuals were being caught nearly 2700 km from the release point (Guillory and Gasaway 1978). Such "explosive" spread of introduced species is commonly observed. Inland silversides *(Menidia beryllina)* were introduced into a central California lake in 1967; by 1972 they were abundant in the outlet stream. By 1975 they were abundant downstream in the Sacramento-San Joaquin estuary, and had spread by 1981 to reservoirs in southern California, a distance of about 1000 km from the point of introduction (Moyle 1976b; J. St. Amant, personal communication). In 1879, a small number of striped bass were introduced into the Sacramento-San Joaquin estuary; by 1914 they had colonized the Coos Bay estuary in Oregon and by the 1930s had probably reached their present range, from northern Mexico to southern British Columbia (Wydoski and Whitney 1979). Similar, if slightly less spectacular, dispersal records exist for many other introduced species.

2.3. Saltwater Introductions

Marine fish assemblages rarely contain introduced species, presumably because marine fishes are more difficult to transport than freshwater fishes and because the general abundance and diversity of marine fishes has discouraged the type of faunal manipulation that has taken place in fresh water. That introduced species can become established even in species-rich marine fish assemblages is illustrated by the success of the introduction into Hawaiian waters of two species of herrings (Clupeidae), two species of snapper (Lutjanidae), and one

species of grouper (Serranidae) to reef assemblages that already contained in excess of 450 species (Maciolek 1984). Hawaiian reefs, however, do contain fewer species than similar reefs in the South Pacific from which the introduced species came. Along the Pacific coast of North America, only two introduced species seem well established, excluding estuarine fishes such as striped bass *(Morone saxatilis)* and American shad *(Alosa sapidissima)* which are often found at sea. The two introduced marine species are the chameleon goby *(Tridentiger trigonocephalus)* and the yellowfin goby *(Acanthogobius flavimanus)*. Both species are euryhaline and apparently arrived in North America by traveling in the ballast water of ships coming from Japan (Moyle 1976b). Similarly, J. Carlton (personal communication) observed threespine sticklebacks *(Gasterosteus aculeatus)* in the ballast water of ships traveling from Europe to North America. This is alarming because coastal populations of sticklebacks are used for the study of evolutionary biology on both continents (Bell 1976). Actually, it is surprising that more non-native marine fishes have not become established in North America, as Carlton (1974) records over 150 species of exotic marine invertebrates that have become established on the Pacific coast alone, mainly through "hitch-hiking rides" on or in ships.

2.4. Estuarine Introductions

The fish fauna of estuaries is a mixture of euryhaline marine and freshwater species, plus anadromous species. There are few "true" estuarine species (Moyle and Cech 1982). Furthermore estuarine fish assemblages show considerable seasonal and year-to-year variation, depending on the amount of freshwater inflow and the abundances of euryhaline species that use the estuaries. It should not be surprising therefore to find that estuaries on the East and Gulf coasts contain few or no introduced species, because such species are uncommon in the associated marine and freshwater environments. In west coast estuaries, introduced species are common and may be the most abundant species. Thus in a 5-year study of a portion of the Sacramento-San Joaquin estuary in California, Moyle et al. (1986) collected 42 species, half of which were introduced species. The most abundant species was striped bass, introduced from the east coast. The importance of introduced species declines in more northern estuaries and in smaller estuaries that are formed by isolated coastal streams. In a small estuary in southern California, 19% of the 32 species were introduced (Allen 1982), whereas only 6% of the 31 species found in a large British Columbia estuary were introduced (Levy et al. 1979). One factor that may contribute to the ease with which introduced species seem to become established in western estuaries is their youth: the Sacramento-San Joaquin estuary is probably only around 10,000 years old, and has only three native species that are estuarine-dependent. Estuaries on the east coast are much older and more numerous and, as a consequence, have a larger number of marine or anadromous species that depend on them for at least part of their life cycle. Such species are much less common in western estuaries.

2.5. Fish Assemblages Containing Introduced Species

In this section I describe three fish assemblages in which introduced species are important constituents: (1) Suisun Marsh, (2) coldwater streams of the Lahontan drainage, California, and (3) Lake Michigan. The object is to see if these assemblages differ significantly in their structure and dynamics from assemblages without introduced species.

2.5.1. Suisun Marsh

Suisun Marsh is a brackish to seasonally freshwater tidal marsh that is part of the Sacramento-San Joaquin estuary. Of the 42 species collected in the marsh over a 5-year period of intense sampling, only about 20 were abundant and/or predictable enough in occurrence to be considered ecologically significant (Moyle et al. 1986). The remaining species, however, do represent the species pool available to invade the marsh should conditions change. Of the important species, half were introduced, including two abundant species (yellowfin goby and inland silverside) that have been present for less than 20 years.

The present fish assemblages can be divided into seasonal species and resident species. The seasonal species move into the marsh in response to salinity; the marsh tends to be fresh in the winter and spring and brackish by late summer and fall. The seasonals are both native and introduced species and seem to respond to environmental conditions independently of one another. The resident species in contrast fall into two distinct groups. One group consists of the four most abundant native species that are characteristically found mainly in small, dead-end sloughs while the other group consists mainly of introduced species that are found throughout the marsh. The populations of fishes within these two groups tend to fluctuate in concordance with one another.

One significant difference between the two resident groups is their response to the annual summer invasion of opossom shrimp *(Neomysis mercedis)*. This small shrimp seems to be a preferred food item for most of the fishes and, when abundant, is important in diets of both native and introduced species. When the shrimp leaves the marsh in the fall and/or its populations are reduced by predation, the native fishes switch to other food items, each species feeding on different items. The resident introduced species, however, continue to feed largely on the shrimp. Consequently, their stomach fullnesses decline, while those of the native species do not (Herbold 1986). An analysis of Lyapunov stability of the total resident assemblage indicates that the native species seem to form a very stable assemblage that becomes increasingly unstable (likely to change) as introduced species are added. Presumably, the introduced fishes are not as well adapted to the particular sets of local conditions as are the native fishes, which may explain in general terms why local fisheries management agencies have problems maintaining adequate populations of introduced gamefishes such as striped bass. Despite the differences between native and introduced species, the structure and dynamics of the fish assemblage of Suisun Marsh still resemble those of estuarine fish assemblages in other parts of the world (Moyle et al. 1986).

2.5.2. Lahontan Drainage Streams

Martis Creek and Sagehen Creek are two small streams on the eastern slope of the Sierra-Nevada that empty into the ancient bed of Lake Lahontan, via the Truckee River. In both streams the native Lahontan cutthroat trout *(Salmo clarki henshawi)* has been replaced by three introduced species of trout: brown trout, rainbow trout, and brook trout. Each stream also contains six species of native nongame fishes. Despite the abundance of the introduced trout, fish assemblages show remarkable persistence in composition over long periods (Gard and Flittner 1974; Moyle and Vondracek 1985), especially among the native fishes. The reason for this seems to be that the three trout species together occupy the same ecological role once occupied by the Lahontan cutthroat trout. All three are aggressive and defend feeding areas from other trout; size is more important than species in determining outcome, although brown trout seem to be the most aggressive of the three species (e.g., Fausch and White 1981). The species segregate partially on the basis of physiological tolerances; brook trout dominate the cold headwaters, while brown trout dominate the warmer downstream areas. Different spawning times also facilitate coexistence. Seegrist and Gard (1972) found that a good year class of brook trout, a fall spawner, depends on the absence of severe winter floods. Under such conditions, the young brook trout are larger and established in feeding territories by the time the rainbow trout emerge from the gravel. The smaller rainbow trout are forced to keep moving and, presumably, eventually are eaten or starve to death. If severe winter floods reduce the survival of brook trout embryos in the gravel, rainbow trout can safely emerge and establish feeding territories.

Overall, despite the replacement of one major species with three equivalent species, there is little reason to suspect that the basic structure of the fish communities in the streams (as described by Moyle and Vondracek 1985) changed following the introductions.

2.5.3. Lake Michigan

Few, if any, aquatic ecosystems have been altered as dramatically through the introduction of new species as the Laurentian Great Lakes (Smith 1972). In Lake Michigan, populations of native fishes collapsed (or in a few cases temporarily increased) following the successive establishments of the rainbow smelt *(Osmerus mordax)* in 1923, the sea lamprey *(Petromyzon marinus)* in 1936, and the alewife *(Alosa pseudoharengus)* in 1949. These three species altered the system so much that even the zooplankton composition changed (Wells 1970). Apparently, the reason the impact of these three introductions was so great was that many of the native fish populations were already being stressed by overexploitation, pollution, and habitat alteration (Smith 1972).

The impact of the lamprey was most dramatic because its preferred prey were the larger species of fish, such as lake trout *(Salvelinus namaycush)*, burbot *(Lota lota)*, and large whitefish *(Coregonus* spp.). The lamprey virtually wiped out these fishes, although predation on native fish larvae by alewife and smelt may also have contributed to the decline of some species (Crowder and Bin-

kowski 1983). While this was happening, the smallest species of whitefish, the bloater *(C. hoyi)*, actually showed a large increase in numbers, presumably due to the greater availability of zooplankton caused by the absence of the larger whitefish species. However, the bloater population also collapsed when the alewife population exploded in the lake, due to competition for zooplankton and/or predation on bloater eggs and larvae by alewife (Crowder and Binkowski 1983). The bloater persisted, however, by showing a shift in habitat and feeding habits; they now have a more benthic orientation, which seems to be reflected in their reduced number of gill rakers (Crowder 1984).

Populations of a number of the species affected by the lamprey, alewife, and smelt are now recovering because of a number of factors, most prominently artificial control of lamprey populations in their spawning streams, planting of hatchery-reared lake trout, and addition of three species of Pacific salmon to the fauna. The salmon are all pelagic piscivores and prey heavily on the alewife; they may be controlling alewife numbers because the alewife has not achieved its former abundance in the lake, following a "natural" collapse of its populations in 1967 (Crowder 1980).

At the present time it is difficult to predict what will happen to the fish communities of Lake Michigan in the future, except that dramatic changes can be expected and that introduced species will continue to play a dominant role.

Why have the Great Lakes been so vulnerable to introduced species? Although environmental change and overfishing were certainly contributing factors, they at best only set the stage for the takeovers. Probably the most important factors were the biological characteristics of the lamprey and alewife. Both, when they invaded the Great Lakes, found an environment filled with prey (fish for lampreys, zooplankton for alewife) that lacked defenses for their particular style of predation. Appropriate predators to control their populations were also absent and conditions for spawning were close to optimal. Pelagic salmon species were then able to become established because of superabundant prey and, perhaps, resistance to lamprey attacks. In contrast to the situation in the Great Lakes, numerous large coldwater lakes that are dominated by salmonids in other parts of the world have not suffered such dramatic consequences of introductions, even though many introduced species have become established in them (see examples in Stevenson 1972).

2.6. Conclusions

Throughout North America, fish faunas have been altered through the introduction of new species. Most of the introductions have been of North American species into North American waters from which they were absent, although some of the more spectacular introductions (common carp, tilapia, etc.) have been of exotic species. Given the rapidity of modern transportation systems and our increased knowledge of fish ecology and physiology, it is possible to supplement the fish fauna of virtually any place in the world, no matter how depauperate or how rich in species. In North America, successful introductions

have been made into environments ranging from fishless lakes to species-rich streams of the Mississippi drainage, to coral reefs. The general effect of introductions is usually to increase the number of species in an area, although extinctions and severe declines of native species often follow the establishment of introduced fishes. The disappearance of native species is often attributed to predation and competition, but unequivocal examples are few (Taylor et al. 1984; Moyle et al. 1986). Usually introductions accompany major environmental changes or other stresses on the native fish community, so it is uncertain whether the declines were caused by the changes, the introduced species, or both working together. In the western United States, many new reservoirs initially developed large populations of native nongame fishes, but as introduced fishes became established through massive planting programs, most species of native fishes declined. This replacement of native with introduced species happened in a number of large natural lakes as well, such as Clear Lake, California (Moyle 1976b). In many cases, the native fishes persisted in the streams feeding the lakes and reservoirs, providing a continuous stream of native recruits to their fish populations. The inability of these recruits to colonize the reservoirs successfully in large numbers, despite past success, indicates that predation and perhaps competition from introduced fishes is keeping them from doing so. Introduced diseases and parasites may also be a factor. On the other hand, it is possible that the limited ability of introduced fishes (except salmonids) to colonize relatively undisturbed streams in the west may be due to competition and predation from native fishes, although it is more likely that they simply cannot survive the extreme floods and droughts to which these streams are subjected (Minckley 1973; Moyle 1976b).

Although fishes have been successfully introduced into a wide variety of environments, the greatest success has been achieved in waters with relatively depauperate fish communities, such as those found in the western United States. The reason for this seems to be related to the greater efforts that have been made to introduce species into such waters and to the greater likelihood that such waters will have been seriously altered by human activities, making them less suitable for native fishes. These may also be the reasons why some environments (reservoirs, coldwater streams) seem more vulnerable to introductions than others (warmwater streams, most marine habitats). However, in the case of most temperate lakes it can be argued that, because of their comparatively short duration in geological time, they are occupied mainly by native fishes adapted for life in streams (Moyle and Cech 1982), increasing the likelihood of the success of introductions of lake-adapted fishes. A similar argument can be made for Pacific coast estuaries (Moyle et al. 1986) and estuarine-adapted fishes.

Fishes that have been successfully introduced have a wide variety of biological features. Aspects of their biology such as spawning habits, degree of parental care, adult size, and feeding habits seem to have little bearing on whether or not a species can be successfully introduced. Most such fishes do seem to have a fairly broad range of environmental conditions under which they survive, have the ability to disperse rapidly, and have the ability to interact

successfully with other fishes and become integrated into local fish communities. However, the mechanisms of how introduced species become established, especially how they manage to "fit" into complex communities, are poorly understood, as are the mechanisms of how they replace native species in depauperate systems. Many communities that are made up of a mixture of native and introduced species seem to be similar in structure to those without introduced species, especially where environmental fluctuations are high. However, mixed communities have existed only for a short time (less than 100 years), so it is not known how stable they are. Presumably their stability is inversely related to the proportion of introduced fishes in the community, the amount of human perturbation of the waterway, and the amount of natural environmental variability. The constant problems fisheries managers have with systems containing a high proportion of introduced species are probably good indications that their degree of stability is generally low. Such communities should consequently continue to provide good opportunities for studying competition, predation, and other ecological processes.

2.7. Acknowledgments

The manuscript was reviewed by D.M. Baltz, B.Herbold, R.A. Daniels, and B. Vondracek, and was "processed" several times by Kimberly Strauch and Donna Raymond. C.R. Gilbert and R.A. Daniels provided information on the fishes of Florida and New York, respectively.

2.8. References

Allen LG (1982) Seasonal abundance, composition, and productivity of the littoral fish assemblages in Upper Newport Bay California. Fishery Bull 80:769–790

Balon EK (1975) Reproductive guilds of fishes: a proposal and definition. J Fish Res Bd Canada 32:821–864

Baxter GT, Simon JR (1970) Wyoming Fishes. Wyoming Game and Fish Department, Cheyenne, Bull 4, 168 p

Becker GC (1983) Fishes of Wisconsin. University of Wisconsin Press, 1052 p

Bell MA (1976) Evolution of phenotypic diversity in *Gasterosteus aculeatus* superspecies of the Pacific coast of North America. Syst Zool 25:211–227

Bond CE (1973) Keys to Oregon Freshwater Fishes. Agricultural Experiment Station, Oregon State University, Technical Bull 58, 42 p

Brown CJD (1971) Fishes of Montana. Big Sky Books, Montana State University, Bozeman Montana, 207 p

Buchanan TM (1973) Key to the Fishes of Arkansas. Arkansas Game and Fish Commission, 170 p

Carlton JT (1974) Introduced intertidal invertebrates. In: Smith RI, Carlton JT (eds), Light's Manual of Intertidal Invertebrates of the Central California Coast. University of California Press, pp 17–25

Christie WJ, Fraser JM, Nepsky SJ (1972) Effects of species introductions on salmonid communities in oligotrophic lakes. J Fish Res Bd Canada 29:969–973

Clay WM (1975) The Fishes of Kentucky. Kentucky Department of Fish and Wildlife Resources, Frankfort, Kentucky, 416 p

Cooper GL (1983) Fishes of Pennsylvania. Pennsylvania State University Press, University Park, 243 p

Courtenay WR Jr, Deacon JE (1983) Fish introductions in the American southwest: a case history of Rogers Spring, Nevada. Southwest Nat 28:221–224

Courtenay WR Jr, Hensley DA, Taylor JN, McCann JA (1984) Distribution of exotic fishes in the continental United States. In: Courtenay WR Jr, Stauffer JR Jr (eds), Distribution, Biology and Management of Exotic Fishes. Johns Hopkins University Press, Baltimore, pp 41–87

Courtenay WR Jr, Sahlman HF, Miley WW, Herrema DJ (1974) Exotic fishes in fresh and brackish waters of Florida. Biol Conserv 6:292–302

Cross FB (1967) Handbook of Fishes of Kansas. Museum of Natural History, University of Kansas, Lawrence, Kansas, 357 p

Crowder LB (1980) Alewife, rainbow smelt and native fishes in Lake Michigan: competition or predation? Environ Biol Fish 5:225–233

Crowder LB (1984) Character displacement and habitat shift in a native cisco in southeastern Lake Michigan: evidence for competition? Copeia 1984:878–883

Crowder LB, Binkowski FP (1983) Foraging behaviors and the interaction of alewife, *Alosa pseudoharengus,* and bloater, *Coregonus hoyi.* Environ Biol Fish 8:105–113

Deacon JE, Williams JE (1984) Annotated list of the fishes of Nevada. Proc Biol Soc Wash 97:103–118

Douglas NH (1974) Freshwater Fishes of Louisiana. Claitor's Publishing Division, Baton Rouge, Louisiana, 443 p

Fausch KD, White RJ (1981) Competition between brook trout *(Salvelinus fontinalis)* and brown trout *(Salmo trutta)* for positions in a Michigan stream. Can J Fish Aquat Sci 38:1220–1227

Gard R, Flittner GA (1974) Distribution and abundance of fishes in Sagehen Creek, California. J Wildlife Mgmt 38:347–358

Grossman GD, Moyle PB, Whitaker JO (1982) Stochasticity in structural and functional characteristics of an Indiana stream assemblage: a test of community theory. Am Nat 120:423–454

Guillory V, Gasaway RD (1978) Zoogeography of the grass carp in the United States. Trans Am Fish Soc 107:105–112

Herbold B (1986) Resource partitioning within a non-coevolved assemblage of fishes. Thesis, University of California Davis

Hubbs C (1976) A checklist of Texas freshwater fishes. Texas Parks, Wildlife Tech Series II, 12 p

Lee DS, Gilbert CR, Hocutt CH, Jenkins RE, McAllister DE, Stauffer JR Jr (1980) Atlas of North American Freshwater Fishes. North Carolina State Museum of Natural History, Raleigh, North Carolina, 867 p

Lee DS, Norden A, Gilbert CR, Franz R (1976) A list of the freshwater fishes of Maryland and Delaware. Chesapeake Sci 17:205–211

Levy DA, Northcote TG, Birch GJ (1979) Juvenile salmon utilization of tidal channels in the Fraser River Estuary, British Columbia. Westwater Research Center University, British Columbia, 20 p

Li HW, Moyle PB (1981) Ecological analysis of species introductions into aquatic systems. Trans Am Fish Soc 110:772–782

Maciolek JA (1984) Exotic fishes in Hawaii and other islands of Oceania. In: Courtenay WR Jr, Stauffer JA Jr (eds), Distribution, Biology, and Management of Exotic Fishes. Johns Hopkins University Press, pp 131–161

Magnuson JJ (1976) Managing with exotics—a game of chance. Trans Am Fish Soc 105:1–9

Mathur D (1977) Food habits and competitive relationships of the bandfin shiner in Halawakee Creek, Alabama. Am Midl Nat 97:89–100

Menhinick EF, Burton TM, Bailey JR (1974) An annotated checklist of the freshwater fishes of North Carolina. J Elisha Mitchell Sci Soc 90:24–50

Miller RJ, Robison HW (1973) The Fishes of Oklahoma. Oklahoma State University Press, Stillwater, Oklahoma, 246 p

Minckley WL (1973) Fishes of Arizona. Arizona Game and Fish Department, 293 p

Morrow JE (1980) The Freshwater Fishes of Alaska. Alaska Northwest Publishing Company, Anchorage, Alaska, 248 p

Moyle PB (1973) Ecological segregation among three species of minnows in a Minnesota lake. Trans Am Fish Soc 103:799–805

Moyle PB (1976a) Fish introductions into California: history and impact on native fishes. Biol Conserv 9:101–118

Moyle PB (1976b) Inland Fishes of California. University of California Press, Berkeley, California, 403 p

Moyle PB (1984) America's carp. Natur Hist 93(9):42–51

Moyle PB, Cech JJ Jr (1982) Fishes: An Introduction to Ichthyology. Prentice-Hall, Englewood Cliffs, New Jersey, 593 p

Moyle PB, Daniels RA (1982) Fishes of the Pit River System, McCloud River System and Surprise Valley Region. University of California Publ Zool 115:1–82

Moyle PB, Daniels RA, Herbold BC, Baltz DM (1986) Annual and long-term patterns in distribution and abundance of a noncoevolved assemblage of estuarine fishes in California. Fisher Bull 84 (in press)

Moyle PB, Li HW, Barton B (1986) The Frankenstein affect: impact of introduced fishes on native fishes in North America. In: Stroud RH (ed), The Role of Fish Culture in Fishery Management. Am Fish Soc, Bethesda, Maryland (in press)

Moyle PB, Nichols R (1974) Decline of the native fish fauna of the Sierra Nevada foothills, Central California. Am Midl Nat 92:72–83

Moyle PB, Vondracek B (1985) Persistence and structure of the fish assemblage in a small California stream. Ecology 66:1–13

Nilsson N A (1972) Effects of introductions of salmonids into barren lakes. J Fish Res Bd Canada 29:693–697

Noble RL, Germany RD, Hall CR (1975) Interactions of the blue tilapia and largemouth bass in a power plant cooling reservoir. Proc 29th Ann Conf Southeast Assoc Game and Fish Comm, pp 247–251

Ono RD, Williams JD, Wagner A (1983) Vanishing Fishes of North America. Stone Wall Press, Washington DC, 257 p

Pflieger WL (1975) The Fishes of Missouri. Missouri Department of Conservation, Jefferson City, Missouri, 342 p

Phillips GL, Schmid WD, Underhill JC (1982) Fishes of the Minnesota Region. University of Minnesota Press, Minneapolis, Minnesota, 248 p

Regier HA, Applegate VC (1972) Historical review of the management approach to exploitation and introduction in SCOL lakes. J Fish Res Bd Canada 29:683–692

Schoenherr AA (1981) The role of competition in the replacement of native fishes by introduced species. In: Naiman RJ, Soltz DL (eds), Fishes in North American Deserts. John Wiley, New York, pp 173–203

Scott WB, Crossman EJ (1973) Freshwater Fishes of Canada. Fisheries Research Board of Canada, Ottawa, 966 p

Seegrist DW, Gard R (1972) Effects of floods on trout in Sagehen Creek, California. Trans Am Fish Soc 101:478–482

Shapovalov L, Cordone AJ, Dill WA (1981) A list of the freshwater and anadromous fishes of California. Calif Fish Game 67:4–38

Sheeley WW (1917) History of the introduction of food and game fishes into the waters of California. Calif Fish Game 3:3–12

Sigler WF, Miller RR (1963) Fishes of Utah. Utah State Department of Fish and Game, Salt Lake City, Utah, 203 p

Simpson JC, Wallace RL (1982) Fishes of Idaho. University of Idaho Press, Moscow, Idaho, 238 p

Smith PW (1979) The Fishes of Illinois. University of Illinois Press, Urbana, Illinois, 314 p

Smith SH (1972) Factors of ecologic succession in oligotrophic fish communities of the Laurentian Great Lakes. J Fish Res Bd Canada 29:717–730

Smith-Vaniz WF (1968) Freshwater Fishes of Alabama. Auburn University Agricultural Experiment Station, Auburn, Alabama, 211 p

Stevenson JC (ed) (1972) Proceedings of international symposium on salmonid communities in oligotrophic lakes. J Fish Res Bd Canada 29:611–986

Taylor JN, Courtenay WR Jr, McCann JA (1984) Known impacts of exotic fishes in the continental United States. In: Courtenay WR Jr, Stauffer JR Jr (eds), Distribution, Biology and Management of Exotic Fishes. Johns Hopkins University Press, Baltimore, pp 322–373

Trautman MB (1981) The Fishes of Ohio. Ohio State University Press, Columbus, Ohio, 782 p

Welcomme RL (1984) International transfers of inland fish species. In: Courtenay WR Jr, Stauffer JR Jr (eds), Distribution, Biology, and Management of Exotic Fishes. Johns Hopkins University Press, Baltimore, pp 22–30

Wells L (1970) Effects of alewife predation on zooplankton populations in Lake Michigan. Limnol, Oceanogr 15:556–565

Werner RG (1980) Freshwater Fishes of New York State. Syracuse University Press, Syracuse, New York, 186 p

Wydoski RS, Whitney RR (1979) Inland Fishes of Washington. University of Washington Press, Seattle, Washington, 220 p

Zaret TM, Paine RT (1973) Species introduction into a tropical lake. Science 182:421–487

3. Patterns of Plant Invasion in North America

H.G. Baker

3.1. Introduction

Twenty years ago, an international conference on *The Genetics of Colonizing Species* was held in Asilomar. The proceedings of the conference were published in the next year (Baker and Stebbins 1965). The plant materials that were discussed then were mainly weeds.* This would seem to be a restriction but it is a fact that most colonizing plants are weeds. In North America, except for the sea shores, the moraines in front of receding glaciers, and the recolonization of devastated areas around Mount St. Helens, there are few opportunities for introduced plants to take part in plant succession. Otherwise, the disturbance in the natural system that almost always seems to be necessary for plant invaders to be successful is caused by human activities.

3.2. Origins and Modes of Spread of Weeds

Plant invaders are weeds even when they dominate regional vegetation as do the Mediterranean annual plants in the grasslands of California.

In North America, habitat alterations by human beings have changed some

---

* A plant is a weed if, in any specified geographical area, its populations grow entirely or predominantly in situations markedly disturbed by man (without, of course, being deliberately cultivated plants) (Baker 1965, p. 147).

native plants that were previously ecologically restricted into weeds with a broader distribution. For example, the Turkey Mullein, the monotypic genus *Eremocarpus setigerus,* has spread from natural occurrences in dry open places in the grassland, foothills, and desert from southern California to Washington (Munz and Keck 1958), to a disturbance along the highways that have been constructed (Frenkel 1970). This is invasion in a technical sense only.

More obviously invaders are those species that have been helped wittingly or unwittingly by man to expand an existing North American distribution to a part of the country where they were unknown previously. Walter Conrad Muenscher's (1955) classic book on weeds gives many examples (mostly for central and northern states) of such invasions. Consider the range expansion of the common sunflower *(Helianthus annuus)* which is "native from Minnesota to Saskatchewan, southward to Missouri and Texas; frequently introduced eastward and westward" (Muenscher 1955, p 472) and is now firmly established here as a roadside weed over much of North America (see also Heiser 1965).

But most invaders have been introduced by transport from a foreign land by man, his domesticated animals, or machinery. Usually the source is climatically similar to the area of introduction, and there is likely to be similarity in the soils and life forms of the vegetation (Baker 1962, 1974).

A listing of some of the ways in which alien flowering plants have reached North America and Hawaii is given in Table 3.1.

Some of the important modes of introduction of the past are not now significant. Thus the emptying of ballast from the holds of ships at east coast and west coast ports was undoubtedly responsible for the introduction of such weeds as the white campion *(Silene alba)* through Philadelphia and the Australian fireweed *(Senecio minimus)* through San Francisco (Baker unpublished).

Such a method of introduction would probably involve many seeds of the invader in the same load of ballast and would produce a group of plants flowering contemporaneously rather than a single example of the species, and this may have been important for a dioecious species such as *Silene alba* where pollen is needed for the pistillate plant to set seed. But this method of introduction is not now operative.

A description of ballast heap introductions is given by Muhlenbach (1979), who also records the spread of introduced plants along railroad tracks in St. Louis, Missouri. Frenkel (1970) studied the effect of road building on the distribution of introduced plants in California.

Contamination of seed of crop plants also was more likely in the past. A striking example where the introduction could be dated and the source geographically located is provided by the camel thorn *(Alhagi camelorum)* (Robbins et al. 1951). This drought-resistant, deep-rooting leguminous bush grows naturally in Asia Minor and adjacent European Russia. It is reported to have come to California as a contaminant of alfalfa seed imported by the Miller and Lux agricultural development company from Turkestan prior to 1915 and also, less convincingly, in the packing around North African date offshoots. Scattered infestations resulted in the San Joaquin valley. Some spread of the seed by ingestion and defecation by cattle and horses has enlarged its range subsequently, but its invasive power seems to be accentuated by the very vigorous

Table 3.1. Modes of entry of weeds into North America

By ballast in ships
By impure crop seed
By adhesion to domesticated animals
In soil surrounding roots of nursery stock
Deliberate introduction as:
  Forage plants
  Fiber plants
  Medicinal plants
  Ornamentals
  Erosion controls
  Timber plantations (e.g., introduction of *Melaleuca* into Florida and *Eucalyptus*
    into California)

---

circular growth of the plants by spreading rootstocks which may reach 25 feet away from the original plant and which put up shoots along their length. The impenetrable infestations may cover several acres as almost pure stands in low places (Robbins et al. 1951).

Deliberate introduction of species from another part of the world to supply forage, or erosion control, or some crop, or for ornamental use, has brought many potential colonizers to North America and to Hawaii (Holm et al. 1977). These introductions have been followed sometimes by escape of the cultivated plants to life as weeds.

3.3. Different Kinds of Invaders

Most frequently successful invaders are the pastoral forage plants—they have had to compete with grass in pastures and they are consequently able to compete as weeds. Good examples are the European clovers (of the genus *Trifolium*). Also successful are the sand-binders such as the European *Ammophila arenaria* which, after deliberate planting, take over whole sand dunes (Barbour et al. 1973).

Parentucellia viscosa and *Bellardia trixago* (both hemiparasitic members of the Scrophulariaceae) are introduced pasture weeds from Europe into western North American grasslands. Peter Atsatt and Donald Strong (1970) reported that, in experiments, the growth and fecundity of *Parentucellia* was improved by attachment to a wider range of potential host plants than were beneficial to *Bellardia*. They correlate this with the wider distribution of *Parentucellia* in northern California and Oregon, whereas *Bellardia* is restricted to several counties around San Francisco Bay.

Equally capable of survival are plants that were introduced to combat soil erosion. The blanketing of the southeast by the leguminous vine Kudzu (*Pueraria lobata*), introduced from Japan, is notorious. One article (Blackwell 1974–1975) described it as "the vine that ate the south." Presumably its vigor is related to leaving behind controlling pests in the Orient (for further interpretation

see Ewel, this volume). Rather less aggressive but vigorously colonizing chaparral, from the roadsides to which it was introduced by the Highways Department in California, is the French Broom, *Cytisus monspessulanus* (Munz
and Keck 1958).

The opposite situation is found in the case of ornamentals. Unlike most of
the other kinds of plants that have been mentioned, the ornamentals are generally
given special care in gardens. In this way they are protected from competition
with wild plants and are certainly never grown in a closed community (where
most of them would be failures).

To begin with, many of them need not set seed. Although in most cases they
are required to flower, very often they are propagated vegetatively. Indeed,
for many purposes it is considered undesirable that an ornamental plant should
be able to seed. Seed pods may detract from the appearance of the plant, the
formation of the pods may reduce further flowering, or the projection of seeds
in an uncontrolled manner around a garden may be deprecated by its owner.
As a consequence of this (and, sometimes, because to a gardener's taste "dou-
ble" flowers are preferable to normal flowers), many ornamental plants have
been deliberately deprived of their ability to reproduce by seed. It is clear that
they have little hope of naturalization as weeds (Baker 1962, 1974).

Even within a single genus, one may see these differences. On the northwestern Pacific coast, *Chrysanthemum leucanthemum* (the oxeye daisy) and
Chrysanthemum segetum (the corn marigold) have become established as quite
common roadside and field weeds (Munz and Keck 1958). They have never
been anything except incidentally beautiful and they are with us as a result of
impurities in grass-seed and grain. On the other hand, the commonly cultivated
garden chrysanthemums—the florists' chrysanthemum and the shasta daisy—
are extremely variable in chromosome number (Darlington and Wylie 1955) and
are highly infertile. Although they may persist for a season or so after having
been removed from a garden together with rubbish, they never become established.

Crop plants that are raised for seed yield have more chance of escaping from
cultivation and life as a weed in disturbed habitats. *Raphanus sativus,* the European radish, is raised for seed as well as the swollen root which is formed
by plants raised from that seed. The assistance that has been given to it by
hybridization with the already weedy *Raphanus raphanistrum* has enabled it
to become a problem weed in waste places in California, and, undoubtedly
elsewhere in North America (Baker 1971, 1974).

But more bizarre is the behavior of the cabbage *(Brassica oleracea)* which
was domesticated long ago from an ancestor found on sea cliffs in Europe (Clapham et al. 1952). On Point Bonita, on the north side of the Golden Gate, in
California, *Brassica oleracea* has returned to its ancestral habitat on the sea
cliffs (Howell 1970; Baker 1972).

In Hawaii, the vigorous spread into grasslands of the guava *(Psidium gua-
java)*, introduced as a fruit tree from tropical America, illustrates what can
happen especially in the reduced floral richness of oceanic islands (Holm et al.
1977).

Medicinal plants (in the wide sense) have a modest record of invasion following introduction. An example is the horehound, *Marrubium vulgare*, which was introduced from Europe to provide stems and leaves that produce an extract that is a constituent of cough medicines. It has spread across the United States and southern Canada by seed and has become particularly troublesome as a weed in the grasslands and hedges of the Pacific coast states. Its softly spiny calyx teeth, persisting around the nutlets, catch in the fur of sheep and other animals and contribute to the dispersal of the seed. The plant has a bitter taste so that it is not grazed or browsed by domesticated animals in pastures (Muenscher 1955; Robbins et al. 1951).

There is a sort of medicinal connection between the introduction and spread in North America of *Plantago indica* (Muenscher 1955). Seeds of this species are imported in large quantity because the mucilage that they exude when wetted is used as a laxative. But the seed survives passage through the human gut and commonly gives rise to plants at old-fashioned sewage farms.

It must not be forgotten that there are species that developed as weeds from long association with man in the Old World before being introduced to North America. An example is *Picris echioides* (Compositae) which developed as an annual, arable land weed (from perennial ancestors) in Europe but is found widely distributed as a roadside and waste place weed in North America (Baker 1974). Mulligan (1965) pointed out that *Matricaria inodora* (Compositae) has been introduced from Europe into Canada from both coasts and consequently has two widely separated distributions. Investigation of the chromosome numbers showed that the more aggressive dry prairie form is tetraploid, while the form from the maritime provinces (in the east) is diploid. However, these two forms are indistinguishable morphologically.

It seems inconceivable that such a plant as the poison hemlock *(Conium maculatum,* Umbelliferae) would be deliberately introduced from Europe, although it has been suggested that it was brought here to serve a medicinal use in view of its historical connections that date back over 2,000 years. It has become widespread in the northeastern and northcentral states and adjacent Canadian provinces, and, separately, in the Pacific coast states, especially California (Robbins et al. 1951; Muenscher 1955). Multitudes of other cases of Old World weeds introduced to North America could be quoted.

3.4. Susceptible Ecosystems

It should be obvious that some ecosystems are more vulnerable to invasion than others. Those most likely to be affected are grasslands (particularly when they are overgrazed), riparian habitats, waterways, roadsides and trodden paths, sand dunes, and some light forests. All of these have relatively frequent breaks in their natural plant cover wherein the invader can establish a foothold. Sometimes this foothold is maintained and expanded by allelopathic influences from the invader. This is believed to be one of the secrets of success by the mustards (*Brassica* spp.) in spreading masses of plants over California hills (Black et al. 1969).

In 1939, Talbot, Biswell, and Hormay estimated that in the San Joaquin Valley of California, introduced plants, mostly annuals, constituted 63% of the herbaceous plants of the grasslands, 66% in the woodlands, and 54% in the chaparral. These proportions must be even greater at the present day.

Some statistics for California based on lists in Robbins et al. (1951) are quite revealing. Although the species concerned differ considerably, it is an interesting fact that for vineyards, orchards, alfalfa fields, range lands, natural lowland meadows, and artificial pastures the proportion of weed species derived from Europe, Asia, and North Africa, taken together, remains roughly constant at 50–65%. Most of the remainder are natives of North America (Baker 1962, 1974).

In the extremely artificial environment of lawns and golf courses, however, the Eurasian element soars to 73% while the North American drops from about 30% to only 12%. Normally, in California, lowland terrestrial summer-green communities are rare, so that the indigenous plants can hardly contribute to the weed flora of moist soils of these continuously watered habitats. By contrast, in California rice fields and wet alkali soils, tule swamp California natives are quite at home and here the balance is tipped in the opposite direction—with only 17 to 20% of Eurasian adventives and 70% of North American species (Baker 1962, 1974).

The variety of geographical sources that may contribute to a weed flora is well illustrated by what I call "treading weeds" (Baker 1974) (Table 3.2.). Thus these morphologically and physiologically very similar (though taxonomically very distantly related) species that make up the assemblage on paths and roadsides in Berkeley come from a wide range of sources. On a broader scale, Muenscher (1955) found the proportions indicated in Table 3.3., for the northern states, where European and North American species are the most numerous.

In contrast, some ecosystems seem to be relatively resistant to invasion, and these include dense forests, high montane ecosystems, salt marshes, and deserts. Here the failure of invasion by introduced species can be related to the closed nature of the vegetation or stressful climatic features that can be withstood only by specialized plants.

3.5. Patterns of Spread

Patterns of spread shown by invaders after they have achieved a foothold may be of two kinds—the steady advance of a population or the scattering of "satellite" populations from an original center of introduction followed by a filling in of the gaps (Baker 1974). The latter pattern has been observed by the amateur British botanists who keep tabs on the spread of exotics in Britain and Ireland (Salisbury 1961). It is also likely that it is the most frequent pattern here in North America, too (Baker 1974).

The Division of Plant Industry of the State of California has produced maps for the benefit of its weed survey officers on a "township" basis. Species probably in the early stages of invasion are restricted to more or less adjacent "townships." These include *Rorippa austriaca* (Cruciferae) in extreme north

Table 3.2. "Treading weeds" in the Berkeley area of California

Species	Family	Geographical Origin
Matricaria matricarioides	Compositae	Northwest North America
Soliva sessilis	Compositae	Chile
Cotula australis	Compositae	Australia or New Zealand
Coronopus didymus	Cruciferae	Europe
Lepidium spp.	Cruciferae	Native
Sagina apetala	Caryophyllaceae	Europe
Polycarpon tetraphyllum	Caryophyllaceae	Europe
Polygonum aviculare	Polygonaceae	Europe

eastern California and *Setaria faberi* (Gramineae) in the Sierra Nevada foothills of north central California. Later stages may be shown by *Aegilops triuncialis*, (Gramineae), which is fairly widespread in north central California; *Chondrilla juncea* (Compositae), which has a focus in north central California and scattered locations in coastal, central, and southern California; and *Salvia aethiopis* (Labiatae), which has spread in northeastern California. But, in *Helianthus ciliaris* (Compositae), with its main occurrence in the Los Angeles area and scattered localities throughout California, the latter may be incipient "satellite populations" (if not separate introductions) at considerable distances from the apparent primary place of introduction. *Agropyron repens* (Gramineae), with two foci (the Los Angeles area and the San Francisco Bay area) is also scattered through northern California. *Alternanthera philoxeroides* (Amaranthaceae), has two distinct foci—the south end of the Central Valley and the Los Angeles area.

In the cases of *Centaurea repens* (Compositae) and *Solanum elaeagnifolium* (Solanaceae), the gaps between populations are well on the way to being filled.

Table 3.3. The source of five hundred weeds of the Northern United States

Origin	Number of Species		Percentage of Species	
Native to North America		196		39.2
Widespread	51		10.2	
Eastern	95		19	
Western	42		8.4	
Southern U.S.	8		1.6	
Tropical America		15		3
North America and Europe		13		2.6
North America and Eurasia		16		3.2
Europe		177		35.4
Asia		12		2.4
Eurasia		66		13.2
Africa and Eurasia		3		.6
Doubtful		2		.4
Total		500		100

From Muenscher WC. In: Weeds, 2nd edit. 1955. Reprinted with permission of Cornell University Press and Macmillan Co.

Italian thistles *(Carduus pycnocephalus* and *Carduus tenuiflorus)* are spreading rapidly over range lands at the present time (T.A. Fuller personal communication).

E.J. Salisbury (1961), in England, postulated that an introduced species would have to reproduce locally to build up an "infection pressure" before spreading widely from a center of introduction. It is difficult to pin down what might be the genetical processes involved in this but the generation of a sufficient measure of genotypic variability to accommodate to habitat variation may be part of the story because the initial introduction may have been by one or a few seeds leading to a founder effect.

3.6. Minor and Major Weeds

I have earlier distinguished "minor" and "major" weeds (Baker 1972, 1974). These differ in that the "minor" weeds have some genetically controlled limit on their climatic, edaphic, or biotic tolerance which restricts them to certain habitat types whereas the "major" weeds are much less inhibited.

This can be illustrated by *Ulex europaeus* (Leguminosae), the gorse, which has an "Atlantic" distribution in Europe, where it is further restricted to acid soils (Tansley 1939). On introduction into Pacific North America it is similarly generally restricted to sandy soils in coastal grasslands (Munz and Keck 1958; Howell 1970; Baker personal observation).

The foxglove, *Digitalis purpurea* (Scrophulariaceae), is a biennial that is also restricted to acid soils in Europe but with the difference that it is a woodland inhabitant (Tansley 1939). Since being introduced to the Pacific states, either as an ornamental or a medicinal plant, it has become naturalized in woodlands near the coast from Santa Barbara to British Columbia (Munz and Keck 1958)—but always in markedly acid soils (Baker personal observation).

These "minor" weeds may be able, after initial growth in a favorable spot, to develop genotypes appropriate to a wider range of habitats (Baker 1972, 1974). Possible cases of this may be the Medusa-head grass *Taeniatherum asperum* (McKell et al. 1962) and *Bromus tectorum* (Mack 1981).

"Major weeds" have preadaptations that very often enable individual plants to thrive more or less in a variety of habitats; their populations contain plants with "general purpose genotypes" (Baker 1965, 1972, 1974) that can build populations immediately. If they stay long enough in a particular habitat they may be selected for closer adaptation which would increase their chance of surviving in the habitat even if the native vegetation is permitted to return (see review in Baker 1974). Thus, they would show ecotype formation. These "opportunistic" major weeds are particularly likely to show disjunct populations, that is, they have not yet filled in the gaps in their ranges.

But there is always the possibility that a similar picture may be produced by multiple introductions. The apparently adaptive variation between populations of *Bromus mollis* in California in length of time spent in the vegetative condition before "heading" (Knowles 1943) could have been brought about by

multiple introductions from European sources with different climatic conditions or by ecotypical differentiation in California (or both).

3.7. Ecotypical Differentiation

Related taxa that segregate out in a new continent in patterns similar to those in the source continent may be illustrated by Shull's (1929, 1937) treatment of what he called *Bursa bursa-pastoris* but which we know now as *Capsella bursa-pastoris* (Cruciferae)—the shepherd's purse. He found large numbers of biotypes in this largely self-pollinated species and their morphological variation on a latitudinal basis was similar in North America to the presumably original pattern in Europe.

In flower gardens in Europe and North America, the herbaceous perennial *Kentranthus ruber* (Valerianaceae) from southern Europe is grown in three different flower colors—red, pink, and white. But Salisbury (1961) noted in Britain that the pink form appeared to be more vigorous than the other colors and is most frequent in the weedy populations of escapers from cultivation. In the San Francisco Bay Area we are finding the same general picture (Baker unpublished)—that weedy populations rarely contain white flowered plants. The white flowered morph is characterized by fewer flowering shoots, fewer flowers per shoot, and lighter seeds. Apparently it is less drought-tolerant than the colored morphs. Particularly striking is the fact that skippers, butterflies, and hummingbirds, which are the major pollinators of this species in California, ignore the white-flowered form which remains a more or less horticultural item in western North America as it is in Europe (Baker unpublished).

But not all patterns of distribution in the source area are repeated in the new continent. In the annual weed *Spergula arvensis* (Caryophyllaceae), in Britain, New (1958, 1959, 1978) found clines in the proportions of plants with papillate seeds compared with plants that have smooth nonpapillate seeds. In Britain this species is autogamous, so that the plants with papillate and nonpapillate seeds each breed true. Intermediates (heterozygotes) are rare. Herbarium specimens showed that the papillate seed form predominated in the north of Europe while the nonpapillate form had a southern distribution. In the British Isles, in mixed populations, there is a cline in the ratios of the two morphs which roughly corresponds to the European picture. New related this to differences in germination temperature requirements of the two morphs.

However, in the Pacific States (where *Spergula arvensis* reached California before 1848), Wagner (1983), who made population analyses over a wide range of localities (mostly in disturbed grasslands), found no geographical order in the population constitutions. Populations pure for one seed type occurred (probably as a result of a founder effect), others were dimorphic but without recognizable patterns in the proportions. It is possible that, in California, in particular, a greater prevalence of sunshine has led to a greater degree of outcrossing and a disruption of the linkage between genes controlling seed morphology and those concerned with germination and growth physiology.

In all cases, it should be emphasized that genetical interactions may take place, not only within species but also between related species.

3.8. Relationships between Species: *Cakile*

A striking case of sequential invasions is provided by the strand and sand dune genus *Cakile* (Cruciferae) on the Pacific coast of North America. Barbour and Rodman (1970) have concluded that historically there was no *Cakile* on the American shores of the Pacific Ocean. They found evidence that *Cakile edentula,* already present on the Atlantic coast of North America (and in the Great Lakes region), became established at a beach on San Francisco Bay (at West Berkeley) in 1881 or 1882, probably being brought in in ship's ballast. It is doubtfully distinct from the east coast material but was given the varietal name *californica*. It then spread north and south at an average rate of 65 km per year, eventually reaching Kodiak Island, Alaska, and the border between the U.S.A. and Mexico by 1936. Barbour and Rodman point out that *Cakile edentula* probably spread in a series of jumps, subsequently filling in the spaces between. Its fruits can float without harm in sea water. By the fourth decade of the 20th century it was common or occasional along the Pacific coast from San Diego to Alaska.

But then a surprising thing happened; *Cakile maritima*, which is a European species (from the Mediterranean region), appeared at Stinson Beach, Marin County, California and was collected in 1935. It also spread rapidly (an average of somewhat over 50 km per year) and became "common" along the entire California coast. It reached into Baja, California. The spread of *maritima* appears to have been accompanied by a great reduction in abundance of *edentula* so that at the time of the publication of their own work (in 1972) Barbour and Rodman found *edentula* to be "nearly extinct" in California. Farther north the replacement does not seem to be as complete, suggesting that *edentula* is better adapted to a cooler climate.

The rapid establishment of populations along a long coastline is striking enough, but the rapid replacement of one species by the other is especially remarkable. It is also puzzling because these plants grow in open communities— seashore strand and sand dunes—so there is unlikely to be serious competition between them, and allelopathic influences also seem unlikely.

Barbour and Rodman mention the possibility that the arrival of *maritima* might have coincided with the advent of a predator or parasite with an affinity to *edentula* over *maritima* but they have seen no evidence to support such an hypothesis.

They also speculate that the slightly larger flowers of *maritima* might give it an edge in competition for pollinators and, although both species are self-compatible, they think that *maritima* may produce more seeds than *edentula* and win in competition by that means. However, the nature of the habitats and the rapidity of the interactions suggest to me that the replacement—which can only be termed "catastrophic"—speaks in favor of a disease carried by *maritima*

but devastating only to *edentula*. The disease might be a virus that might not produce very obvious symptoms but that is effective in preventing reproduction.

As Barbour and Rodman pointed out, it might be instructive to investigate other regions where the two species coexist—for example, the northwest coast of South America and south coast of Australia, and they are doing this (J.E. Rodman personal communication).

3.9. Conclusions

The available facts about successful invaders suggest:

1. Climate of source and reception area must be similar.
2. Life forms of the vegetation should be similar.
3. Soils should not be significantly different.
4. The invader should have a generalized pollination system—wind, generalized insect, or self pollination.
5. There should be a seed-dispersal system appropriate not only to bring the immigrant to the new area but to spread it around in the new habitat.
6. The breeding system is extremely important. It should allow seed reproduction of the immigrant, while at the same time providing for genetical recombination. Facultative apomixis appears ideal, particularly as the amphimictic reproduction can produce new genotypes by outcrossing, while the agamospermy allows any genotypes that are favorable to be reproduced although highly heterozygous. *Poa pratensis*, (Kentucky bluegrass) is believed to be a native of Europe, but it has invaded successfully grasslands all over the world. In addition to facultative apomixis, it has vigorous vegetative reproduction which is favorable to it in spreading in closed grassland (Clausen 1953, 1954).

Obligate apomixis can also establish a weed in a new area if the original genotype is appropriate. In coastal communities in California the pampus grass *Cortaderia jubata*, from Ecuador (it is an obligate agamospermous apomict), has successfully invaded disturbed grasslands and even logged-over redwood forests in California (Costas-Lipmann 1979). This is in marked contrast to the lack of weediness in another species of pampas grass, *Cortaderia selloana*, that has been cultivated even longer, but is dioecious. This species needs staminate and pistillate plants close enough together to form seed. But even *C. jubata* has some restrictions on its success as an invader (climatic restriction to coastal areas, inability to germinate successfully in closed grassland with its tiny seeds in which the absence of sexual recombination does not permit variation), though it has a very broad tolerance of soil differences.

Self-compatibility and an absence of strong inbreeding depression will help establish an invader from a small introduction. Dioecism can be overcome if it is "leaky" with some seeds being set from hermaphrodite flowers that make their appearance occasionally (Baker and Cox 1984). Dispersal of multiseeded diaspores also could help a dioecious species in establishment after long-distance dispersal, and extreme vegetative reproduction can help a plant of one sex wait out the arrival of an individual of the opposite sex. All of these features may

have functioned in generating the high proportion of dioecious species (27.5%) in the Hawaiian flora after transoceanic distribution from their source areas (Baker and Cox 1984).

7. Vegetative reproduction helps in the invasion of communities where very few of the native species reproduce frequently by seed. This is well-demonstrated by the Kikuyu grass *(Pennisetum clandestinum)* in lawns and pastures (Baker 1974, 1978) and by water hyacinth *(Eichhornia crassipes)* (Pontederiaceae) in the waters of the Delta region of Central California (Bock 1968, Barrett 1982) where abundant vegetative growth of stolons is the key to the success of these invaders. Probably the same applies to the infestation of quiet spots on the Hudson River, in New York, by floating *Trapa natans* (Baker personal observation).

3.10. Epilogue

I have only scratched the surface of the subject of patterns in North American plant invasions. There is a vast amount of information that is needed but not yet available about the plants that have invaded new geographical areas and new ecosystems. When available and analyzed this may provide the information needed to prevent further invasions. In this context it is admirable that C.F. Reed (1977) has produced (for the U.S.D.A.) a most useful volume in which weed species that are not yet successfully established in the United States are named and described, on the basis of their foreign performances, as warning to observers here.

3.11. Acknowledgments

I am deeply indebted to Irene Baker who has devoted much time and skill to the preparation of this manuscript and in general support of this research.

3.12. References

Atsatt PR, Strong DR (1970) The population biology of annual grassland hemiparasites. I. The host environment. Evolution 24:278–291
Baker HG (1962) Weeds—native and introduced. J Calif Hort Soc 23:97–104
Baker HG (1965) Characteristics and modes of origin of weeds. In: Baker HG, Stebbins GL (eds), The Genetics of Colonizing Species. Academic Press, New York, pp 147–172
Baker HG (1972) Migrations of weeds. In: Valentine DH (ed), Taxonomy, Phytogeography and Evolution. Academic Press, London, pp 327–547
Baker HG (1974) The evolution of weeds. Annu Rev Ecol Syst 5:1–24
Baker HG (1978) Invasions and replacement in Californian and neotropical grasslands. In: Wilson JR (ed), Plant Relations in Pastures. CSIRO, East Melbourne, Chapter 24, pp 367–384.

Baker HG, Cox PA (1984) Further thoughts on dioecism and islands. Ann Missouri Bot Gard 71:230–239

Baker HG, Stebbins GL (eds) (1965) The Genetics of Colonizing Species. Academic Press, New York

Barbour MG, Craig RB, Drysdale FR, Ghiselin MT (1973) Coastal Ecology, Bodega Head. University of California Press, Berkeley

Barbour MG, Rodman JE (1970) Saga of the West Coast sea-rockets: Cakile edentula subsp californica and C maritima. Rhodora 72:370–386

Barrett SCH (1982) Genetic variation in weeds. In: Charudattan R, Walker HL (eds), Biological Control of Weeds with Plant Pathogens. John Wiley, New York

Black CC, Chen TM, Brown RH (1969) Biochemical basis of plant competition. Weed Sci 17:338–344

Blackwell J (1974-1975) The vine that ate the South. Brooklyn Bot Gard Record 30:29–30

Bock JH (1968) The water hyacinth in California. Madroño 197:281–283

Clapham AR, Tutin TG, Warburg EF (1952) Flora of the British Isles. Cambridge University Press, Cambridge

Clausen JC (1953) New bluegrasses by combining and rearranging genomes of contrasting *Poa* species. Proceedings of the 6th International Grassland Congress, University Park, Pennsylvania, pp 216–221

Clausen JC (1954) Partial apomixis as an equilibrium system in evolution. Caryologia 6 (Suppl):469–479

Costas-Lipmann M (1979) Embryology of *Cortaderia selloana* and *C. jubata* (Gramineae). Bot Gaz 140:393–397

Darlington CD, Wylie AP (1955) Chromosome atlas of flowering plants. Allen and Unwin, London

Frenkel RE (1970) Ruderal vegetation along some California roadsides. University of California Publications in Geography 21:1–163

Heiser CB (1965) Sunflowers, weeds and cultivated plants. In: Baker HG, Stebbins GL (eds), The Genetics of Colonizing Species. Academic Press, New York, pp 391–403

Holm LG, Plucknett DL, Pancho JV, Herberger HP (1977) The World's Worst Weeds: Distribution and Biology. University Press of Hawaii, Honolulu

Howell JT (1970) Marin Flora. University of California Press, Berkeley

Knowles PJ (1943) Improving an annual brome grass, *Bromus mollis* L, for range purposes. J Am Soc Agron 35:584–594

Mack RN (1981) Invasion of *Bromus tectorum* L into Western North America. Agro-Ecosystems 7:145–165

McKell CM, Robison JP, Major J (1962) Ecotypic variation in medusahead, an introduced annual grass. Ecology 43:686–698

Muenscher WC (1955) Weeds, 2nd edition. Cornell University Press, Ithaca

Muhlenbach V (1979) Contributions to the synanthropic (adventive) flora of the railroads in St Louis, Missouri, USA. Ann Missouri Bot Gard 66:1–108

Mulligan GA (1965) Recent colonization by herbaceous plants in Canada. In: Baker HG, Stebbins GL (eds), The Genetics of Colonizing Species. Academic Press, New York, pp 127–146

Munz P, Keck DD (1958) A California Flora. University of California Press, Berkeley

New JK (1958) A population study of *Spergula arvensis*. I. Two clines and their significance. Ann Bot 22:457–477

New JK (1959) A population study of *Spergula arvensis*. II. Genetics and breeding behaviour. Ann Bot 23:23–33

New JK (1978) Change and stability of clines in *Spergula arvensis* (Corn Spurrey) after twenty years. Watsonia 12:137–143

Reed CF (1977) Economically important foreign weeds. Potential problems in the United States. USDA Agriculture Handbook no 498, Washington

Robbins WW, Bellue MK, Ball WS (1951) Weeds of California. State Printing Division, Sacramento

Salisbury EJ (1961) Weeds and Aliens. Collins, London

Shull GH (1929) Species hybridizations among old and new species of shepherd's purse. Proceedings of the International Congress of Plant Science 1:837-888

Shull GH (1937) The geographical distribution of the diploid and double-diploid species of shepherd's purse. In: Youngken HK (ed), Nelson Fithian Davis Birthday Volume. Published privately, Boston, pp 1–8

Talbot MW, Biswell HH, Hormay AL (1939) Fluctuations in the annual vegetation of California. Ecology 20:394–402

Tansley AG (1939) The British Islands and Their Vegetation. Cambridge University Press, Cambridge

Wagner LK (1983) The population biology of an introduced weedy annual: *Spergula arvensis*. PhD Thesis, Botany, University of California, Berkeley

4. Patterns of Invasions by Pathogens and Parasites

A.P. Dobson and R.M. May

4.1. Introduction

For thirty years I have read publications about this spate of invasions; and many of them preserve the atmosphere of first-hand reporting by people who have actually seen them happening, and give a feeling of urgency and scale that is absent from the drier summaries of text-books. We must make no mistake: we are seeing one of the great historical convulsions in the world's fauna and flora. We might say, with Professor Challenger, standing on Conan Doyle's 'Lost World', with his black beard jutting out: 'We have been privileged to be present at one of the typical decisive battles of history—the battles which have determined the fate of the world.' But how will it be decisive? Will it be a Lost World? These are questions that ecologists ought to try to answer.

Elton (1958 pp. 31-32)

A lot has been written about invasions by pathogens and parasites, and about their effects on native populations of plants and animals. For instance, Elton's (1958) work, *The Ecology of Invasions*, remains fresh and full of telling case histories, while Soule and Wilcox (1980) give good accounts of more recent studies. The present chapter therefore does not offer an encyclopedic review of the subject matter embraced by its title, much of which would repeat material presented elsewhere, but rather concentrates on a few ideas that may be relatively unfamiliar.

The chapter is organized as follows. First, we consider invasions by parasites with direct life cycles (DLC), briefly sketching some examples and then dis-

cussing factors that bear upon the establishment and the spread of such invaders. Second, we give a parallel discussion of invasions by parasites with indirect life cycles (ILC), where one or more species of intermediate hosts are necessary for the parasite to be able to complete its transmission cycle. We present such sparse evidence as is available (from studies of parasites of fish) in support of our expectation that DLC parasites are likely to be relatively more successful invaders than are ILC parasites. Next we turn to some of the more complicated situations that can arise with parasites and two or more host species: it may be that an introduced alternative host is necessary for the invading parasite to persist and continue to have a serious effect on a native host population; or it may be that the invading parasite affects the outcome of competition among endemic or introduced host species; or it may even be that an invading species benefits by leaving its usual parasites behind. We conclude with a brief discussion of the possible trajectories and time scales for coevolution between hosts and parasites.

In all this, we define parasite broadly to embrace a range of pathogenic organisms from viruses, bacteria, protozoans, and fungi through to the more conventionally defined helminth and arthropod parasites. Such distinctions as we do make among these parasites are based on population dynamics rather than on taxonomy. We use the term *microparasite* in those cases where the host population may reasonably be regarded as made up of relatively few distinct classes (susceptible, infected but latent, infectious, recovered, and immune). Most viral, bacterial, and many protozoan infections are of this kind. In contrast, for *macroparasites* the pathogenic effects upon the host, the egg output per parasite, such immune responses as may be elicited in the host, and other factors all depend on the number of parasites harbored by the host individual in question; macroparasites are, moreover, almost invariably distributed among the host population in an aggregrated or clumped fashion, with relatively few of the hosts often harboring most of the parasites. Whereas host–*microparasite* associations can be modeled by studying the flows among the relatively small number of different classes of hosts, the modelling of host–*macroparasite* systems requires a full description of the distribution of parasites among hosts. For a more detailed discussion of these ideas, with emphasis on the ways in which such a dynamically based classification does and does not correlate with more familiar classification schemes based on biology and taxonomy, see Anderson and May (1979; May and Anderson 1979).

4.2. Invasions by Parasites with Direct Life Cycles

DLC parasites are communicated from one host to the next by contact, or via droplets, or—in the dynamically most complicated case—by free-living infective stages (which range from encapsulated viruses to the larval stages of many DLC helminth parasites).

For any invading organism it is useful to discriminate among three distinct phases of the invasion process: getting there; becoming established; and

spreading. For most plant and animal species, it is very hard to decide exactly what factors determine the minimum critical size for the population to become established and to maintain itself, much less to estimate the magnitude of this critical population. For parasites, however, establishment and maintenance can be related to a *threshold* magnitude of the host population in a way that is fairly clear in principle and sometimes (e.g., for many human host–parasite associations) measurable in practice. In this section we first indicate some examples of DLC parasite invasions, and then concentrate on how threshold host densities can affect parasite establishment and spread.

The classic example of a DLC parasite invader is rinderpest in Africa. Endemic in Asia, rinderpest first arrived in sub-Saharan Africa in 1889, where it was brought into Somaliland by cattle imported to feed the Italian Army. The infection spread rapidly among indigenous ruminants; south of the Zambesi the disease is estimated to have attained mortality levels of 90%, killing over 5 million buffalo, antelopes, and other ruminants in 2 years. When the epidemic eventually died out in the Serengeti, the annual survival probability of wildebeest doubled from 0.25 to 0.50 or higher. Indeed, it is argued that rinderpest is the main determinant of ruminant biogeography in East Africa over the past century (Spinage 1962; Sinclair and Norton-Griffiths 1979). For North America, the most notable example of a DLC parasite invasion is probably the fungus *Endothia parasitica*, which effectively removed the chestnut from eastern deciduous forest. Other examples are discussed by Elton (1958) and Soule and Wilcox (1980).

4.2.1. Threshold Host Densities for Parasite Establishment

For a DLC microparasite, the threshold density of hosts, N_T, is given in the simplest case by (Anderson and May 1979)

$$N_T = (\alpha + b + v)/\beta. \tag{1}$$

Here α is the disease-induced death rate, b is the per capita death rate from all other causes, v is the recovery rate, and β is a parameter measuring the intensity of transmission. If the host population N exceeds the threshold density, $N > N_T$, the parasite in effect has a basic reproductive rate greater than unity, and can maintain itself. Conversely, if $N < N_T$, the basic reproductive rate of the parasite is below unity; the host population is too small for the parasite to become established within it. The formula of Eq. (1) holds whether or not recovered hosts have acquired immunity, and is also independent of whether or not the parasite affects the reproductive ability of infected hosts. We note that if the microparasitic infection is highly virulent (large α), or if recovery is rapid (large v), then N_T will tend to be large: as discussed elsewhere, DLC microparasites are typically infections of organisms that are found in dense aggregations (social insects, colonial birds, herding ungulates, human populations only since the Agricultural Revolution some 10,000 years ago: May and Anderson 1979).

A variety of realistic refinements can alter Eq. (1) in particular situations. For example, if "vertical transmission" results in congenital infection in a fraction f of all births from infected females, it can be easier to establish and maintain the infection:

$$N_T = (\alpha + b + v - fa)/\beta. \tag{2}$$

Here a is the per capita birth rate. Other things being equal, N_T is lower than would be the case in the absence of vertical transmission ($f = O$); indeed, if $fa > \alpha + b + v$, the infection can persist in an indefinitely small population of hosts. On the other hand, a long latent period (in which hosts are infected but not infectious) results in N_T being higher than estimated by Eq. (1). For a full discussion, see Anderson and May (1979; summarized in Table 1 of that paper).

For DLC macroparasites, the expression for the host threshold density is usually messier than for microparasites (May and Anderson 1978, 1979). To a rough approximation, N_T for a typical macroparasite can be written

$$N_T \cong (\alpha + b + \mu)/\beta. \tag{3}$$

Here α is the contribution to the per capita host deaths per worm (assuming the overall host death rate rises roughly linearly with worm burden), b as before is the per capita death rate of hosts from all other causes, μ is the worm death rate, and β measures the transmission rate. The transmission parameter β can be a lot higher than for microparasites, by virtue of the free-living transmission stages that characterize many helminth and anthropod infections. Conversely, α will typically be a lot smaller than the disease-induced death rate α of virulent microparasites. Finally, μ is typically much smaller than the recovery rate v of Eq. (1); many helminths live for years in their hosts, whereas recovery times for most microparasitic infections are measured in days or at most weeks. In combination, these three factors result in N_T characteristically being significantly lower for macroparasites than for microparasites (Anderson and May 1979).

On these grounds, we expect macroparasites to be more common as invaders of low-density populations of hosts than are microparasites.

4.2.2. Disturbance and Invasion by DLC Microparasites

It has been shown that many kinds of environmental disturbance—ranging from toxic pollutants to "enrichment" with nitrogen or phosphorus, or with heat effluent from power stations or the like—tend to disrupt pristine patterns of relative abundance of species, producing new patterns that resemble those in the early stages of succession in that relatively few species are highly abundant. That is, man-made disturbance often results in "outbreaks," with one or two species rising to levels of abundance higher than any found in the undisturbed ecosystem (for a review, see May 1981, Chap. 9).

It has been observed that invasions by pathogens and other organisms often follow disturbances. Clearly, this is very commonly a correlation without causation: humans made the disturbance, and humans brought the invaders. In the case of many DLC microparasites, however, there may indeed be a causal chain. The factors noted in the preceding paragraph can be considered alongside the earlier observation that DLC microparasites often require high threshold densities of hosts to maintain themselves, to suggest that man-made disturbance actually creates the circumstances under which such microparasites can become established.

4.2.3. Diffusive Spread of Parasites

Once a parasite has been introduced and has established itself at the point of introduction, its spread can depend on many factors. As reviewed elsewhere in this volume by Roughgarden, it is often sensible to make the simple assumption of diffusive spread. In particular, there is a substantial mathematical literature dealing in a rather abstract way with the diffusion of DLC microparasites. Kallen et al. (1985) have recently shown how this analysis can be used to give practical estimates bearing on the spatial spread and possible control of rabies, following its introduction into fox populations.

Kallen et al. use a simple model for the diffusive spread of a DLC microparasite in a spatially distributed host population, focusing on the essentials and ignoring most of the fine details of fox behavior and family structure. They show that the velocity of spread, c, of the invading infection depends on the fox mortality rate caused by rabies, μ, on the basic reproductive rate of rabies infections within the fox population, R_o; and on a diffusion coefficient, D. The explicit relationship between the velocity of spread and these three parameters is

$$c = 2[D \mu (R_o - 1)]^{1/2}. \qquad (4)$$

The parameter μ can be fairly easily determined, and R_o can be estimated from serological data (or, more crudely, by knowing the fraction of the fox population infected at any one time, in the endemic state) by established techniques (Anderson et al. 1981). The diffusion coefficient has dimensions $(\text{length})^2/(\text{time})$, and is estimated by Kallen et al. as the ratio between the average area of one host (fox) territory and the average time that elapses before an infected fox leaves its territory. This estimate of D ignores the much larger distances typically travelled by juvenile foxes when they leave the parent group to seek their own territory, but this event represents a small fraction of the fox's life cycle, and it is correspondingly less likely that such a fox will actually be rabid; we estimate that this complication is likely to have relatively little effect on the effective value of D. More accurate estimates of D could ideally be obtained from field observations of infective foxes.

In Eq. (4), notice that both D and R_o will in general vary with the average density of fox populations. Thus a map of the probable rate of spread of rabies

in Britain subsequent to introduction at, say, Southampton will not simply be a pattern of concentric circles, but rather will have spurts and lags in regions where fox densities are known to be high and low, respectively.

The same analysis can be used to calculate how wide a "firebreak" (within which foxes are killed) must be, in order to halt the spread of infection. Kallen et al. show the critical width, L, is given by

$$L \cong \# [D/\mu R_o]^{1/2}. \tag{5}$$

Here $\#$ represents a numerical constant (characteristically around 5 to 10) that depends on the exact value of R_o and on assumptions about how large a fraction of all foxes are killed in the "firebreak": for $R_o = 2$ and 80% of foxes killed, $\# = 8$ (for further details, see Kallen et al. 1985).

Although developed in the specific context of rabies and foxes, this analysis can be applied to the spread of most DLC microparasites. The analysis gives rough but useful estimates of rates of spread and other quantities upon which control programs can be based.

For the fox-rabies case, Kallen et al. estimate that $\mu \sim 10$/year (corresponding to rabies killing in 30 to 40 days) and $R_0 \sim 2$ (Anderson et al. 1981). Fox territories range in size from 2.5 to 16 km^2; Kallen et al. use an average figure of 5 km^2, and assume infected foxes move on a typical time scale of 1 month, to arrive at $D \sim 60$ km^2/year. It follows that the velocity of spread is around $c \sim 50$ km/year, and the characteristic width for a firebreak (assuming an 80% kill within it) is $L \sim 14$ km. These estimates match observed values of c, and the fact that firebreaks 20 km wide halted rabies spread in Jutland. This case study illustrates techniques that are more generally applicable.

4.3. Invasions by Parasites with Indirect Life Cycles

ILC parasites must pass through one or more species of intermediate hosts, or vectors, to complete their transmission cycle from one primary host to the next. Successful invasion by such ILC parasites is obviously more complicated than for DLC parasites; the ILC parasite not only must find a suitable primary host (in sufficient abundance to exceed the transmission threshold), but also needs to be accompanied by its customary intermediate host species or to find a substitute such vector in its new environment.

Examples of invasion by ILC parasites frequently revolve around the population biology of an intermediate host. A notable example of this kind is presented by Warner (1968), who argues that the extinction of roughly half the endemic land birds of the Hawaiian Islands since their discovery by Europeans in 1778 is due mainly to introduced diseases. In particular, he argues that avian malaria, birdpox, and other unidentified diseases "swept through the lowland bird populations, causing a widespread disappearance of birds from even those forests where vegetation had not been disturbed by man," following the accidental introduction of the mosquito vector *Culex pipiens fatigans* in 1826.

This tropical subspecies of the night mosquito is restricted to elevations below 600 m; the extant Drepaniidae species are found only above 600 m elevation, despite the existence of apparently habitable regions at lower elevations. Warner observes that the temperate-zone subspecies of the night mosquito, *C. pipiens pipiens,* is not restricted to lowland elevations, and warns that its accidental introduction to the Hawaiian Islands could well allow avian malaria and birdpox to extinguish the remaining drepaniids.

The helminth parasite *Cyathocotyle bushiensis* provides another illustration where invasion by an ILC parasite depends on an intermediate host, the snail *Bithynia tentaculata.* This parasite causes substantial mortality in duck populations, and has been steadily moving up the St. Lawrence river in recent years, correlated directly with the introduction and spread of its intermediate snail host (Gibson et al. 1972). Elton (1958) discusses the elm—bark beetle *(Hylurgopinus rufipes)*—fungus *(Cerastomella ulmi)* system whose workings have altered the landscape of many parts of Britain and North America; as documented by Elton, this example has the twist that the vector appears to have invaded and spread in advance of the fungus. To come nearer home for an example, it is estimated that 200,000 or more inhabitants of New York City harbor schistosomes (brought with them from other places), but the infection is not transmitted because the snail vector is absent.

Bauer and Hoffman (1976) present an interesting overview of invasions by ILC helminth parasites (digeneans, acanthocephalans, nematodes, cestodes). Such parasites usually do not establish themselves if they are introduced on a primary host, because intermediate hosts are usually absent and the parasites are unable to use other species. Exceptions sometimes come from among the nematodes and cestodes, some of which tend to be less specific in their choice of an intermediate host. When introduced in conjunction with an intermediate host species, such as a snail or dipteran, these parasites can have devasting effects; they commonly are nonspecific with respect to primary hosts, and can thus attack a range of species that lack the resistance often found in regions where the parasite is long-established.

Expressions for the threshold magnitude of the primary host population needed to maintain an ILC parasite are essentially similar to those given in Section 4.2.1., except that now the transmission parameter β involves the population dynamics of the intermediate host. The resulting expressions can become quite complicated (Dietz 1975; Anderson and May 1978, 1979; for a simple overview, see May 1984a).

In particular, the relationship between the threshold density of the primary host, N_T, and the population density of the intermediate host, M, depends on the details of the life history of the vector. If the passage of infection from primary to intermediate host, and from intermediate host back to primary host, depends essentially on the frequency of encounters between such hosts, then we require

$$N > N_T \sim (\text{constant})/M. \tag{6}$$

That is, a high density of vectors reduces the density of primary hosts required for successful extablishment of the parasite. This is the case for most molluscan vectors. A less obvious relationship between N_T and M arises when the intermediate host transmits or acquires infection in the course of taking a bloodmeal from a primary host; such vectors usually have some roughly fixed biting rate, and the important thing now is the average number of vectors per primary host, rather than simply the absolute abundance of vectors. This circumstance, which arises for malaria and many other ILC infections transmitted by dipterans, leads to Eq. (6) being replaced by

$$N < N_T \sim (\text{constant})M.$$

That is, the host population needs to be *below* some threshold, which is roughly proportional to intermediate host density. Although at first sight surprising, this result is sensible enough: if a mosquito vector makes a fixed number of bites on humans per day, there may not be enough mosquitoes to ensure that infected humans are bitten sufficiently often to spread malaria if the human population is too large in relation to the number of available mosquitoes.

Likewise, the spread of ILC parasites is a more complicated story than for DLC ones. It will often be that one factor predominates, so that the spread depends only on the diffusive association between primary hosts and parasites (as in Section 4.2.3.), or only on the diffusive population dynamics of the intermediate vector (as in some of the examples discussed above). More generally, however, the complex spatial dynamics of the association among three or more populations will need to be considered.

4.4. Comparisons between Parasites with Direct and Indirect Life Cycles

The above discussion suggests that DLC parasites will in general find it easier to become established, following introduction to a new region, than will ILC parasites. The comparison in many ways parallels that between specialist and generalist pollination systems in the establishment of invading plants, as discussed by Orians elsewhere in this book; Darwin and Wallace were among the first to note the relative preponderance of generalist pollinators among plants on remote islands (see Carlquist 1974).

These ideas about the life cycles of successfully invading parasites may in principle be tested for any specified group, by calculating the number of invading species with DLCs and with ILCs that were successful (by some objective criterion), as a ratio to the total pool of possible invaders in the respective groups. This is easier said than done. The only systematic compilation of information that we have been able to find is for parasites of fish.

Table 4.1. summarizes the data compiled by Hoffman (1970) for 48 species of protozoan and helminth parasites of freshwater fish; these parasites have

Table 4.1. Direct versus indirect life cycles among successful invaders

Life Cycle	Parasite Group	A: Number of Successfully Invading Species	B: Total Number of Species in Britain	C: Total Number of Species in Canada
Direct	Protozoans	5	22	N/A
(DLC)	Monogeneans	31	35	166
	Subtotal	36	57	166
	Protozoans	0	24	N/A
	Digeneans	5	29	184
Indirect	Nematodes	3	18	58
(IDL)	Acanthocepyhalans	1	7	30
	Cestodes	0	24	102
	Subtotal	9	102	374
Indirect or Direct	Copepods	3	N/A	N/A
Total		48	159	540

This table shows number of species of protozoan and helminth parasites of freshwater fish, distinguishing DLC parasites from ILC parasites. Column A shows parasite species that have successfully invaded other continents through the transfer of infected fish (from Hoffmann 1970). Columns B and C serve to calibrate this information about invasions, by showing total numbers of parasites of freshwater fish in Britain and in Canada, respectively.
From Kennedy 1974 and Margolis and Arthur 1979.

successfully invaded new continents as a result of the transfer of infected live fish (or even, arguably, frozen fish). Of these 48 successful invaders, 36 are DLC and 9 are ILC (while 3 can be either).

Of itself, this preponderance of DLC over ILC in Table 4.1. does not confirm our expectation. We need also to know how many of the total of all unsuccessful such invaders were DLC, and how many ILC. We can find no information that directly addresses this need. We can, however, approach the problem indirectly by asking how many of the endemic parasites of freshwater fish are DLC versus ILC. Here information is available for Britain and for Canada: the numbers and kinds of species of parasites are different in these two regions, but in both regions there are roughly twice as many ILC parasite species as DLC species. Against this background, the predominance of DLC over ILC (in the rough proportions 4 to 1) among successful invaders in Table 4.1. appears significant; DLC parasite species do better, by about an order of magnitude, than we may expect from chance alone. We are aware of the crudities inherent in the above analysis, but the available information is such that it is easier to perceive the shortcomings than to remedy them.

4.5. Parasites and Two or More Hosts

Up to this point, we have dealt only with associations between a parasite species and a single species of primary host. We now explore some of the additional complications that can enter when two or more primary host species are in-

volved. First, we consider the circumstance where the adverse effects of an introduced parasite upon a native host species could not be sustained were it not for the presence of an alternative, introduced host species (upon which the parasite has relatively minor effects); some possible examples are suggested, but the discussion here is mainly theoretical. Second, we note how the introduction of a parasite may alter the competitive relationships between two species; we support this discussion by outlining some case studies and theory. Third, we tabulate evidence that two successful invaders of North America (starlings and sparrows) have left behind a significant number of their endemic parasites, and we speculate that such a measure of release from parasitism may on occasion facilitate invasion by the liberated hosts.

4.5.1. Invading Host Species Facilitates Extinction of Native Host Species

In the absence of an alternative host or some other mechanism for continual reintroduction, it is usually difficult for a parasite to drive its host population to extinction. This is because once the host population has been driven to sufficiently low levels by the depredations of the parasite it will be below threshold for parasite maintenance, and the infection is likely to die out before the host population is extinguished. A variety of effects can override this broad generalization in particular cases: demographic stochasticity (possibly compounded by behavioral factors) may extinguish the host population once it is carried to low levels; long-lived transmission stages of the parasite, or long-lived and infected intermediate hosts, or vertical transmission may result in the threshold host density being effectively zero. But for highly virulent microparasites that have DLC with short-lived transmission stages or ILC with short-lived intermediate hosts, extinction of the host population is not easy.

Possible illustrations of these ideas are frankly anecdotal. It seems to us that (following the introduction of the appropriate mosquito vector) the extinction of much of the lowland avifauna of the Hawaiian Islands by avian malaria and other infections probably required the annual migratory passage of infected shorebirds and ducks from the nearctic and palearctic mainlands, in order to sustain the final phases of the extinction process. Within our own species, it is likely that the dramatic reductions (by factors of 20 or more; May 1984b) in the native populations of the New World and Oceania following the introduction of microparasites such as measles, smallpox, tuberculosis, and the like required the continuing presence of the invading European hosts for continued maintenance of the infections.

Implicit here is the notion that the long-established associations between invading parasites and invading alternative hosts results in the parasites having less effect on such hosts than on the newly exposed native hosts.

These ideas may be given mathematical shape, as follows. Suppose the interaction between the parasite and the native population of hosts is characterized by the parameters of Section 4.2.: α is the disease-induced death rate (assumed high in this instance); a and b are the birth and "natural" death rates, respectively; v is the recovery rate; and β is the transmission parameter. Suppose

further that $\alpha > (a - b)(1 + v/b)$ (so that the infection is capable of regulating this host population in the absence of other density-dependent factors; Anderson and May 1979), but that other ecological factors keep this host population to a level too low for the parasite to be maintained ($N_T > K$, where N_T is the threshold host density of Eq. (1) and K the host population density as determined by resource limitations or predation or other factors). From Eq. (1) we see that this scenario requires

$$K < (\alpha + b + v)/\beta. \tag{8}$$

This is not an unreasonable assumption if we are dealing with a highly virulent pathogen (α very large).

Now introduce an alternative host species (labelled 2), within which the infection is maintained endemically (which is relatively easy if the parasite is comparatively avirulent for species 2; that is, if α_2 is comparatively small). Let the density of infected hosts of species 2 be Y_2. Then, sustained by the presence of alternative hosts of species 2, the parasite will be able to extinguish the native host population provided

$$\beta \, Y_2 > \frac{r \, [\alpha + b + v]}{\alpha - r \, [1 + (v/b)]}. \tag{9}$$

Here r is the per capita population growth rate of the native population in the absence of the parasite, $r = a - b$; we have already assumed that α is large enough to make the denominator on the right-hand side of Eq. (9) positive.

The result of Eq. (9) is new, and is derived in the Appendix. It is reminiscent of a result derived by Anderson and May (1981) for the rate at which a microparasite (typically a virus or micorspordian protozoan) must be released into the environment, in order to eradicate an insect pest (for which $v = 0$): the critical such rate is $A_c = r(\alpha + b)/(\alpha - r)$.

Some intuitive appreciation of the above results can be gained by considering their approximate form when the pathogen is highly virulent to the native host: $\alpha \gg r, b, v$. In this case, Eq. (8) reduces to

$$K < \alpha/\beta; \tag{10}$$

and Eq. (9) reduces to

$$Y_2 > (\alpha/\beta)(r/\alpha). \tag{11}$$

By assumption, $r/\alpha \ll 1$. Thus the number of invading, alternative hosts that are infectious (Y_2) can be very small relative to the size of the original population of native hosts (K), $Y_2 \ll K$, yet Y_2 can nonetheless be large enough to ensure extinction of the native hosts, $Y_2 > (r/\alpha)K$, provided only that r/α is very small (as we have assumed it will be).

4.5.2. Parasite Affects Competition between Species

Several recent field studies suggest that introduced parasites can affect the relative abundances of species in a community. Burrough and Kennedy (1979) show that, following the introduction of the helminth parasite *Ligula intestinalis* into Slapton Ley in southern England in 1974, the population density of roach has systematically declined and that of rudd has concomitantly increased; the observed prevalence and intensity of infection in the roach population increased over the span 1974–1979, and parasitic infection has been shown to depress growth rates in individual roach (*Ligula* does not appear to infect rudd). In Canada, the meningeal helminth, *Paralaphostrongylus tenuis,* kills moose and caribou but has little affect on white-tailed deer; moose and caribou are being excluded from an expanding area within which *P. tenuis* is transmitted from the invading deer (Embree 1979). Pickering and Gutierrez (1985) suggest that recent changes in the observed densities of two sympatric species of aphids may be explained in quantitative detail by the different mortality rates induced by a fungus, *Erynia neaphidis.* These are among the better-documented of many possible examples, some of which are discussed by Elton (1958).

Holt and Pickering (1985) have used models which combine elements of Lotka-Volterra models with elements of the kind outlined in the Appendix (see also Anderson and May 1979, 1981), to explore the effects that microparasites can have on competitive coexistence between host species. In these models, the host species have no direct interactions; the interactions enter indirectly because the dynamics of each host population can be affected by infection, and the level of infection depends in turn on the dynamics of both host populations. Holt and Pickering show that the outcomes of these indirect interactions between the two host species are very similar to those in conventional Lotka-Volterra models: "when an infectious disease is the only factor regulating population size, one host species can exclude another by means of a shared infectious disease. This is true even though both host species, in isolation from each other, can coexist with the infectious disease. The model analyzed here suggests that apparent competitive dominance can result if individuals of one species, as compared to individuals of the other species, have a higher per capita growth rate when uninfected, are less susceptible to becoming infected, or have a higher tolerance to disease. The higher tolerance to disease of individuals of one species may result from their faster recovery, lower death rates, or higher reproductive rates."

4.5.3. Invading Hosts Leaving Parasites Behind

It is plausible that some invading species may flourish because they are free from debilitating parasites that afflict them in their original habitats. Such liberation could derive from the invaders happening to be uninfected, or from the absence of intermediate hosts necessary for the maintenance of ILC parasites. Such a suggestion parallels the notion, which is subjected to critical examination

by Simberloff elsewhere in this book, that many introduced crop pests attain their pest status by virtue of the absence of their customary natural enemies. This is an appealing idea, but there is little evidence to support it.

For birds, there is evidence to show that the two most conspicuously successful species to have invaded North America—the house sparrow and the European starling—have fewer parasites in North America than in their homelands in Europe.

Since the first release of 60 individuals in Central Park in New York City in 1890 (and another 40 in 1891) the European starling, *Sturnus vulgaris,* has spread throughout North America. Table 4.2. is compiled from data given by Hair and Forrester (1970); the Table shows that the numbers of genera and of species of helminth parasites associated with starlings in Europe are both roughly three times the corresponding numbers in North America. Table 4.2. also indicates that roughly half the parasite genera and species associated with starlings in North America are found also in Europe, with the other half being parasite genera and species freshly acquired in the New World. There is no detailed understanding of the mechanisms whereby many parasite genera and species have been lost: in some cases the absence of intermediate hosts may be responsible, and other cases may be correlated with changes in diet (starlings in North America appear to eat fewer snails than in Europe). More importantly, no advantage has been demonstrated to accrue from this diminished list of parasites.

Table 4.3. is compiled from data presented by Brown and Wilson (1975), and it similarly shows the numbers of genera and of species of ectoparasites found on the house sparrow, *Passer domesticus,* in Europe and in North America. Here the numbers of genera and of species associated with house sparrows in Europe are almost twice the corresponding numbers found in North America (with about 1/3 of the genera and species found in North America also being found in Europe, and the remaining 2/3 being newly acquired parasite genera and species). Several of the European ectoparasites that are missing in North America are known to be vectors for blood parasites, which may explain the observed fact (Manwell 1957) that North American house sparrows have relatively few blood parasites. Since many of these blood parasites induce mortality

Table 4.2. Helminth parasites of European starlings in Europe and in North America

| Helminth Group | Number of Genera (Species) | | | |
	In Europe	In North America	Total Number	Number in Common
Trematoda	17 (26)	4 (4)	18 (28)	3 (2)
Cestoda	9 (12)	4 (5)	10 (14)	3 (3)
Nematoda	14 (26)	6 (10)	17 (30)	3 (6)
Acanthocephala	4 (6)	2 (3)	6 (9)	0 (0)
Total	44 (70)	16 (22)	51 (81)	9 (11)

From data in Hair JD., Forrester DJ. AM Mdld Natur 83:555–564, 1970.

Table 4.3. Ectoparasites of house sparrows in Europe and in North America

Ectoparasite Group	Number of Genera (Species)			
	In Europe	In North America	Total Number	Number in Common
Acarina	8 (35)	5 (24)	10 (43)	3 (16)
Mallophaga	8 (18)	4 (9)	9 (22)	3 (5)
Siphonoptera	1 (7)	1 (4)	1 (9)	1 (2)
Total	17 (60)	10 (37)	20 (74)	7 (23)

From data in Brown NS, Wilson GI. Am Mdld Natur 94:154–165, 1975.

or morbidity, house sparrows invading North America have probably benefitted from their lower burden of ectoparasite genera and species. There are, however, no quantitative studies of the magnitude of this benefit.

In brief, the idea that invading species may enjoy a comparative advantage because some of their usual parasite species have been left behind is an appealing one. But even in those few cases where such a reduction in associated parasite species has been documented for invaders, ensuing benefits such as higher birth rates or lower death rates have not yet been demonstrated.

4.6. Coevolution of Hosts and Parasites

As mentioned above, invading parasites are often much more harmful to newly acquired host species than they are to hosts with which a long-evolved relationship has likely been established. This tendency has, indeed, been invoked to support the generalization that "successful" or long-established parasites inflict little harm on their hosts. Levin et al. (1982), Levin and Pimentel (1981), May and Anderson (1983a,b), and others, however, have shown that such a generalization is not true, either in theory or in practice; the evolutionary endpoint of a particular host-parasite association depends on the detailed interplay among parasite virulence, transmissibility, and the costs of resistance to the host. In particular, the myxoma virus-Australian rabbit association provides an example where the changing virulence of an introduced virus has been studied in detail, and where the system appears to have settled fairly quickly to a state in which the preponderant grade of virus found in the field is one of intermediate virulence. For a full discussion of these points, see the above references.

A subclass of the above questions concerns the time taken for a native host population to evolve a significant degree of resistance to an introduced pathogen or parasite. This is a topic that has understandably received more attention for insect pests evolving resistance to introduced pesticides than for hosts and parasites. As summarized in a recent review of the evolution of pesticide resistance (May and Dobson 1986), the characteristic time taken for resistance to appear, T_R, in the simplest case is

$$T_R \cong T_g \, [\ln \, (p_f/p_o)/\ln \, (w_{RS}/w_{SS})]. \tag{12}$$

Here T_g is the cohort generation time of the host or pest species; p_p is the frequency of the resistance gene in the pristine population; p_f is the resistance gene frequency when resistance is first recognized; and w_{SS} and w_{RS} are the fitnesses of susceptible (SS) and heterozygous resistant (RS) genotypes, respectively. The essential feature of Eq. (12) is that T_R depends directly on T_g, but only logarithmically on such factors as the initial frequency and degree of dominance of the resistance gene, and the strength of selection; this probably explains why resistance to pesticides (and antibotics) typically appears in 5 to 50 pest (or pathogen) generations, despite enormous variation (ranging over 10 or more orders of magnitude) in p_o and w_{RS}/w_{SS}. These points are elaborated and documented in May and Dobson (1986, and references therein).

The basic message from Eq. (12) bears repeating: the characteristic time for resistance to appear depends linearly on the generation time, and only very weakly on other factors. This, of course, is the main reason why resistance takes so much longer, when measured in absolute time, to appear in trees and mammals than in weeds and insects. Elton (1958) recounts the story of "a fungus disease of asparagus in Europe, where it does not develop epidemics, [that] got carried to the United States where it swept over the entire country and virtually destroyed the entire asparagus industry. Gradually, however, the rust has become less important until now asparagus growing has become rehabilitated and the disease is of minor importance. This happy result seems to have taken place by genetic changes in the populations of fungus or asparagus or in both, not by chemical sprays." No such spontaneous solution to the fungal afflictions of elm or chestnut have appeared in North America, partly because the generation times of these trees are so much longer than that of asparagus.

4.7. Conclusions

Table 4.4. attempts to summarize some of the scattered comments we have made about invasions by microparasites and macroparasites with DLCs and ILCs.

In particular, we would emphasize the expectation that—other things being equal—DLC parasites are likely to be more successful invaders than ILC parasites. Table 4.1. represents an unsatisfactory attempt to demonstrate this for protozoan and helminth parasites of freshwater fish. There is much room for further work here.

We also emphasize that man-made disturbances often produce changes in the relative abundance of species, whereby one or two species become unusually abundant; such high population densities may then be above the threshold required to maintain a DLC microparasite. In this way human disturbances can not only introduce parasites, but at the same time create the circumstances that permit them to become established and to spread.

In Section 4.5. we aired speculations about ways in which the presence or absence of parasites could shift the balance of interaction among species, often in subtle ways. Tables 4.2. and 4.3., for example, showed that starlings and sparrows in North America are relatively free from parasites, but offered no

Table 4.4. Some rough generalizations about invading host and parasite species

Life Cycle of Parasite	Invading Host Species	Invading Parasite Species
Direct	New parasite species may be picked up, because they are often not very specific. Conversely, parasites may invade with invading hosts and spread to native species.	*Microparasites:* May spread rapidly to new hosts (particularly among the high density of hosts often found —for a few species— following disturbance). Many of the most spectacular and devasting parasitic invasions are in this class. *Macroparasites:* Able to spread to new hosts if sufficiently nonspecific (as is the case for some nematodes).
Indirect	Often existing parasites cannot persist, and few new parasites are acquired, because specific intermediate vectors are absent (e.g., few blood parasites in house sparrows in North America).	Invading species cannot maintain themselves in the absence of appropriate intermediate hosts (e.g., ILC versus DLC parasites of freshwater fish).
Vectors	Absence of vector often reduces parasite burdens in invading hosts. Conversely, introduction of vectors can have dramatic effects (e.g., avian malaria in Hawaii; *Cyathocotyle* in ducks along the St. Lawrence).	

evidence at all that such freedom was responsible for the great success of these avian invaders. It would be very nice to have more complete information about any one such system.

4.8. Acknowledgments

We are indebted to R. M. Anderson, D. Simberloff, D. Wilcove, and others for helpful discussions. This work was supported in part by NSF grant BSR83-03772 (R.M.M.) and by a NATO Postdoctoral Fellowship (A.P.D.).

4.9. References

Anderson RM, Jackson H, May RM, Smith T (1981) The population dynamics of fox rabies in Europe. Nature 289:765–771

Anderson RM, May RM (1978) Regulation and stability of host-parasite population interactions: I, regulatory processes. J Anim Ecol 47:219–247

Anderson RM, May RM (1979) Population biology of infectious diseases: Part I. Nature 280:361–367

Anderson RM, May RM (1981) The population dynamics of microparasites and their invertebrate hosts. Philos Trans R Soc B 291:451–524

Bauer ON, Hoffman GL (1976) Helminth range extension by translocation of fish. In: Wildlife Diseases (Proceedings of the 3rd International Wildlife Disease Conference). Plenum Press, New York, pp 163–172

Brown NS, Wilson GI (1975) A comparison of the ectoparasites of the House Sparrow (*Passer domesticus*) from North America and Europe. Am Mdld Natur 94:154–165

Burrough RJ, Kennedy CR (1979) The occurrence and natural alleviation of stunting in a population of roach, *Rutilus rutilus* (L.) J Fish Biol 15:93–109

Carlquist S (1974) Island Biology. Columbia University Press, New York

Dietz K (1975) Transmission and control of arbovirus diseases. In: Ludwig D, Cooke KL (eds) Epidemiology. Society for Industrial and Applied Mathematics, Philadelphia, pp 104–121

Elton CS (1958) The Ecology of Invasions by Animals and Plants. Methuen and Company, London

Embree DG (1979) The ecology of colonizing species, with special reference on animal behavior invaders. In: Horn DJ, Stairs GR, Mitchell RD (eds), Analysis of Ecological Systems. Ohio University Press, Columbus, pp 51–65

Gibson GG, Broughton E, Choquette LPE (1972) Waterfowl mortality caused by *Cyathocotyle bushiensis* Khan 1962 (Trematoda: Cythocotylidae) St. Lawrence River, Quebec. Can J Zool 50:1351–1356

Hair JD, Forrester DJ (1970) The helminth parasites of the starling (Starnus vulgaris L.): a checklist and analysis. Am Mdld Natur 83:555–564

Hoffman GL (1970) Intercontinental and transcontinental dissemination and transfaunation of fish parasites with emphasis on whirling disease *(Myxosoma cerebrals)*. In: Snieszko SF (ed), A Symposium on Diseases of Fishes and Shellfishes. American Fisheries Society, Washington, DC, pp 69–81

Holt RD, Pickering J (1985) Infectious disease and species coexistence: a model of Lotka-Volterra form. Am Natur 126:196–211

Kallen A, Arcuri P, Murray JD (1985) A simple model for the spatial spread of rabies. J Theor Biol 116:377–394

Kennedy CR (1974) A checklist of British and Irish freshwater fish parasites with notes on their distribution. J Fish Biol 6:613–644

Levin BR, Allison AC, Bremermann HJ, Cavalli-Sforze LL, Clarke BC, Frentzel-Beyme R, Hamilton WD, Levin SA, May RM, Thieme HR (1982) Evolution of parasites and hosts (group report). In: Anderson RM, May RM (eds), Population Biology of Infectious Diseases. Springer-Verlag, New York, pp 212–243

Levin SA, Pimentel D (1981) Selection of intermediate rates of increase in parasite-host systems. Am Natur 117:308–315

Manwell RD (1957) Blood parasitism in the English sparrow with certain biological implications. J Parasitol 43:428–433

Margolis L, Arthur JR (1979) Synopsis of the parasites of the fishes of Canada. Bulletin of the Fisheries Research Board of Canada, Ottawa, 199:269

May RM (ed) 1981 Theoretical Ecology: Principles and Applications, 2nd edit. Blackwell, Oxford and Sinauer, Sunderland, Massachusetts

May RM (1984a) Ecology and population biology of parasites. In: Warren KS, Mahmoud AF (eds), Tropical and Geographical Medicine. McGraw-Hill, New York, pp 152–166

May RM (1984b) Prehistory of Amazonian Indians. Nature 312:19–20

May RM, Anderson RM (1978) Regulation and stability of host-parasite population interactions: II, destabilizing processes. J Anim Ecol 47:249–267

May RM, Anderson RM (1979) Population biology of infectious diseases: II. Nature 280:455–461

May RM, Anderson RM (1983a) Coevolution of parasites and hosts. In: Futuyma DJ, Slatkin M (eds), Coevolution. Sinauer, Sunderland, Massachusetts, pp 186–206

May RM, Anderson RM (1983b) Epidemiology and genetics in the coevolution of parasites and hosts. Proc R Soc Land B 219:281–313

May RM, Dobson AP (1986) Population dynamics and the rate of evolution of pesticide resistance. In: Pesticide Resistance Management. NAS-NRC Publications

Pickering J, Gutierrez AP (1985) Impact of a fungus on community composition of *Acyrthosiphon* aphids

Sinclair ARE, Norton-Griffiths M (1979) Serengeti: Dynamics of an Ecosystem. The University of Chicago Press, Chicago

Soule ME, Wilcox BA (eds) (1980) Conservation Biology: An Evolutionary-Ecological Perspective. Blackwell, Oxford and Sinauer, Sunderland, Massachusetts

Spinage CA (1962) Rinderpest and faunal distribution patterns. African Wildlife 16:55–60

Warner RE (1968) The role of introduced diseases in the extinction of the endemic Hawaiian avifauna. Condor 70:101–120

4.10. Appendix

In this Appendix we derive Eq. (8).

Following lines laid down in detail elsewhere (see, e.g., Anderson and May 1979, 1981), we consider a DLC microparasite and let the number of susceptible, infected, and recovered and immune hosts of species 1 be denoted by X_1, Y_1, and Z_1, respectively. The total population of native hosts is then $N_1 = X_1 + Y_1 + Z_1$. We further let Y_2 denote the number of infectious hosts of species 2. The parameters α, b, v, β, and $r = a - b$ all refer to species 1 and have their usual meanings (as set out in Sections 4.2. and 4.5., or in Anderson and May 1979, 1981). Then, ignoring other possible density dependent limitations, the dynamical behavior of species 1 is given by the usual set of nonlinear, first-order differential equations:

$$dX_1/dt = aN_1 - bX_1 - \beta X_1 (Y_1 + Y_2) \qquad (A1)$$

$$dY_1/dt = \beta X_1 (Y_1 + Y_2) - (\alpha + b + v) Y_1 \qquad (A2)$$

$$dN_1/dt = rN_1 - \alpha Y_1 \qquad (A3)$$

The dynamics of Z_1 follow from the constraining relation among X_1, Y_1, Z_1 and N_1.

If we ignore the presence of species 2 ($Y_2 = 0$), but include other density-dependent limitations that lead to $N_1 = K$ in the absence of infection ($Y_1 = Z_1 = 0$), the relation $K > (\alpha + b + v)/\beta$ is elsewhere obtained as the criterion for the infection to maintain itself (Anderson and May 1979, 1981). Conversely, Eq. (8) is the criterion for the microparasitic infection to be unable to persist within this host population.

If species 2 is indeed present, under what circumstances will the parasite drive species 1 to extinction? This question may be answered by considering Eqs. (A1) to (A3) in the limit when N_1 (and consequently X_1, Y_1, Z_1) is very small. In this limit we may treat Y_2 as constant (species 1 will be at densities

too low to affect the level of infection within species 2), and also we may neglect second-order terms such as $X_1 Y_1$, to reduce Eqs. (A1) to (A3) to the set of linear equations:

$$dX_1 (t)/dt = aN_1 (t) - [b + \beta Y_2] X_1 (t) \tag{A4}$$

$$dY_1 (t)/dt = \beta Y_2 X_1 (t) - [\alpha + b + v] Y_1 (t) \tag{A5}$$

$$dN_1 (t)/dt = rN_1 (t) - \alpha Y_1 (t) \tag{A6}$$

In these linear equations, the time dependence of $X_1 (t)$, $Y_1 (t)$, $N_1 (t)$ may be factored out as $\exp(\lambda)$ in the usual way. Thus when species 1 is at low population levels its dynamical behavior is characterized by the eigenvalues λ, and the species will eventually become extinct if all these eigenvalues λ have negative real parts (corresponding to ever decreasing values of N_1). These eigenvalues are determined by the requirement that

$$\det \begin{vmatrix} -(b + \vartheta) - \lambda & 0 & a \\ \vartheta & -(\alpha + b + v) - \lambda & 0 \\ 0 & -\alpha & r - \lambda \end{vmatrix} = 0. \tag{A7}$$

Here ϑ has been defined as $\vartheta = \beta Y_2$. This leads to a cubic equation for λ:

$$\lambda^3 + A\lambda^2 + B\lambda + C = 0. \tag{A8}$$

The coefficients have the values:

$$A = \vartheta + \alpha + v + 2b - r \tag{A9}$$

$$B = (\alpha + b + v)(\vartheta + b - r) - r(\vartheta + b) \tag{A10}$$

$$C = a\alpha \vartheta - r(\vartheta + b)(\alpha + b + v) \tag{A11}$$

All three roots will have negative real parts if, and only if, $A > 0$, $B > 0$, $C > 0$, and $AB > C$. The requirement that $C > 0$ leads directly, via Eq. (A11), to Eq. (9) in the main text. If Eq. (9) is indeed satisfied, then it is a routine (though algebraically messy) exercise to show that it implies $A > 0$, $B > 0$, and $AB > C$. Details of the proofs are available upon request.

2. Attributes of Invaders

5. Which Animal Will Invade?

P.R. Ehrlich

5.1. Introduction

One of the most persistent puzzles for ecologists and population geneticists is why some animals are extremely successful invaders (colonists), while close relatives are not. Thus the European cabbage butterfly *(Pieris rapae)* has, in the last century, invaded Bermuda, North America, Australia, New Zealand, Hawaii, and other Pacific islands, while other European *Pieris* species such as *P. mannii* and *P. ergane* have not yet crossed the Atlantic, and the large white, *P. brassicae,* has only recently established a toehold in South America (Feltwell 1978). Furthermore, *Pieris* species established in North America for many millennia, such as *P. napi* and members of the *P. protodice* complex, have not yet managed to establish themselves in Hawaii.

Only one of six species of serranid fishes (groupers, sea trout, and their relatives) introduced purposely to Hawaiian waters (where the native serranid fauna is depauperate) became successfully established. Similarly, only two of four introductions of snappers (Lutjanidae) to Hawaii were successful (Randall and Kanayama 1972; Maciolek 1984), even though relatively few native lutjanids are present.

The house sparrow *(Passer domesticus)* occupied the entire United States, with the help of additional releases by *Homo sapiens,* in a little over 50 years after it was first successfully introduced. Other introductions have led to extensive ranges in South America, southern Africa, Australia, and New Zealand.

The closely related tree sparrow *(Passer montanus)* has a range in Eurasia almost as large as that of the house sparrow. It was successfully introduced into North America at St. Louis in 1870, but until 1960 it was still largely confined to the St Louis area. Since then the tree sparrow has spread a little further into central Illinois, but it has not colonized with anything like the vigor of its close relative. Indeed, in some places it appears to have suffered some competitive displacement by *P. domesticus.* The tree sparrow was introduced into Australia at about the same time as the house sparrow, but also has a much more restricted range there as well. It did not successfully colonize New Zealand.

Two extraordinarily successful invaders are the black or roof rat, *Rattus (Rattus) rattus,* and the Norway or brown rat, *R.R. norvegicus.* Both species are found throughout the world generally in association with human beings. But some 43 other species of the subgenus *Rattus,* a few of which are commensals with *Homo sapiens* in limited areas, have not become ubiquitous.

Gray and red wolves *(Canis lupus* and the closely related *C. rufus)* have undergone a dramatic decline in North America as a result of depredation by human beings (Nowak and Paradiso 1983). With the exception of a gray wolf population in Minnesota and scattered reds in the coastal swamps of Louisiana and Texas, wolves are essentially gone from the coterminous United States. Their congener, the coyote *(Canis latrans),* with which they hybridize easily even though the species have had separate evolutionary histories since the late Pliocene (Nowak 1978), has proven to be a successful invader. The range of the coyote expanded dramatically while those of its relatives were shrinking.

As a final example, *Homo sapiens* has been the most successful invader of all, while its closest living relatives, *Pan troglodytes* and *Gorilla gorilla,* have been pushed almost to the brink of extinction.

5.2. What Is an Invader?

What are the biological attributes of the species just discussed that make some much better invaders than others? Are there any general rules that would permit an ecologist to predict which of an array of exotic species might be a potential invader (or perhaps more importantly, an economically damaging invader)? Before considering these questions, let me say what I mean by an "invading" or "colonizing" species. I use the two terms interchangeably for a species that easily crosses barriers (with or without the help of human beings) and rapidly establishes itself and then expands its range and numbers relatively rapidly in new habitats on the other side. Most of the really successful animal invaders are ones that, for a variety of reasons, are able to cross major barriers because of their relationship with *Homo sapiens* (Elton 1958). The principal exceptions are beasts such as tardigrades that are passively wind dispersed and have become ubiquitous. No larger animals, however, approached ubiquity until humanity did.

The examples of successful-unsuccessful pairs that I have given are, clearly, dissimilar. At one extreme there are organisms such as the cabbage butterfies and rats that invaded without any *deliberate* aid from humanity. In the middle

are the sparrows, which were purposely introduced, but without systematic thought as to consequences (many birds have been introduced around the world by colonists from Western nations who wanted familiar attractive animals around them). At the other extreme of the examples are the reef fishes, which were transplanted by biologists with considerable knowledge of the organisms to be moved and the communities into which they were being introduced.

It seems obvious that the characteristics of successful invaders in these three groups might be very different. An additional category of invaders in which I have done no paired comparisons are organisms, especially insects, introduced as biological control agents. It may, indeed, be stretching the definition of "invader" to include them. After all, they are introduced specifically because it is believed, sometimes after very careful investigation, that an especially suitable environment—an empty niche—is awaiting. Furthermore, there are statistical reasons why comparisons of successes and failures in this category are difficult. Often introductions of different potential control agents are tried until some level of success is achieved, and then the same organism is introduced repeatedly. And when more than one agent actually become established, but one is primarily responsible for the reduction of the host, the latter is counted as the "success," even though the others might become very widespread.

How one should deal with such issues is beyond the scope of this paper. So is the problem of the partitioning of the invasion process. Ability to cross barriers unaided is different from attractiveness for introduction. Both are in turn different from the ability to establish successfully on the other side. And finally, establishment does not necessarily precede expansion; different attributes are often involved. But all of these capabilities clearly will enter into decisions about which organisms are good invaders and which are not—and will eventually have to be addressed.

5.3. Why One and Not Another?

Starting with first principles, what sort of ecological, genetic, and physiological attributes might one expect *a priori* to characterize successful and unsuccessful invaders? Table 5.1 indicates some possible distinctions. I will return to these after considering specific cases.

Table 5.1. Possible concomitants of invasion potential

Successful Invaders	Unsuccessful Invaders
Abundant in original range	Rare in original range
Polyphagous	Monophagous or oligophagous
Short generation times	Long generation times
Much genetic variability	Little genetic variability
Fertilized female able to colonize alone	Fertilized female not able to colonize alone
Larger than most relatives	Smaller than most relatives
Associated with *H. sapiens*	Not associated with *H. sapiens*
Able to function in a wide range of physical conditions	Able to function only in a narrow range of physical conditions

5.3.1. Pierid Butterflies

The large white butterfly, *Pieris brassicae,* is a less successful colonist than *P. rapae*, even though like the former it is a migrant (Williams 1936a,b; Feltwell 1977; Baker 1978) and has an extremely broad diet, feeding on a wide variety of crucifers as well as Capparidaceae, Fabaceae, Resedaceae, Tropaeolaceae, and, secondarily, species in seven other families (Feltwell 1982). One possible reason is that *P. rapae* is less susceptible to parasites and pesticides (Feltwell 1982). *P. rapae* lays its eggs singly, and thus they may be less easily discovered by parasitic Hymenoptera. Its larvae are solitary and feed mostly on internal leaves of the host plant; those of *P. brassicae* feed gregariously on the outer leaves where they are more vulnerable to both poisons and parasites.

On the other hand, however, *P. brassicae* is able to maintain itself over a broad geographical range that includes much of Eurasia, North Africa, Madeira, the Canary Islands, and Chile, so it seems unlikely to me that the differences mentioned above account for its absence where *P. rapae* has penetrated. The success of the latter had often been credited to its ability to outcompete indigenous pierines, but recent evidence suggests that this is not a crucial factor in its success (Chew 1981).

It is not clear why, of the North American endemics, at least *P. protodice* has not reached Hawaii. It is occasionally a pest, but not as commonly so as *P. rapae*. Perhaps its chances of being accidentally transported in produce (which presumably is how *P. rapae* got there at the end of the last century—Zimmerman 1958) have just not been as high.

5.3.2. Reef Fishes

It is not clear why groupers of the genera *Cephalopholis* and *Epinephelus,* and snappers of the genus *Lutjanus,* were originally absent from the Hawaiian Islands. Other groups of reef fishes are well represented there, all presumably having arrived as drifting larvae (long-distance dispersal is not known to occur in post-larval stages). It seems likely that the great isolation of the islands, perhaps combined with a relatively short larval life span (Randall and Kanayama 1972), was responsible for the absent genera.

That, however, does not explain why the serranid *Cephalopholis guttatus* (=*argus*) became established after the introduction of 2285 individuals whereas, for example, *C. urodelus* did not "take" after the release of more than 1800 individuals at the site. Similarly, it does not explain why *Lutjanus kasmira* became established and expanded to commercial significance (Oda and Parrish 1981) after more than 3100 individuals were introduced, while *L. guttatus* did not after 3400 were liberated (Maciolek 1984). Both of the *Cephalopholis* species are extremely widely distributed, occurring as far from the Society and Marquesas Islands (the home of the source populations) as Africa. *L. kasmira* (from the Marquesas) is also widespread, reaching South Africa, and *L. guttatus* occurs from the Gulf of California (the source site) to Ecuador. It thus would appear that all of these species are able to disperse widely and adapt to a variety

of habitats (including a variety of competitors), and there is no obvious intrinsic reason why one of each pair should succeed and the other fail. All occur in locations where the sea temperatures are both warmer and cooler than in the Hawaiian islands. Possibly, as Randall and Kanayama (1972) suggest, some factor influencing the pelagic larval stage of the life history explains the differential success.

5.3.3. Sparrows

The house sparrow as a colonizer and companion of humanity has been the subject of intensive study (e.g., Summers-Smith 1963; Johnston and Selander 1964; Johnston and Klitz 1977). It is thought to have become a commensal of human beings shortly after sedentary agricultural communities raising wheat and barley were first established. This notion is based on the fossil record of *Passer domesticus* (Tchernov 1962; Bar-Yosef and Tchernov 1966) indicating that the stock from which house sparrows evolved was present in the appropriate places in the Near East, and from observations of the evolutionary adaptability of modern house sparrows. For example, in little over a century North American populations have differentiated from one another considerably, following adaptive trends such as that exemplified by Bergmann's rule (Calhoun 1947; Johnston and Selander 1971, 1973).

Their relationship with human beings accounts for both the spread of the sparrows (through deliberate introductions and "hitch-hiking" on ships—Barrows 1889) and for their success in a wide variety of habitats. The birds presumably were originally migratory (Johnston and Klitz 1977), but have lost that behavior and overwinter in the vicinity of *Homo sapiens,* living around garbage dumps, barns, grain elevators, and other places where they can obtain food indirectly from humanity. They were even more common when horse droppings provided an abundant source of partially digested vegetable matter.

The first introduction of *Passer domesticus* involved the release of eight pairs in Brooklyn in 1850, and it failed. The introduction of about 100 birds in the same general area succeeded a few years later (Long 1981). Its success, however, does not explain the relative lack of colonizing success of other species of *Passer,* such as *P. montanus,* which presumably shared a common ancestor with *P. domesticus* that was preadapted to human commensalism (Johnston and Klitz 1977), and whose North American populations trace to some dozen pairs.

It is possible, of course, that the smaller number of tree sparrows represented a genetically depauperate sample. But no information is available on such crucial things as the comparative sizes of the bottlenecks that each population probably went through soon after introduction. It is also possible that pure luck—such as more parasites being introduced with the *P. montanus* than with *P. domesticus*—controlled which species was successful. But the pattern of greater success of the latter in both Australia and New Zealand argues against either of these hypothesis.

5.3.4. Rats

One reason the two ubiquitous species of rats are so successful as colonizers may be their highly developed sense of taste, combined with a cautious ability to add virtually anything to their diets. Rats, confronted with a novel food, will sniff it very thoroughly. They then will eat a very small amount and wait for a full day before trying it again. If they develop any distressing symptoms during that period, they will shun that food in the future. If no symptoms follow, they will sample more the next time (Roper 1984). This behavior makes the rats difficult to poison with substances causing acute toxicity and has led to a dependence on anticoagulant rodenticides such as warfarin. The appearance of warfarin-resistant strains of *R. norvegicus* (e.g., Greaves and Rennison 1973) suggests that that invader may once again increase.

All rats, however, are omnivores (Nowak and Paradiso 1983), and it is not clear why other species have not spread to the extent of *R. rattus* and *R. norvegicus*. These two are, however, relatively large (Lim 1970; Lekagul and McNeely 1977) and fecund (Tamarin and Malecha 1972; Wirtz 1972, 1973), and (perhaps more important) they may have arrived in most places first and exclude other *Rattus* from the human commensal niche. *Mus musculus,* which shares that niche, is only about a third of the weight of the smallest *Rattus* and I would assume coexists in many situations because it can subsist on more limited resources than the rats, and occupy refuges and use pathways too small for the latter.

R. rattus prefers higher, more open habitats, and presumably was originally arboreal in its Indian homeland. The ancestors of *R. norvegicus* were presumably fossorial dwellers in the steppes of central Asia. The latter often displaces the former from basements, sewers, and so on. These species, especially *R. rattus,* like the house sparrow, have differentiated locally in many parts of their range (Ford 1975; Nowak and Paradiso 1983).

5.3.5. Canids

The reason coyotes are better invaders than wolves is easier to explain. It seems to be closely related to their diets and their social behavior. The wolf is almost exclusively a predator of large ungulates (Mech 1970), whereas the coyote is an opportunistic scavenger and predator (Hilton 1978). Coyote populations appear to be limited primarily by environmental factors, especially food supply, rather than by predator control programs (Connolly 1978). In short, humanity is increasing the food supply of coyotes while generally failing in its attempts to control them. It has reduced the food supply for wolves, while their larger size and social nature, including greater dependence on hunting in packs (Kleiman and Brady 1978), made them easier targets for extirpation. Wolf packs and dens are relatively easy to find, and lone wolves would generally have more difficulty surviving than lone coyotes. Interestingly the other large *Canis* that is a successful "invader," *Canis familiaris* (including *C.f. dingo)* is not highly social (Nowak and Paradiso 1983).

The coyote has been divided into 19 subspecies (Hall 1984), a very crude indication that it has plenty of genetic variability. But there is no sign that wolves are any less diverse. Twenty-four subspecies of *Canis lupus* are recognized in North America, and three of *Canis rufus,* which is allopatric with *C. lupus* (Hall 1984) and interfertile with it.

5.3.6. Hominoids

Homo sapiens is, of course, the ultimate invader. Like other animal "weeds," it tends to have a very broad diet. It also is able to use cultural evolution to adapt to very diverse habitats, an ability that complements the loss of estrus which permits it to breed all year round. These attributes give it a very distinct advantage over its less aggressive and less destructive relatives, for whose approaching demise it is exclusively responsible. *Homo sapiens* is, of course, extremely genetically variable.

5.4. Noninvaders: Some Lessons from Checkerspot Butterflies

Checkerspot butterflies of the genus *Euphydryas* (Nymphalidae: Nymphalinae) seem typical of animals that are not thought of as invaders. They tend to occur in restricted colonies and not to disperse great distances (Ehrlich 1961, 1965; Ehrlich et al. 1975), although patterns of movement vary among species (Brown and Ehrlich 1980), among populations of the same species (Gilbert and Singer 1973), and among years in individual populations (White and Levin 1981; Murphy and White 1984). *Euphydryas editha* is the least vagile checkerspot butterfly species in the western United States, although it disperses readily, especially if conditions in its habitat are suboptimal (Ehrlich et al. 1984). It occurs as isolated populations grouped into an array of ecotypes that specialize on different (but usually closely related) larval food plants, and often show other ecological, behavioral, and morphological differences (Ehrlich et al. unpublished).

Some populations feed on native species of the genus *Plantago,* whereas a few colonies have become established on a ubiquitous introduced weed from Europe, *Plantago lanceolata* (Ehrlich et al. 1975; Ehrlich et al. unpublished). But in general *E. editha* has not been able to take advantage of the presence of the abundant introduced *Plantago* to spread significantly beyond its original range or into previously uninhabited interstitial areas. The reasons for this are not clear, particularly because all ecotypes tolerate *P. lanceolata* in the laboratory and rather disparate ecotypes use it in the field. The reasons may well be related to the sedentary behavior of *E. editha,* combined with a tendency to avoid, or at least to delay (Singer 1983), oviposition on novel food plants.

Additionally its habits of laying eggs in clusters of several hundred and of the prediapause larvae living communally in a web may also be factors. A dispersing, fecundated female thus quite literally places all of her eggs in a single basket, or, at most, in a very few baskets. Several hundred scattered eggs in an environment relatively free of competitors and predators and rich in resources

might have a high chance of yielding a few adult individuals of each sex. In contrast, when a *Euphydryas* egg mass is discovered by predatory mites or the hatching larvae are attacked by mirid bugs, the result most often is the loss of the entire clutch (Ehrlich and C. Holdren unpublished). This species also is prone to stochastic extinction and even suitable sites in the immediate vicinity of those still maintaining populations tend to be recolonized very slowly (Ehrlich et al. 1980; Murphy et al. 1983).

The probability of a butterfly species or any similar insect colonizing a new area can be viewed as a product of separate probabilities (Ehrlich 1958). They include the probabilities of departure of a fertilized female; of her surviving in transit, finding a suitable habitat, and successfully ovipositing; of at least one individual of each sex surviving and maturing simultaneously, of them finding each other and mating; and of the cycle of probabilities persisting for several generations until a thriving daughter colony exists. In *E. editha,* enough of these probabilities are small to make their product extremely small, and the species is not a "good invader."

This view is reinforced by studies of another sedentary *Euphydryas* in western North America, *E. gillettii* (Williams 1986; Williams et al. 1984). This species occurs naturally in the northern Rocky Mountains, north of the Wyoming basin gap. Its larvae feed on *Lonicera involucrata,* a caprifoliaceous shrub that also grows abundantly south of the gap in the Colorado Rocky Mountains. A transplant experiment was done (Holdren and Ehrlich 1981) to determine whether *E. gillettii* had not invaded Colorado because it could not successfully cross a barrier of some 160 km from which its food plant was absent, or whether it was unable to survive in Colorado in spite of the presence of its larval food plant and suitable adult nectar resource.

The equivalent of the complete reproductive output of 10 to 20 females was introduced to one transplant site, and the output of 8 to 16 females plus 17 inseminated individuals were introduced to another. The second introduction failed essentially immediately—a small first generation, perhaps only a single pair, flew the next year, and the transplant colony was extinct the second year following the introduction. The first transplant colony, started in 1977, decreased to perhaps 16 individuals in 1978 and may have declined to as few as four (two females) in 1979. The colony rebounded to perhaps two dozen in 1980, and in subsequent years through 1985 rose to the vicinity of 100 (Holdren and Ehrlich 1981; Ehrlich unpublished).

Both the successful and unsuccessful colonies had, as far as could be told, abundant resources, and the site of the *unsuccessful* introduction was climatically superior. And although suitable habitat extends in all directions from the site of the successful introduction, in eight generations the colony has spread only slightly from the precise area of less than a hectare in which the original eggs and larvae were placed on food plants—the most distant egg mass having been found a few hundred meters away. No adult has been captured out of sight of the area of original release.

This experiment illustrates that even a relatively massive introduction—the

equivalent of the arrival of a small flock of fresh, fertilized females into a suitable area—is no guarantee of establishment. The sedentary behavior of the adults has doubtless helped to maintain that colony, but has also militated against any spread. Parasites and predators placed considerable pressure on the one colony that could be studied and may still be a major factor keeping it from "breaking out" and becoming securely established. The impression given by these experiments is that colonization is a difficult and chancy process for *E. gillettii;* and this conforms to the general conclusions from studies of *E. editha,* and for colonizing species in general (Lewontin 1965).

Of three other western *Euphydryas, E. anicia* may be somewhat more vagile than *E. editha* (Cullenward et al. 1979; Ehrlich et al. unpublished), whereas *E. chalcedona* is clearly more vagile (Brown and Ehrlich 1980) and generally appears to be a better colonizer. Nothing is known of the vagility of a third species, *E. colon.* But these three appear to be classic semispecies *(E. chalcedona s.l.),* and their combined geographical distribution is much greater than that of *E. editha. E. chalcedona s.l.* has penetrated habitats such as the deserts of California where *E. editha* does not occur, and its range extends to Alaska, far to the north of any populations of *E. editha.* This appears to be largely due to the greater breadth of diet of the former. *E. chalcedona s.l.* feeds largely on perennials, but can also use annuals; *E. editha* is largely confined to annuals and *Castilleja.* For instance, in laboratory experiments larvae from a California *E. chalcedona* population were able to develop satisfactorily on all of 10 scrophulariaceous and plantaginaceous food plants offered, including not only hosts novel for the population, but some not used elsewhere by *E. chalcedona s.l.* In contrast, *E. editha* survived on only six of the ten and only on ones used by the population or *E. editha* populations from other ecotypes. In short, it may well be that population parameters, such as the average distance and individual moves, may be much less important in determining the dispersal ability of these organisms than their physiological ability to feed on different host plants.

5.5. The Question of Genetics

Whether or not there are significant genetic differences between species that invade easily and those that do not has attracted considerable attention from evolutionary biologists (e.g., Baker and Stebbins 1965; Parsons 1983). Lewontin (1965) raised most of the pertinent questions two decades ago. Answers (which were not already obvious then) have been slow in coming. For instance, Lewontin's conclusion that there should be constant interdeme selection in invaders still seems logically valid, and it seems obvious that a reasonable degree of genetic variability is helpful to invaders. But there is very little evidence to support them or any other generalizations. Unfortunately, nothing is known of the genomes of most animals, and the fragmentary knowledge available about a limited sample of species yields no obvious answers to the dilemma. In fact, it is proving extremely difficult to sort out the genetics of populations in situ,

as the "neutrality controversy" shows. It is hardly surprising that we cannot list in detail genetic attributes, present in some species but absent in close relatives, that have provided advantages to to the former in novel environments.

I am tempted to predict that no relatively simple genetic measure, such as level of structural heterozygosity, will have much predictive value for invasion success or failure, although some general guidelines may emerge. For example, successful invaders may be more likely to come from marginal populations which, at least in *Drosophila* (Brussard 1984, 1986), have more genic (and less structural) heterozygosity and a greater tendency to disperse. But key answers seem more likely to be found in the as-yet tangled domain of genetics and development and to involve such things as the relationship between phenotypic plasticity and the capacity for rapid evolutionary change (e.g., Douglas and Grula 1978). In sum, at the moment we do not even know whether a particular constellation of genetic attributes normally characterizes a "good invader." My guess is that there is not, unless some vague statement such as "good invaders are not adapted to too narrow a niche" is counted as a genetic characteristic.

5.6. Are There Any Rules?

Comparisons of species that are excellent colonists with closely related ones that are not give little support to the principles suggested in Table 5.1. But comparisons of members of different groups tend to support them. For example, it is clear why *Pieris rapae* invades easily while the lycaenid *Sandia macfarlandi,* which is oligophagous on a few closely related desert plants in the southwestern United States and northern Mexico, is not. Obviously, the ability to subsist on a relatively wide variety of foods is often a prerequisite of being a successful colonist. Herbivores that are monophagous or oligophagous on a few species with relatively limited distributions will normally be handicapped relative to polyphagous species. Certainly the ubiquity of the imported cabbage butterfly, house sparrow, and the two rats could not have been achieved by organisms with narrow diet specialization. Of the birds that have been successful colonists, a substantial proportion have been omnivores, often ones in which seeds make up a substantial portion of the diet (Mayr 1965, Long 1981), but that alone is clearly not sufficient to guarantee success.

There are abundant exceptions to this generalization among specialist insect herbivores and predators, but they are found in the very special cases of planned introductions for the purposes of biological control. In these cases, of course, the food plant(s) or prey species are abundant and widespread in the area invaded, and the specialist often can thrive. The cottony cushion scale, *Icerya purchasi,* became an abundant pest on citrus in California and elsewhere outside of its native Australia. This successful invader was quite local and rare in its native land (DeBach 1974). Subsequent introductions of the Australian coccinellid beetle, *Rodolia cardinalis,* which attacks the scale have uniformly resulted in the beetle becoming common enough to reduce the impact of the scale greatly.

In a reverse exchange, the small moth *Cactoblastis cactorum* was a very successful invader in Australia when introduced to areas in which its host, *Opuntia* cactus, had become an extremely serious and widespread pest (DeBach 1974).

But, in contrast to these successes, many attempts to establish biological control of pests by introducing their predators have failed. Of about 50 species feeding on *Opuntia* that were sent to Australia, only about a quarter had become established when the enormous success of *Cactoblastis* ended the experiment (DeBach 1974). And only less than a third of the predators that have been introduced to control coniferous forest pests (largely in North America) have "taken;" even under the most favorable conditions of release, which were preceded by very careful studies, a quarter still failed (Turnock et al. 1976).

Successful colonists do tend to be animals that are relatively abundant and widely distributed where they are endemic. Certainly also, characteristics lending themselves to transport by *Homo sapiens* (purposely or accidentally) and to adaptation to habitats created by them, have been important attributes of these species and many others we think of as successful colonists. Most of the truly widespread terrestrial animals today are ones considered pests or commensals of human beings, and are dependent on *Homo sapiens* for more than transport.

Invading fishes are also mostly human commensals—many of them deliberately distributed by *H. sapiens*. For example, a great many species of exotic aquarium fishes have been released repeatedly in the United States and have become established (Courtenay et al 1984). Most of these are tropical in origin, and southern states (especially Florida) and California and Nevada have been the main sites of invasion. Some of these fishes have spread quite rapidly both through natural dispersal and by repeated transplants (e.g., Hubbs and Deacon 1964; Courtenay and Robins 1975; Moyle 1976; Courtenay and Hensley 1979).

One of the most successful of all piscatorial invaders is the walking catfish, *Clarias batrachus,* which has rapidly occupied much of Florida after escaping from the aquarium trade in the mid-1960s (Courtenay 1978, 1979; Courtenay et al 1984. This species has the great advantage (indicated by its name) of being able to travel considerable distances overland—obviously a great advantage to an invading freshwater fish. It also has apparently been able to evolve rapidly a degree of cold resistance (Courtenay, 1978). Such a species is *a priori* clearly likely to be a more successful invader than a species such as the neon tetra *(Paracheirodon innesi)*, which can reproduce only in warm acid waters and cannot travel overland. Another invading fish that has spread quite rapidly is the oriental goby, *Acanthogobius flavimanus,* which occupied much of the San Francisco Bay-Delta region within a few years in the 1960s. Like *Clarius,* it is "tough and resilient" (Brittan et al. 1970), and it is also euryhaline.

Invaders that are unlikely to be transported in numbers by human beings face numerous problems in crossing substantial barriers. Single female birds, for example, unlike single female lizards, ordinarily cannot travel long distances and then reproduce. Therefore birds that normally travel in flocks are much more likely to be successful colonists—starlings are much better at it than woodpeckers (Mayr 1965). Further, one would expect in all sexually reproducing

invaders an optimum tendency to disperse. Those species that were too sedentary would not explore new habitats; those that dispersed too readily would
find themselves in new habitats without mates. This, for example, would even
be true of organisms such as the checkerspot butterflies, where fecundated
females alone in theory can establish new colonies. Unless, as in the *E. gillettii*
experiments, first immigrant generations remain in a restricted area, it may be
impossible for many females to get mated and for a population to reach a size
at which it is reasonably safe from stochastic extinction. It may be that subtle
differences in patterns of mating behavior, dispersal, and habitat selection hold
the key to differences between relatively successful and unsuccessful colonists.

One of the presumed characteristics of invaders that is supported by some
comparisons among *close* relatives is large size. For instance, *Passer domesticus*
is larger than *P. montanus,* which may account for the ability of the first to
outcompete the latter. And *Rattus rattus* and *R. norvegicus* are among the
largest members of the genus *Rattus,* with the latter being somewhat larger and
having a somewhat broader distribution. Larger size should be an advantage
in predator-prey and competitive interactions and could increase a species'
ability to penetrate cooler areas (where *R. norvegicus* is more successful).

5.7. Chance and Dispersal

What is one to make of all this? One certainty is that population biologists are
a long way from any comprehensive quantitative theory of what determines
the potential for becoming a successful invader. But perhaps this is one area
where no comprehensive theory is either possible or desirable, because the
factors controlling invasion ability are very different in different taxonomic
groups. Most organisms obviously have the potential for expanding their population sizes quite rapidly when introduced into a particularly favorable habitat,
or when their environment becomes more benign. Even elephant populations
(hardly "r-strategists") have, under some circumstances, become pests (Eltringham 1979). But beyond this (which makes me doubt the advantage often
imputed to r-strategists) there are few rules that seem to extend widely over
taxonomic groups. Elephants (or even wolves) are unlikely to evolve the ability
to be dispersed concealed in produce, as the larvae of some butterflies have.
And few insects will be deliberately moved around because people find them
attractive, as many birds have been.

While a few detailed comparisons of migration and dispersal of close relatives
with different histories (e.g. Derr et al. 1981, on plant bugs attacking Malvales)
have revealed important differences between species with high and low dispersal
abilities, I doubt that the results will be transferable to species pairs of *Passer*
or even *Pieris*. And, of course, those general attributes of being "tough" and
"adaptable"—hard to quantify but clearly descriptive of creatures such as coyotes, walking catfishes, and the marine toad, *Bufo marinus* (Sabath et al. 1981)—
certainly enhance the chances of an animal invading successfully.

Obviously the key factor in invasion success is the "fit" of the animal into
the recipient environment. An invader must be to some degree preadapted; to

Table 5.2. Relative success of invasions and reintroductions

	Percent Definitely Established	Percent Possibly Established	Number of Attempts
New Zealand	25/86[a]	2/0[a]	126/7[a]
Hawaii	27/67	17/0	159/3
Australia	10/28	14/40	71/25
North America	25/67	12/24	98/21
South America	47/56	13/22	15/9
Europe	28/56	14/19	42/27
Africa-Arabia	42/50	32/25	19/12

[a] In each pair of numbers nonnatives are on the left and reintroductions on the right.

take an extreme case, the tough and adaptable walking catfish is not threatening to invade the Sahara desert. One situation where this factor can be seen on a finer scale is in the relatively higher success of bird species that are reintroduced into environments from which they have been extirpated as opposed to non-natives introduced into the same environments, as shown in Table 5.2 (modified from Long 1981, p. 15).

One must also suspect that sometimes the good invader in a taxonomic group actually has no attributes that especially suit it as a colonist—only good luck. For example, if the tree sparrow had been introduced into North America and the house sparrow had not, perhaps the former (in the absence of competition from the latter) would have become widespread and abundant. Similarly, many invaders may just happen to be preadapted to some aspect of environments modified by *Homo sapiens*. If this is the case, then we can expect invasions to increase as humanity converts more and more of earth's habitats—indeed the main survivors of the biota may increasingly be "invaders" as the "non-invaders" become extinct.

Furthermore, paired comparisons are unlikely to answer that most fundamental of questions in ecological physiology: Why, when presented with an obvious evolutionary opportunity such as a vast area to invade, do some species take advantage of it while their close relatives cannot? Why, for example, are some butterfly species constrained to migrate north into the cooler regions of the United States each spring and be killed by frosts each fall, while their close relatives have successfully evolved the ability to diapause and persist year round (Ehrlich 1984)? Why doesn't *Pieris brassicae* evolve populations that lay eggs singly and produce larvae that feed inside of cabbage heads? Here a "Spandrels of San Marcos" type of argument (Gould and Lewontin 1979) would not seem to apply. I think research should be focused on the question of what sets evolutionary limits; if it can be answered, many of the mysteries of invasion may be cleared up also.

5.8. Conclusions

The inability of ecologists to predict which of two closely related species is more likely to be a successful invader has been interpreted by some as indicating

a failure of ecology as a science. I think this is incorrect, and that ecologists tend to be too conservative when it comes to evaluating their science. Such precise prediction is rare indeed in science, and does not necessarily represent a lack of adequate theory or a failure of the discipline. Physicists cannot predict which of two identical radioactive nuclei will decay first or which of a series of nearly identical missiles launched from the same silo will come closest to the target. But these are not taken to indicate failures of physics, only that at a certain point stochasticity becomes overriding. In dealing with invasions, ecologists are, in contrast, attempting to predict the fate of diverse, often little-known organisms launched at diverse, complex, usually barely studied environments. And they are attempting to do so even though natural systems are so complex that it is usually very difficult to predict population size changes in well-studied organisms in well-studied environments. In spite of this, it is clear that ecologists can say a great deal about both the probability of invasion success of different organisms and the possible consequences of that success. Indeed, most of the a priori assumptions that I listed, representing what might fairly be called an approximation of a consensus of trained ecologists (or the "folk knowledge" of the field) seem to be valid in many cases. Clearly ecologists can develop better predictive tools, including quite sophisticated mathematical models for the behavior of invaders in some groups (and most successful models will almost certainly be group-specific). But, considering the enormous complexity of the problem, what can already be predicted is far from trivial.

5.9. Acknowledgments

I am grateful to D.S. Dobkin, A.H. Ehrlich, R.W. Holm, H.A. Mooney, D.D. Murphy, P.M. Vitousek, and B.A. Wilcox of the Department of Biological Sciences, Stanford University, for helpful comments on the manuscript. This work was supported in part by a series of grants from the National Science Foundation (the most recent DEB82-069611), and a grant from the Koret Foundation of San Francisco.

5.10. References

Baker HG, Stebbins GL (eds) (1965) The Genetics of Colonizing Species. Academic Press, New York
Baker RR (1978) The Evolutionary Ecology of Animal Migration. Holmes and Meier, New York
Barrows WB (1889) The English Sparrow (*Passer domesticus)* in North America. USDA, Div Econ Ornith and Mammal Bull no 1
Bar-Yosef O, Tchernov E (1966) Archeological finds and the fossil faunas of the Natufian and microlithic industries at Hayonim Cave (western Galillee, Israel). Isr J Zool 15:104–140
Brittan MR, Hopkirk JD, Conners JD, Martin M (1970) Explosive spread of the oriental goby *Acanthogobius flavimanus* in the San Francisco Bay-Delta region of California. Proc Calif Acad Sci 38:207–214

Brown IL, Ehrlich PR (1980) Population biology of the checkerspot butterfly, *Euphydryas chalcedona:* structure of the Jasper Ridge colony. Oecologia (Berl) 47:239–251

Brussard PF (1984) Geographic patterns and environmental gradients: the "central-marginal" model in *Drosophila* revisited. Annu Rev Ecol Syst 15:25–64

Brussard PF (1986) Structural heterozygosity, population topology, and genetic management. Zoo Biol (submitted)

Calhoun JB (1947) The role of temperature and natural selection in relation to the variations in the size of the English sparrow in the United States. Am Nat 81:203–228

Chew FS (1981) Coexistence and local extinction in two pierid butterflies. Am Nat 118:655–672

Connolly GE (1978) Predator control and coyote populations: a review of simulation models. In: Bekoff M (ed), Coyotes: Biology, Behavior, and Management. Academic Press, New York, pp 327–345

Courtenay WR (1978) Additional range expansion in Florida of the introduced walking catfish. Environ Conserv 5:273–276

Courtenay WR (1979) Continued range expansion in Florida of the walking catfish. Environ Conserv 6:1

Courtenay WR, Hensley DA (1979) Range expansion in southern Florida of the introduced spotted tilapia, with comments on its environmental impress. Environ Conserv 6:149–151

Courtenay WR, Hensley DA, Taylor JN, McCann JA (1984) Distribution of exotic fishes in the continental United States. In: Courtenay WR, Stauffer JR (eds), Distribution, Biology, and Management of Exotic Fishes. Johns Hopkins University Press, Baltimore, pp 41–77

Courtenay WR, Robins CR (1975) Exotic organisms: an unsolved, complex problem. Bioscience 25:306–313

Cullenward MJ, Ehrlich PR, White RR, Holdren CE (1979) The ecology and population genetics of an alpine checkerspot butterfly, *Euphydryas anicia.* Oecologia (Berl) 38:1–12

DeBach P (1974) Biological Control by Natural Enemies. Cambridge University Press, London

Derr JA, Alden B, Dingle H (1981) Insect life histories in relation to migration, body size, and host plant array: a comparative study of *Dysdercus.* J Anim Ecol 50:181–193

Douglas MM, Grula JW (1978) Thermoregulatory adaptations allowing ecological range expansion by the pierid butterfly *Nathalis iole* Boisduval. Evolution 32:776–783

Ehrlich PR (1958) Problems of arctic-alpine insect distribution as illustrated by the butterfly genus *Erebia* (Satyridae). Proc X Intl Congr Entomol I:683–686

Ehrlich PR (1961) Intrinsic barriers to dispersal in the checkerspot butterfly, *Euphydryas editha.* Science 134:108– 109

Ehrlich PR (1965) The population biology of the butterfly, *Euphydryas editha.* II. The structure of the Jasper Ridge colony. Evolution 19:327–336

Ehrlich PR (1984) The structure and dynamics of butterfly populations. In: Vane-Wright RI, Ackery PR (eds), The Biology of Butterflies. Academic Press, London

Ehrlich PR, Launer AE, Murphy DD (1984) Can sex ratio be defined or determined? The case of a population of checkerspot butterflies. Am Nat 124:527–539

Ehrlich PR, Murphy DD, Singer MC, Sherwood CB, White RR, Brown IL (1980) Extinction, reduction, stability and increase: the responses of checkerspot butterfly populations to the California drought. Oecologia (Berl) 46:101–105

Ehrlich PR, White RR, Singer MC, McKechnie, SW, Gilbert LE (1975) Checkerspot butterflies: a historical perspective. Science 188:221–228

Elton CS (1958) The Ecology of Invasions by Animals and Plants. Methuen, London

Eltringham SK (1979) The Ecology and Conservation of Large African Mammals. MacMillan, London

Feltwell JSE (1977) Migration of whites. Entomol Mon Mag 112:88

Feltwell JSE (1978) *Pieris brassicae* in South America. Entomol Rec J Var 90:330

Feltwell JSE (1982) Large White Butterfly. W Junk, The Hague

Ford EB (1975) Ecological Genetics. Chapman and Hall, London

Gilbert LE, Singer MC (1973) Dispersal and gene flow in a butterfly species. Am Nat 107:58–72

Gould SJ, Lewontin RC (1979) The spandrels of San Marco and the Panglossian paradigm: a critique of the adaptionist programme. Proc R Soc Lond B 205:581–598

Greaves JH, Rennison BD (1973) Population aspects of warfarin resistance in the brown rat, *Rattus norvegicus*. Mammal Rev 3:27–29

Hall ER (1984) The Mammals of North America. John Wiley, New York

Hilton H (1978) Systematics and ecology of the eastern coyote. In: Bekoff M (ed), Coyotes: Biology, Behavior, and Management. Academic Press, New York, pp 209–228

Holdren CE, Ehrlich PR (1981) Long range dispersal in checkerspot butterflies: transplant experiments with *Euphydryas gillettii*. Oecologia (Berl) 50:125–129

Hubbs CH, Deacon JE (1964) Additional introductions of tropical fishes in southern Nevada. Southwest Natur 9:249–251

Johnston RF, Klitz WJ (1977) Variation and evolution in a granivorous bird: the house sparrow. In: Pinowski J, Kendeigh SC (eds), Granivorous Birds in Ecosystems. Cambridge University Press, Cambridge, pp 15–51

Johnston RF, Selander RK (1964) House sparrows: rapid evolution of races in North America. Science 144:548–550

Johnston RF, Selander RK (1971) Evolution in the house sparrow. II. Adaptive differentiation in North American populations. Evolution 25:1–28

Johnston RF, Selander RK (1973) Evolution in the house sparrow. III. Variation in size and sexual dimorphism in Europe and North and South America. Am Nat 107:373–390

Kleiman DG, Brady CA (1978) Coyote behavior in the context of recent canid research. In: Bekoff M (ed), Coyotes: Biology, Behavior, and Management. Academic Press, New York, pp 163–188

Lekagul B, McNeely JA (1977) Mammals of Thailand. Sahakarnbhat, Bangkok, 758 pp

Lewontin RC (1965) Selection for colonizing ability In: Baker HG, Stebbins GL (eds), The Genetics of Colonizing Species. Academic Press, New York

Lim BL (1970) Distribution, relative abundance, food habits, and parasite patterns of giant rats *(Rattus)* in West Malaysia. J Mammal 51:730–740

Long JL (1981) Introduced Birds of the World. Universe Books, New York

Maciolek JA (1984) Exotic fishes in Hawaii and other islands of Oceania. In: Courtenay WR, Stauffer JR (eds), Distribution, Biology, and Management of Exotic Fishes. Johns Hopkins University Press, Baltimore

Mayr E (1965) The nature of colonizations in birds. In: Baker HG, Stebbins GL (eds), The Genetics of Colonizing Species. Academic Press, New York, pp 29–43

Mech LD (1970) The Wolf: The Ecology and Behavior of an Endangered Species. Natural History Press, Garden City

Moyle PB (1976) Fish introductions in California: history and impact on native fishes. Biol Conserv 9:101–118

Murphy DD, Launer AE, Ehrlich PR (1983) The role of adult feeding in egg production and population dynamics of the checkerspot butterfly *Euphydryas editha*. Oecologia (Berl) 56:257–263

Murphy DD, White RR (1984) Rainfall, resources, and dispersal in southern populations of *Euphydryas editha* (Lepidoptera: Nymphalidae). Pan-Pacific Entomol 60:350–354

Musser GG (1973) Zoogeographical significance of the ricefield rat, *Rattus argentiventer*, on Celebes and New Guinea and the identity of *Rattus pesticulus*. Am Mus Novit no 2511, pp 1–29

Musser GG (1977) *Epimys benguetensis,* a composite, and one zoogeographic view of rat and mouse faunas in the Philippines and Celebes. Am Mus Novit no 2624, pp 1–15

Nowak RM (1978) Evolution and taxonomy of coyotes and related *Canis*. In: Bekoff M (ed), Coyotes: Biology, Behavior, and Management. Academic Press, New York, pp 3–16

Nowak RM, Paradiso JL (1983) Walker's Mammals of the World, 4th edit. Johns Hopkins Press, Baltimore

Oda DK, Parrish JD (1981) Ecology of commercial snappers and groupers introduced to Hawaiian reefs. In: Proc 4th Int Symp Coral Reefs, Manila, 1981

Parsons PA (1983) The Evolutionary Biology of Colonizing Species. Cambridge University Press, Cambridge

Randall JE, Kanayama RK (1972) Hawaiian fish immigrants. Sea Frontiers 18:144–153

Roper T (1984) A question of taste. New Scientist, 29 March, pp 30–33

Sabath MD, Boughton WC, Easteal S (1981) Expansion of the range of the introduced toad *Bufo marinus* in Australia from 1935 to 1974. Copeia 1981:676–680

Singer MC (1983) Determinants of multiple host use by a phytophagous insect population. Evolution 37:389–403

Summers-Smith D (1963) The House Sparrow. Collins, London

Tamarin RH, Malecha SR (1972) Reproductive parameters in *Rattus rattus* and *R. exulans* of Hawaii, 1968 to 1970. J Mammal 53:518–528

Tchernov E (1962) Paleolithic avifauna in Palestine. Bull Res Counc Isr 11:95–131

Turnock WJ, Taylor KL, Schro''der D, Dahlsten DL (1976) Biological control of pests in coniferous forests. In: Huffaker CB, Messenger PS (eds), Theory and Practice of Biological Control. Academic Press, New York, pp 289–311

White RR, Levin MP (1981) Temporal variation in insect vagility: implications for evolutionary studies. Am Midl Nat 105:348–357

Williams CB (1936a) Our butterfly visitors from abroad. Country- side. Spring, No 2

Williams CB (1936b) Collecting records relating to insect migration. 3rd series. Proc R Entomol Soc Lond A 11:6–10

Williams EH (1986) Population structure of *Euphydryas gillettii*. Oecologia (Berl) (submitted)

Williams EH, Holdren CE, Ehrlich PR (1984) The life history and ecology of *Euphydryas gillettii* Barnes (Nymphalidae). J Lepidopt Soc 30:1–12

Wirtz WO II (1972) Population ecology of the Polynesian rat, *Rattus exulans*, on Kure Atoll, Hawaii. Pacific Sci 26:433– 464

Wirtz WO II (1973) Growth and development of *Rattus exulans*. J Mammal 54:189–202

Zimmerman EC (1958) Insects of Hawaii. Vol 7. Macrolepidoptera. University of Hawaii Press, Honolulu

6. Life History of Colonizing Plants: Some Demographic, Genetic, and Physiological Features

F.A. Bazzaz

6.1. Introduction

The term "colonizer" has come to mean different things to different biologists. In a general sense all species are colonizers, as they all must become established in sites in which to grow and reproduce. Agriculturalists have equated colonizers with undesirable and, in many cases, nonnative species that affect agroecosystems detrimentally by reducing the growth and yield of the desired species. In this agronomic sense the terms colonizers, "weeds," and often "aliens," are synonymous. Theoretical ecologists usually think of colonizers as those species whose disseminules travel relatively long distances, and arrive in unoccupied or presumably incompletely occupied habitats where they subsequently interact with other species present or become locally extinct [e.g., the island biogeographic model of MacArthur and Wilson (1967)].

The contributors to this book are concerned with species that are introduced intentionally or unintentionally into ecosystems where the species have never been before, and cause changes in the structure and function of these ecosystems. Such species often spread relatively quickly. They may modify the resident community, and in some cases may usurp much of the available supply of resources of the habitat. Prominent examples in North America are several Eurasian grasses [e.g., *Bromus* (Mack 1981), *Hordeum,* and *Avena*] that have been introduced in the American west and a number of exotics, usually of tropical or subtropical origin, that have been introduced in the southeastern states, California, and Hawaii (e.g., *Schinus, Melaleuca,* and *Eucalyptus*).

Ecologists and population geneticists have been interested in the biology of colonizing species because of the economic and aesthetic impact of these species and their role in community dynamics. A central question has been: why are some species successful colonizers while others are not?

6.2. Classification and Definition

Species may enter unoccupied sites and initiate primary succession, enter disturbed habitats, or enter relatively intact vegetation. Species that enter unoccupied or sparsely occupied habitats may be called *"colonizers."* Those species that enter relatively intact vegetation and strongly dominate it or even displace it altogether may be called *"invaders."* The integration of an invader into an ecosystem may require a long period in which there may be extinctions as well as coevolution resulting in character displacement, niche shifts, and better combining ability (Fig. 6.1). Species that enter disturbed habitats are likely to have life history features similar to those of plants in early succession (Table 6.3).

Species that do not displace or markedly depress the resident populations and become integrated into the communities they enter may be called *"immigrants."* Their life history features must mesh well with those of the resident species and their entry is prevented mainly by geographical barriers or the lack of dispersal agents. *"Alien"* species may be defined as those species that are economically or aesthetically undesirable. They come into an area from other geographical regions, usually across continental or international boundaries. *"Weeds"* are plant pests of agricultural and other managed ecosystems, regardless of their geographical origin. They reduce the value of these systems. *"Aliens"* and *"weeds"* are anthropomorphically based terms. Their use should be minimized or altogether avoided in ecological literature.

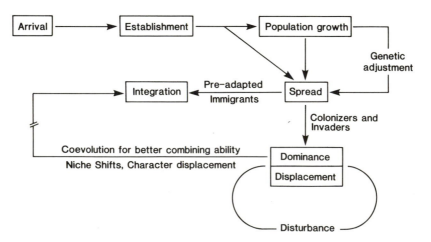

Figure 6.1. Diagramatic representation of population spread of colonizing species.

Here I shall discuss some aspects of demography, genetics, ecophysiology, and niche relationships of colonizing species (sensu lato) as defined ecologically. It is recognized that the generalizations that emerge from this discussion are founded on our limited knowledge of the biology of colonizing species and therefore may not stand the test of time.

6.3. Population Growth Rates and Geographical Spread of Colonizing Plant Species

6.3.1. Population Growth

The first step in colonization is the arrival of disseminules in a new site (Fig. 6.1). In many cases this involves long-distance dispersal. Though in the majority of plant species dispersal is by wind and/or animals, man appears to be the primary agent of long distance dispersal of colonizing species. There are many examples of colonizing species that have been introduced intentionally (e.g., as ornamental exotics), or unintentionally (e.g., as contaminants of desirable plant material) into new habitats (see chapters by Mack and Ewel, this volume). Introductions may occur in one location or more commonly in many locations, at one time or at different times, from single disseminules or from many disseminules. Some of these introductions will act as sources from which colonizers spread in the new habitat. The speed and direction of spread will depend on a large number of intrinsic and extrinsic factors, but first the population has to become established at the site of introduction. It is likely that many colonies will be eliminated because their initial size is too small or the early coincidence of factors causing high mortality hinders their establishment. Although seed germination of many colonizing species is high under laboratory or culture conditions, the percentage of seed that germinates in nature may be low. Field experiments in which seeds are introduced into relatively intact vegetation clearly indicate that for successful establishment of colonizers a large quantity of seed must be introduced. For example, Martins and Jain (1979) found that only about 12% of the sown colonies of *Trifolium hirtum* survived the following season; some of these disappeared a year later. Seedling death caused by drought was especially high during the second year of the experiment. Repeated introductions of a large number of disseminules may be necessary for initial establishment of colonizers. The origin and the number of successful introductions will influence the genetic structure of the populations and therefore their growth, spread, and success.

Populations of colonizing species are often characterized by exponential growth, but at differing rates determined by intrinsic properties and by competitors, herbivores, and pathogens. Lag times before new dispersal and spread undoubtedly differ among colonizing species and the limited available literature, which is largely based on historical records, does not seem to suggest any general trends. Furthermore, the fact that increase in numbers and increase in area may be more or less correlated in different populations complicates the situation.

Mathematical modeling has been little applied to colonizing plant species,

though such modeling has been used for other organisms, especially pathogens (May, this volume). Neither have diffusion equations, which have been used in population genetics (Roughgarden, this volume) to study the spread of genes, been extensively used for colonizing plant populations. Auld and Coote (1980) used a simple model (incorporating population growth rate at the introduction site, fractional loss by dispersal, and the distance this fraction can move) to study the rate of spread of the weed *Nassella trichotoma* in eastern Australia. Their model predicts that the rates of total population growth tend to be exponential and rates of spread linear. Analysis of the spread of *Bromus tectorum* in the western United States shows that population growth and spread are faster from several small introduction foci than from one large focus (R.N. Mack, personal communication). Resistance to pathogens and pests may enhance the rate of spread of a colonizer in new habitats but is not absolutely necessary. For example *Bromus tectorum* is a very successful invader, but suffers up to 48% annual mortality among adults because of infection by smut (Mack and Pyke 1984).

A crucial life history feature of colonizing species is efficient dispersal which may be aided by humans. A species may possess several physiological and demographic features that would make it a highly successful colonizer, but its success may be limited by inefficient dispersal. For example, *Ambrosia trifida*, the giant ragweed, is a highly competitive annual in disturbed areas of the eastern and midwestern United States. It decisively dominates communities that it enters, yielding up to 98% of annual net production and lowers species diversity to 12%. It possesses a number of physiological and demographic features that make it so strongly dominant in these communities (for details see Bazzaz 1984). Except in very dry sites, the species is undoubtedly capable of dominating any annually disturbed habitat in the entire region. The one attribute that is preventing this regional dominance is inefficient dispersal. Seeds of *Ambrosia* are the largest among these herbaceous plants and therefore dispersal is severely limited.

6.3.2. Geographic Spread

The pattern of spread by species that depend on disturbance for establishment will likely differ from those that do not. The first group should spread, at least initially, as many isolated small populations while the second group should spread in an expanding front. The rate of spread of the first group will undoubtedly depend heavily on the size, frequency, and the spatial and temporal distribution of disturbance (Bazzaz 1983). Furthermore, colonizing populations that do not require genetic adjustment (which occurs by several means, see Baker 1967) in the new habitat very likely will spread more quickly initially than those that do. The initial extended lag phase in population spread as well as plateaus in that spread observed for some species may indicate the need for genetic adjustment before further spread occurs. New genotypes, each adapted to a certain range of habitats, may be necessary for rapid spread and establishment. Much experimental and theoretical work needs to be carried out before we can improve our understanding of the process of growth and spread of col-

onizing species. The work will necessarily require an understanding of properties of the invaded ecosystem and an identification of their attributes that make them invasable by certain species.

6.4. The Genetics of Colonizing Species

The genetics of colonizing species has received much theoretical treatment in the literature. A prominent concept that emerged from the book *Genetics of Colonizing Species* (Baker and Stebbins 1965) is that of the "general purpose genotype" (Baker 1965). Colonizing species with such genomes are capable of succeeding in a wide range of environmental circumstances. Jain (1983), and Jain and Rice (1985) have reviewed the genetic and demographic attributes of colonizing species. Brown and Marshall (1981 in Jain 1983) have identified several genetic characteristics of colonizers. These include propagation by self-fertilization, greater phenotypic plasticity, and fixed heterozygosity through polyploidy. Furthermore, colonizers were predicted to show substantial between-population differentiation as a result of founder effects and patchy environments.

6.4.1. Breeding Systems

Colonizers may increase the level of selfing to avoid segregational load; on the contrary, they may increase the level of outcrossing which produces high genetic variability which promotes exploitation of a patchy environment (Jain and Rice 1985). One advantage of selfing for colonizers is that it preserves coadapted gene complexes. Baker (1974) proposed that selfing may insure reproduction in habitats with a scarcity of potential pollinators. It has also been suggested that selfing would permit the establishment of a population from one individual, but the chances of colonizer populations being started from only one individual are slim. In some species small population size may not be detrimental to their future success. These species can generate a reasonable level of genetic diversity after a bottleneck period caused by low population numbers (Clegg and Allard 1972; Jain 1975).

Most colonizing species that have been examined have some degree of outcrossing. The advantage of this outcrossing is that it could generate some adaptive genotypic diversity in the colonizing populations. The flexibility in the degree of outcrossing observed in some populations (e.g., Jain 1983; Jain and Martins 1979) may itself be important in the colonizing ability. These populations could match the relative benefits of selfing versus outcrossing to the environments they encounter.

6.4.2. Levels of Heterozygosity

Experimental work on the role of genetic variation of colonizers in their success has been limited to only a few species and the results have been equivocal.

Populations of the colonizing species *Lupinus succulentus* have higher levels of heterozygosity than expected whereas populations of the noncolonizing *L. nanus* do not (Harding and Barnes 1977). Jain (1983) found that isolated stands of *Avena barbata* on roadsides showed lower genetic variation than did central populations in the grasslands. The percentage of polymorphic loci in the roadside populations was about one half that of the grassland populations (Jain et al. 1981; Jain 1983). These authors attributed this to size bottlenecks and random drift during colonization. Martins and Jain (1979) examined populations of *Trifolium hirtum* from older extensive plantings and smaller naturally colonized roadside sites. Seedlings in the older populations had higher survivorship and the mature individuals had a lower number of heads per plant than did roadside populations. There were no significant differences in seed size between the older and younger populations (Table 6.1.). Furthermore, most of these populations were highly polymorphic and had similar genotypic frequencies except those found in marginal and stressful habitats.

Martins and Jain also planted colonies from mixed seed such that the colonies had either low, medium, or high amount of genetic diversity (based on four morphological markers and four allozyme loci). As previously mentioned (cf. p.183) only a small proportion of colonies got started during the first year of the experiment and some of them disappeared in the following year. New colonies appeared in the second year and there were differences in seedling death rates between populations during the two years. The role of genetic polymorphism during the first year was unimportant, but in the second year new colonies that started from dormant seed came up only in locations where the seed source was of high or medium genetic diversity (Table 6.2.). Large, expanding patches of roadside populations showed a significant increase in genetic polymorphisms. These patches apparently had higher outcrossing rates and higher levels of heterozygosity. Colonization success of these populations seemed to involve rapid genetic change (Jain and Rice 1985).

In another study Jain and Martins (1979) found correlations between some life history features and size and location of *Trifolium* colonies. Roadside colonies showed higher reproductive efforts, lower seedling survivorship, and earlier flowering. They concluded that colonizing success of *Trifolium* seems to be determined largely by a few, rapid morphological changes and by the retention of some outbreeding characteristics and genetic variability. Long-term studies

Table 6.1. Some genetic and demographic variables of small roadside populations and large range populations of *Trifolium hirtum*

	Outcrossing Rate	Genetic Diversity	Seed Size	Seedling Survivorship
Rangeland	3.8	1.17	0.255	60.3
Roadside	5.1	1.36	0.243	46.7
Mann-Whitney	NS	NS	NS	NS

From Jain 1983.

Table 6.2. Relative success of colonies of *Trifolium hirtum* with different degrees of polymorphism in seed source

Polymorphism Index	No. Colonies Planted	First Year		Second Year	
		No. Colonies Established	Success (%)	No. Additional Colonies Established	Success (%)
Low	21	3	14.2	0	0
Medium	89	8	8.9	3	3.7
High	25	3	12.0	4	18.1
Total	135	14	10.3	7	5.7

Polymorphism index was based on four morphological markers and four allozyme loci.
From Martins PS, Jain SK, Am Nat 113:591–595.

of this nature on several colonizers are clearly necessary before we know what role genetic diversity plays in colonizing success.

Brown and Marshall (1981) suggested that colonizing species are somewhat depauperate in allozyme variation. But there are conflicting theories and results about trends in allozyme variation in central versus marginal populations and other geographical gradients, and along successional gradients (see review in Zangerl et al. 1977). Whether colonizing species have high levels of ecotypic differentiation is not clear. Some populations of these species show ectoypic differentiation both locally (Jain 1977) and over a large geographical scale (Hancock and Bringhurst 1978). Furthermore, circumstantial evidence suggests that many of the most successful worldwide invaders have numerous ecotypes.

The preceding discussion shows quite clearly that colonizing species may adopt a variety of breeding strategies and no one genetic attribute or group of attributes could ensure their success. Furthermore, it may be undesirable or even impossible to characterize a set of features common to most colonizing species (Jain and Rice 1985). Both a thorough analysis of the available literature and more research based on the proposed definitions of colonizers, invaders, and immigrants (see above) may clarify the situation.

6.5. Physiological Ecology of Colonizing Species

Except for our knowledge of the physiological ecology of early successional versus late successional plants (see reviews in Bazzaz 1979; Bazzaz and Pickett 1980) the physiological ecology of colonizing species is also poorly understood. Generally, early successional plants (which are one class of colonizing species) have high photosynthetic, respiration, and transpiration rates. They have high stomatal and mesophyll conductances, high individual growth rates, rapid responses to changes in environmental resources, and a high acclimation potential: they behave opportunistically in response to disturbance (Bazzaz 1983, 1984). However, when we consider colonizing plants in general, our knowledge of their physiological ecology is very limited. Here I discuss the little that is known

about the physiological ecology of colonizing species, first with respect to recruitment and then with respect to carbon gain.

6.5.1. Recruitment

Germination and emergence polymorphisms may aid colonizers in their ability to invade. For example, seed germination of *Bromus tectorum* in the field occurs over a long period of time during the year and results in several seedling cohorts established at different times in the same site. The germination pattern and the establishment of seedling cohorts usually follow, and very likely are caused by, the pattern of intermittent rain (Mack and Pyke 1984; Fig. 6.2.). Despite much mortality caused by drought, frost, and disease, some cohorts escape to produce seed and contribute to future generations. *Setaria faberii,* another Eurasian colonizer of agricultural field and disturbed areas, shows a similar pattern in recruitment (Raynal and Bazzaz 1973) and repeated disturbances during the season will not prevent its reestablishment from the persistent seed bank and growth to maturity (Perozzi and Bazzaz 1978).

The flexibility of life history behavior caused by the time of recruitment and resource availability patterns is also an important feature of some colonizers. For example, *Bromus tectorum* individuals in western Washington can behave as biennial monocarpic, delayed "biennial" monocarpic (Harris 1967), ephemeral monocarpic, annual monocarpic, and winter annual monocarpic plants (Mack and Pyke 1983).

6.5.2. Carbon Gain and Allocation

The physiological ecologies of two species of *Agropyron* with regard to carbon and water use have been compared by M. M. Caldwell and his associates.

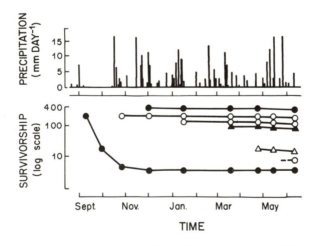

Figure 6.2. Composite portrayal of the physical environment and the survival of constituent cohorts of the *Bromus tectorum* population at a mesic site in eastern Washington, U.S.A., during 1979-1980. (From Mack RN, Pyke DA, J. Ecol 72:731–748.)

Agropyron spicatum is a native grass in the Great Basin Desert and *A. desertorum* (though not strictly a colonizing species) was introduced from Eurasia around the turn of the century. Both are bunchgrasses, but *A. desertorum* is more tolerant of grazing than is *A. spicatum*. The two species have comparable rates of photosynthesis but *A. desertorum* has a slightly lower transpiration rate and higher photosynthesis/transpiration ratio (Fig. 6.3.). On a whole-plant basis, however, the photosynthesis/transpiration ratios under field conditions do not differ between the two species (Caldwell et al. 1983). The two species, despite some differences in architecture and quantity of shaded foliage, intercept about the same amount of solar radiation. Following partial defoliation, the colonizer *A. desertorum* rapidly establishes a canopy with three to five times the photosynthetic surfaces as the native *A. spicatum*. Furthermore, leaf blades of the regrowing tillers of *A. desertorum* have higher photosynthetic capacity than blades or unclipped plants of both species but the relative increase was greater for *A. desertorum* than for *A. spicatum*. *Agropyron desertorum* had a lower investment of nitrogen per unit area of photosynthetic tissue. It showed a greater flexibility than *A. spicatum* in allocation of carbohydrates and nitrogen (Caldwell et al. 1981). Net photosynthesis of the inflorescence and the stem that sustains it was significantly greater on clipped than unclipped *A. desertorum* plants but not on those of *A. spicatum* (Nowak and Caldwell 1984). A major factor in the success of *A. desertorum* is its ability to extract moisture from deeper parts of the soil profile relative to *A. spicatum* (Thorgeirsson and Richards 1983).

Some characteristics of one class of colonizing species are presented in Table 6.3.. For those colonizing species that do not require disturbance there may be many syndromes and a summary of their attributes is not possible at this time.

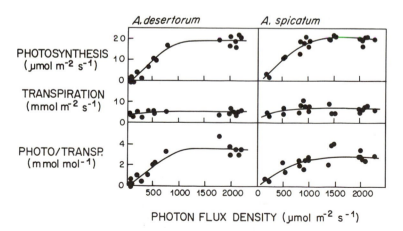

Figure 6.3. Photosynthesis, transpiration rates, and photosynthesis/transpiration ratios for *Agropyron desertorum,* an introduced species and *Agropyron spicatum,* a native species. (From Caldwell MM et al., Oecologia (Berl) 59:178–184, 1983.)

Table 6.3. Some characteristics of colonizing species that depend on some disturbance to enter into new habitats

1. High population growth rate
2. Relatively short life cycle
3. Early reproductive maturity
4. High reproductive allocation
5. Autogamous, or wind-pollinated, or serviced by generalist pollinators
6. Long-range seed dispersal capability
7. Generalists (broad-niched) in resource use
8. High acclimation capabilities
9. Rapid response to resource availability
10. High rates of photosynthesis, respiration, transpiration, and growth

6.6. Niche Relations and Competitive Abilities of Colonizing Species

The niche characteristics of colonizing species probably vary with the status of the invaded community. Species that require disturbance for establishment may have broad niches because very likely in those relatively open habitats fluctuations in the physical environment may be unpredictably high (Bazzaz 1983; Canham and Marks 1985). Thus these species (a) may or may not be competitively competent; (b) may or may not be genetically polymorphic; but (c) they must have plastic response to the environment; (d) tolerate shortages and excesses of resources; and (e) behave opportunistically. In contrast, colonizing species that are independent of disturbance ought to be competitively superior *or* competitively competent and have complementary behavior such as the ability to acquire resources when and where the native species are less active. Broad niches, which may be associated with lower competitive ability, could be disadvantageous in these situations.

Preadaptation to coexistence with members of the resident community represents a major mechanism of entry and establishment of some colonizing species. In California the colonizing species Medusahead grass *(Taeniatherum asperum)* continues to grow well in the dry season after the native grasses have stopped growth (McKell et al. 1962). It uses what is left of soil moisture that is not fully utilized by the early maturing native vegetation (McKell et al. 1959). *Poa pratensis,* the European bluegrass, has invaded and become integrated into sections of the North American grassland. The species grows well in cool temperatures. In the tall and mixed grasslands it reaches a peak in growth and flowers in May before the native grasses are very active. Its growth declines during the hot summer months while the native species (mostly C_4 grasses) are growing fast. After mowing or the decline of the native grasses in autumn, *Poa* resumes active growth until the time of heavy frost (Weaver and Albertson 1956). Thus *Poa* entered the grasslands and is maintained by temporally displaced growth behavior (Bazzaz and Parrish 1982) (Fig. 6.4.).

Another example of complementary behavior that may help the invasion, establishment, and integration of plants is that of *Lactuca scariola*. This species

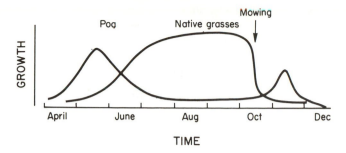

Figure 6.4. Representation of growth phenology of *Poa pratensis* and native grasses of the Tall Grass Prairies of the U.S. (From Bazzaz FA, Parrish JAD, 1982, based on Weaver JE, Albertson FW, 1954.)

and *Erigeron annuus* flower in midsummer in oldfields in the Midwest, and they usually attract the same insect visitors (Parrish and Bazzaz 1978, 1979). *Erigeron,* the more common of the two species, begins to flower in midmorning when the insects are quite active. *Lactuca* flowers open earlier in the morning and close just before *Erigeron* flowers open. Pollinators captured early in the morning have pollen loads mostly of *Lactuca;* those captured in mid and late morning mostly have *Erigeron* pollen (Fig. 6.5.). Thus the two species do not compete for the pollinators. In its native habitat in Europe *Lactuca* co-occurs with *Senecio* and other plants similar to *Erigeron* in floral timing (see discussion in Parrish and Bazzaz 1979). In these communities coevolutionary pollinator niche differentiation between *Lactuca* and its neighbors may have taken place. Therefore, *Lactuca* may have been preadapted to fit into the winter annual community in the midwestern and eastern United States.

Another poorly understood issue in the biology of colonizing species is trade-offs in allocation of carbon and other resources to various functions and in plant architecture: how much is allocated toward reproduction and growth and how best to be built. The trade-offs between colonizing ability and competitive

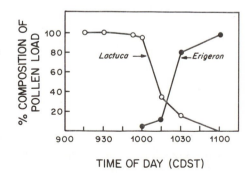

Figure 6.5. Shifts in percent pollen load on Helictid pollinators shared by *Lactuca scariola* and *Erigeron annus* occuring in the same field. (Data from Parrish JAD, Bazzaz FA, Oecologia 35:133–140, 1978.)

ability are also unknown. Since effective colonization may require rapid and efficient dispersal it is conceivable that competitive ability would be sacrificed as the plants would produce very large numbers of small, highly dispersable seeds (r-strategy). There is virtually no direct experimental examination of this trade-off for colonizing species. Our work with populations of *Antennaria parlinii* may shed some light on this question though the species may not be considered a colonizer. The apomictic individuals produce more inflorescences and more seeds per individual than do sexuals. Seeds of the apomicts are smaller and remain aloft in air longer, and the resulting seedlings suffer higher mortality rates than do those of the sexuals. Apomictic and sexual populations also differ in the patterns of allocation to ramet production. Sexuals produce longer stolons which have lower mortality than do those of asexuals (Michaels and Bazzaz 1986). Thus asexuals have a higher potential for dispersal while sexuals should do better in competitive situations. The sexuals generally have a higher between-phenotype and a lower within-phenotype response breadth suggesting that apomicts have a high level of plasticity while sexuals have a high level of genetic diversity (H. Michaels and F.A. Bazzaz, unpublished data). These results agree with Baker's (1974) prediction that phenotypic plasticity is a common feature of colonizing species. The results are also consistent with the findings of Jain (1975) in *Avena fatua* and *A. barbata* and of Wu and Jain (1979) in *Bromus rubens* and *B. mollis*. In both of these cases the less genetically variable species of a pair of congeners was the more phenotypically plastic.

Carbon allocation and the trade-offs in competitive and rapid dispersal ability may not be optimally fixed in colonizing species. Instead, flexibility of response to changing environmental circumstances may be the more common life history design.

Colonizers in general are likely to have some but not necessarily all of these attributes. Successful colonization, however, will depend largely on the appropriate fitting of life history features of the invading species and the ecological attributes of the invaded habitat.

6.7. Conclusions

1. Colonizing species (*sensu lato*) are plants that are introduced into ecosystems where they have not been before. The species spread quickly and may cause changes in structure, productivity, and functioning of these ecosystems.
2. Species that enter unoccupied or sparsely occupied sites and may initiate successions may be called colonizers. Species that enter relatively intact vegetation, and strongly dominate it or altogether displace it, may be called invaders. Species that do not displace or markedly depress the resident populations, but become integrated into the communities they enter may be called immigrants. "Aliens" and "weeds" are strongly anthropomorphic terms and should be avoided.
3. The integration of new arrivals into the resident ecosystems may require a long period in which coevolutionary interactions occur resulting in character displacement, niche shifts, and better combining ability.

4. A crucial life history feature of colonizing plant species is efficient dispersal.

5. Colonizing species are likely to become established after repeated introductions of large numbers of disseminules and very rarely from single introductions of a small number of disseminules.

6. Mathematical modeling of invasion by colonizing plants has not been extensively used and this is an area of much promise in understanding the biology of these species.

7. Colonizing species that depend on disturbance for establishment spread as isolated patches, whereas those species that do not require disturbance are likely to spread more or less as an advancing front.

8. The genetics of colonizing species has received considerable attention but only a few generalizations have emerged. Greater phenotypic plasticity seems to be a common feature, but the species examined so far differ in the degree of outcrossing, levels of heterozygosity in populations, and ecotypic differentiation. The species may adopt a variety of strategies and no one genetic attribute or group of attributes could ensure their success.

9. Carbon and other resource allocation and the trade-offs in competitive and rapid dispersal ability may not be optimally fixed in colonizing species. Instead, flexibility of response to changing environmental circumstances may be the more common life history design.

10. Early successional plants (which are one class of colonizing species) have high photosynthetic, respiration and transpiration rates. They have high stomatal and mesophyll conductances, high individual growth rate, rapid responses to changes in environmental resources, and a high acclimation potential.

11. Colonizing species that require disturbance for establishment may have broad niches. These species may or may not be competitively competent, may or may not be genetically polymorphic but they must have plastic response to the environment, tolerate shortages and excesses of resources, and behave opportunistically. In contrast, colonizing species that are independent of disturbance ought to be competitively superior or competitively competent and have complementary behavior in resource acquisition relative to native species of the resident community. Preadaptation to coexistence with the members of the resident community is a major mechanism for colonization.

12. Further examination of the diverse literature on colonizers and much experimental and theoretical work need to be done before we can improve our understanding of the growth and spread of colonizing species. The work should also involve the invaded systems and the identification of their properties that make them invasable by species with certain life history features. The colonizer and the colonized are partners in the process.

6.8. References

Auld BA, Coote BG (1980) A model of a spreading plant population. Oikos 34:287–292
Baker HG (1965) Characteristics and modes of origin of weeds. In: Baker HG, Stebbins GL (eds), Genetics of Colonizing Species. Academic Press, New York, pp 147–172

Baker HG (1967) The evolution of weedy taxa in the *Eupatorium microstemon* species aggregate. Taxon 16:293–300

Baker HG (1974) The evolution of weeds. Annu Rev Ecol Syst 5:1–24

Baker HG, Stebbins GL (eds) (1965) The Genetics of Colonizing Species. Academic Press, New York

Bazzaz FA (1979) The physiological ecology of plant succession. Annu Rev Ecol Syst 10:351–371

Bazzaz FA (1983) Characteristics of populations in relation to disturbance in natural and man-modified ecosystems. In: Mooney HA, Godron M (eds), Disturbance and Ecosystems—components and response. Springer-Verlag, Heidelberg, pp 259–275

Bazzaz FA (1984) Demographic consequences of plant physiological traits: some case studies. In: Dirzo R, Sarukhán J (eds), Perspectives in Plant Population Ecology. Sinaur Publishers, Sunderland, Massachusetts

Bazzaz FA, Parrish JAD (1982) Organization of grassland communities. In: Estes J, Tyrl RJ, Brunken JN (eds), Grasses and Grasslands. University of Oklahoma Press, Oklahoma

Bazzaz FA, Pickett STA (1980) The physiological ecology of tropical succession: a comparative review. Ann Rev Ecol Syst 11:287–310

Brown AHD, Marshall DR (1981) The evolutionary genetics of colonizing plants. Paper presented at the IInd Int Congr Syst Evol Biol, Vancouver

Caldwell MM, Richards JH, Johnson DA, Nowak RS, Dzurek RS (1981) Coping with herbivory: photosynthetic capacity and resource allocation in two semiarid *Agropyron* bunchgrasses. Oecologia (Berl) 50:14–24

Caldwell MM, Dean TJ, Nowak RS, Dzurec RS, Richards JH (1983) Bunchgrass architecture, light interception and water use efficiency: assessment by fiber optic point quadrats and gas exchange. Oecologia (Berl) 59:178–184

Canham CD, Marks PL (1985) The response of woody plants to disturbance: patterns of establishment and growth. In: Pickett STA, White PS (eds), The Ecology of Natural Disturbance and Patch Dynamics. Academic Press, Orlando, Florida and London

Clegg MT, Allard RW (1972) Patterns of genetic differentiation in the slender wild oat species, *Avena barbata*. Proc Natl Acad Sci USA 69:1820–1824

Hancock JF, Bringhurst RS (1978) Interpopulational differentiation and adaptation in the perennial, diploid species of *Fragaria vesca* L. Am J Bot 65:795–803

Harding JA, Barnes K (1977) Genetics of *Lupinus*. X Genetic variability, heterozygosity and outcrossing in colonial populations of *Lupinus succulentus*. Evolution 31:247–255

Harris GA (1967) Some competitive relationships between *Agropyron spicatum* and *Bromus tectorum*. Ecol Monogr 37:89–111

Jain SK (1975) Patterns of survival and microevolution in plant populations. In: Karlin S, Nevo E (eds), Population Genetics and Ecology. Academic Press, New York

Jain SK (1977) Genetic diversity of weedy rye in California. Crop Sci 17:480–482

Jain SK (1983) Genetic characteristics of populations. In: Mooney HA, Godron M (eds), Disturbance and Ecosystems. Components of Response. Springer-Verlag, Heidelberg

Jain SK, Martins PS (1979) Ecological genetics of the colonizing ability of rose clover (*Trifolium hirtum* All). Am J Bot 66:361–366

Jain SK, Rai KN, Singh RS (1981) Population biology of *Avena XI*. Variation in peripheral isolates of *A barbata*. Genetica 56:213–215

Jain SK, Rice K (1985) Plant population genetics and evolution in disturbed environments. In: Pickett STA, White PS (eds), The Ecology of Natural Disturbance and Patch Dynamics. Academic Press, Orlando, Florida and London

MacArthur RH, Wilson EO (1967) The Theory of Island Biogeography. Princeton University Press, Princeton, New Jersey

Mack RN (1981) The invasion of *Bromus tectorum* L into western North America: an ecological chronicle. Agro-Ecosystems 7:145–165

Mack RN, Pyke DA (1983) The demography of *Bromus tectorum:* variation in time and space. J Ecol 71:69–93

Mack RN, Pyke DA (1984) The demography of *Bromus tectorum:* the role of micro-climate, grazing and disease. J Ecol 72:731–748

Martins PS, Jain SK (1979) The role of genetic variation in the colonizing ability of rose clover (*Trifolium hirtum* All). Am Nat 113:591–595

McKell CM, Major J, Perrier ER (1959) Annual range fertilization in relation to soil moisture depletion. J Range Mgmt 12:189–193

McKell CM, Robison JP, Major J (1962) Ecotypic variation in Medusahead, an introduced annual grass. Ecology 43:686–698

Michaels HJ, Bazzaz FA (1986) Demography and allocation patterns of sexual and apomictic *Antennaria parlinii.* Ecology 67:27–36

Nowak RS, Caldwell MM (1984) A test of compensatory photosynthesis in the field: implications for herbivory tolerance. Oecologia (Berl) 61:311–318

Parrish JAD, Bazzaz FA (1978) Pollination niche separation in a winter annual community. Oecologia (Berl) 35:133–140

Parrish JAD, Bazzaz FA (1979) Difference in pollination niche relationships in early and late successional plant communities. Ecology 60:597–610

Perozzi RE, Bazzaz FA (1978) The response of an early successional community to shortened growing season. Oikos 31:89–93

Raynal DJ, Bazzaz FA (1973) Establishment of early successional plant populations on forest and prairie soil. Ecology 54:1335–1343

Thorgeirsson H, Richards JH (1983) Root elongation rate and water extraction by two aridland bunchgrasses, *Agropyron spicatum* and *Agropyron desertorum.* Bull Ecol Soc Am 64:159

Weaver JE, Albertson FW (1956) Grasslands of the Great Plains. Johnsen Publishing, Lincoln, Nebraska

Wu KK, Jain SK (1979) Population regulation in *Bromus rubens* and *B. mollis.* Life cycle components and competition. Oecologia 39:337–357

Zangerl AR, Pickett STA, Bazzaz FA (1977) Some hypotheses on variation in plant populations and an experimental approach. Biologist 59:113–122

7. Models of Genetically Engineered Organisms and Their Ecological Impact

P.J. Regal

7.1. Introduction

A new category of reproducing organisms will be introduced into the environment. Advances in molecular biology have created a diverse, multibillion dollar, worldwide research effort and an industry that is actively combining genes from widely unrelated taxa. Recombinant DNA techniques are creating a wide range of unicellular and multicellular organisms in the laboratory. It is intended that many of these will multiply in the open environment. Moreover, some that are not *intended* to be released may well escape once they are "upscaled" from the research test tube to the large systems of industrial production.

The industry has an exciting potential not only to promote advances in basic biological research, but also to provide useful products in medicine, agriculture, pollution management, chemical engineering, mining, and other areas. On the other hand there is concern that some free-living recombinant forms might possibly multiply out of control and injure human health, damage crops and natural areas, or otherwise become pests. Society hopes to avoid the experiment that produces the equivalent of "killer bees" or "chestnut blight." Chapters in this volume (and Elton 1958) detail the ecological and economic disaster that runaway species can cause, as well as how difficult and expensive they can be to contain.

This chapter explores the possible ecological effects of free-living genetically engineered organisms. Will they be comparable to domesticated species? Will

they be comparable to wild native species? Will they be comparable to the sorts of wild alien species that we have known before? Can the accumulated knowledge from the history of introduced species, both benign and destructive, be used to help make intelligent decisions about the questions that will inevitably arise in the new era? Will the new forms have novel genetic or ecological properties that society has not had to deal with before? Such questions present a challenge to knowledge in basic biology at all levels.

7.2. A Brief History of the Controversy

Concern over certain laboratory experiments with recombinant DNA began in 1971, leading to the well-known Asilomar meeting at which the community of molecular biologists decided to police their own science. There followed a wave of public anxiety, political activity, and regulation that made research in this field expensive, reputedly difficult, and for 8 years often quite frustrating. But during that time certain of the earlier questions were answered and cautious laboratory work was agreed to be safe (but see Krimsky 1982). The public fear that some sort of "Andromeda strain" could be made by accident, escape from the laboratory, and wipe out all human life is not considered to be realistic. The irrational notion that a new organism that did not previously exist is probably intrinsically dangerous has passed. Regulations have been relaxed. But the tensions created made public discussion more difficult.

During the past decade a general impression was developed that all recombinant organisms would be essentially "crippled" and unable to survive outside of the laboratory. Thus, it came as a surprise to many outside of science and industry, and small segments of government, that "uncrippled" recombinant organisms were being made and some were intended to multiply in the environment. Many products will be conventional crops and livestock modified by recombinant techniques. Others are expected to be wild species modified for food production or environmental manipulation, such as microbes modified to help clean up chemical spills. Robust growth in nature may be quite desirable in such cases.

I initiated and chaired a workshop on the ecological and short-term microevolutionary implications of genetic engineering held in August 1984 at the Banbury Center, Cold Spring Harbor Laboratories (Brown et al. 1984). In background discussions with scientists around the country I had found that many university scientists were uninformed of progress in genetic engineering. In part this was because much of the research had moved off campus.

The issue, though, is no longer only the safety of laboratory experiments. Scientific discussion must now logically include an analysis of the potential ecological roles of a wide variety of free-living engineered organisms.

The discussions in Sharples (1982) and Gore (1984), built carefully on evidence, and cautiously explore the use of introduced species as a model for free-living recombinant organisms. The Gore Report, with a focus on unacceptable hazards such as the costly gypsy moth outbreak, concludes that genetically

engineered organisms in nature will present a "low probability of high consequence risks." However, much of the public and casual discussion, even among scientists, has been superficial, unsystematic, and contradictory. One common generic safety argument is that recombinant forms will be so minimally modified that they will be essentially normal species and will be necessarily harmless. Alternatively, defenders of the generic safety premise also argue that engineered species will be so highly *abnormal* that they cannot possibly survive on their own except in situations where they are "targeted" for survival. Another common argument is that the balance of nature has been so finely tuned by millions of years of evolution that it will purge itself of any artificial novelties. In contrast, we also hear that nature is not so delicately balanced and that disturbance and extinction are natural phenomena. Hence any disturbance or destruction done by novel organisms should not come as a surprise.

As these examples suggest, the general discussion on this issue has been contradictory and confused. It has too often been polemical, and certainly it has been inadequate. Points that may be valid in some circumscribed context have been used unjustifiably in sweeping generic defenses of all potential recombinant organisms. Much of this state of affairs seems to have resulted from a lack of communication between molecular biologists and evolutionary ecologists, and from the fact that adequate public discussion would necessarily have to draw on complex technical understanding in both fields. Much of the lack of communication results from (1) a confusion of professional ecology with philosophical/political "environmentalism;" (2) philosophical differences over scientific methods between molecular biologists and evolutionary ecologists; and (3) considerable differences in the nature of the knowledge basic to the two fields of molecular biology and evolutionary ecology. Systematic discussion and dialogue is needed to address not only the scientific issues but disciplinary issues as well.

7.3. Levels of Risk

The views in futurist fiction have always been divided between the utopian and the dystopian. People tend first to see large new possibilities in terms of their hopes or fears. Similarly, speculation on the issue of recombinant techniques has generated both what some would call euphoria and others hysteria.

Before going on with a discussion of the scientific issues, we must examine the nature of some of the confusion over terms such as "high-probability risk" and "low-probability risk." Various groups have their own ideas of what each of these means.

In discussions of a specific case one should distinguish between the magnitude of the potential *hazard* and its probability of occurrence, and risk would be their product. But magnitude can be treated as constant in this brief and general chapter.

Suppose a researcher *knows* that the probability is one in a hundred that his project will cause a "disaster." Then he may feel that this is a low risk and

that one should not be concerned about it. The public may be more concerned that the risk would be imposed on them *involuntarily* than with the probability of occurrence. A government agency may regard this specific risk as too high since several hundred researchers and companies may be intending to release organisms within a short period. A *collective* industry view might be similar, since a single dramatic mishap could damage public trust for all.

At this hypothetical risk level a small number of "incidents" would almost certainly occur, without some screening. If the product does not have a "track record" in the situation where it will actually be used, and risk estimates are based only on extrapolation from laboratory studies, then there is even more room for disagreement. For example, some will claim that any estimation of risks for humans based on toxicity studies on mice is invalid. We know how difficult such issues are to define and judge since they are seldom all black or white—animal models are useful, but they are not perfect; some projections can be taken very seriously and others not.

I would guess that at some point in the future the public will regard risks of 10^{-2} to 10^{-3} or perhaps 10^{-4} as too high and will demand screening attempts to identify and avoid potential problem organisms and situations. We obviously cannot have a risk-free society. But we demand greater safety margins than this when we buy an airplane ticket, or a steak, and I imagine that society would want the decision makers that scientists advise to be nearly as cautious in taking risks with the future of its public health, economy, and environment. Introduced species that have become pests have caused huge economic damage.

For the sake of this preliminary discussion, I will assume a danger with a rough range of probability between 10^{-2} and 10^{-4}. At least 99 out of every 100 genetically engineered forms may be safe—many will not even be able to establish themselves in nature, and most that can do so will have no destructive effects. If the engineers learn to establish reliably vigorous free-living populations, then without precautions the risk estimate for an advanced generation of creations should obviously be higher.

The above estimates of potential risk are necessarily crude and are based on the record of attempted introductions of alien species given in Sharples (1982) and Simberloff (1981). Often it clearly has been difficult to establish species in new areas, though exact figures are lacking. In one data set of species that *actually became established* in nature 71 out of 854 caused the extinction of a resident species through habitat alteration, predation, or competition (Simberloff 1981). Extinction was a convenient criterion for tabulation, but it is only one possible ill-effect (Elton 1958).

7.4. Models Proposed to Predict Ecological Consequences of Genetically Engineered Organisms

Scientific discussion can proceed only if the multifaceted arguments of potential consequences of introduced recombinant organisms are catalogued. One then can begin systematic analysis of the underlying assumptions and of the claims for generality of each of the various models that in effect have been proposed.

7.4.1. The Domesticated Species Model

It is commonly pointed out that mankind has been manipulating genes for centuries. This is "the old biotechnology" and it has caused few problems, only benefits. Indeed, it is argued, domesticated species do not establish themselves in nature (though one may reply that feral cats, dogs, pigs, goats, horses, burros, water buffalo, and even the "killer" bees have done more damage than is generally known). So, it is argued, the "new biotechnology" is essentially nothing new after all, and it will be generically safe.

We can quickly reject this model. *All forms of manipulation by humans are not equivalent*. Domesticated species are bred for particular features and in the course of this many of their wild traits may be lost. That is a likely reason that their breeding populations do not often escape farms and they may not do well in competition with native species. If, though, one splices genes into a weed to breed a population that will extract and concentrate pesticides from the soil, increase its drought tolerance, and so on, this process will not eliminate the wild traits of the weed. The "hand of man" can produce creatures with very different biological potentials. The range of biotechnology projects even today is dynamic and domesticated organisms cannot stand as a model for *all* the varied projects in biotechnology research and industry.

The common claim that all that genetic engineering involves is the insertion of only a few characterized genes into organisms is, moreover, little more than an ideal. This *can* be done, and there is no doubt that gene-splicing *can be more selective* than traditional breeding. But in practice genes are difficult to characterize completely with regard to both structure and function. Much work is being done by "shotgunning" into organisms large uncharacterized restriction fragments made up of many genes. Vectors also usually remain in the host along with the other new genetic material (see Sect. 7.6).

The papers by Brill (1985) and by Hardy and Glass (1985) argue that *virtually all* released recombinant organisms will be safe based on an idealized model in which only a *few characterized* genes are inserted into highly inbred, *domesticated* organisms. This idealized argument does not, moreover, explore potential secondary effects of the transfer vectors that may be used even in genetically "clean" and conservative modification of highly inbred domesticated organisms.

Finally, the phylogenetic range of experimentation that can be done using breeding methods is minuscule compared to what is even now being tried with recombinant techniques. Traditional breeding does nothing as ambitious as taking the genes for human interferon and placing them in corn. We also read of attempts to place cattle genes in tomatoes, and other laboratory wonders. To introduce highly unusual traits into *wild* species certainly exceeds the model utilized by Brill (1985) and Hardy and Glass (1985).

7.4.2. The Ordinary Sexual Species Model

Similarly, it has been claimed that recombination by genetic engineering is not different from ordinary sexual recombination in the many harmless bisexual

species around us. But this is only superficially true. In ordinary sexual re-combination the genes from remotely related creatures are not mixed as they may be through gene-splicing. Humans and grasses simply do not interbreed. The engineered combinations of genes can dramatically bypass natural genetic barriers. The next section begins to explore the implications of this.

7.4.3. The Unintended Super-Species Model

Many recombinant organisms may be expected to have novel genetic options in the arena of natural selection compared to domesticated or wild-type indi-viduals. This could create completely new survival modes and advantages at once; for example, plant species belonging to groups that are completely mesic may be made drought-resistant for the first time in their evolutionary history. Similarly, it is sometimes of advantage to one player in chess to gain a new piece that allows new moves. A shift in habits would expose the population to new natural selection pressures possibly leading in time, moreover, to a more general genetic reorganization. In either case ecological release by increase in numbers, or shifts in habits, could result.

One may be tempted to counter that hybridization of distantly related taxa often produces sterile or otherwise less fit individuals; so gene-splicing can only decrease fitness. But recombinant techniques are not, as is often implied, comparable to "wide hybridization" in plant breeding. Possible chromosomal mismatching aside, in wide hybridization the organism has *given up* one set of chromosomes in gaining another. The recombinant organism has foreign genes added to a *stable* genome. In wide hybridization the organism has two sets of noncoadapted chromosomes that originate from long-separated gene pools. This is why back-crossing to a parental form to produce a polyploid may often sta-bilize the hybrid genome and capture the benefits of the added genetic material (Wet 1971).

Could recombinant organisms rapidly evolve into ecologically disruptive "super species" with new specific advantages, overall increased vigor, or both? Since microevolutionary, population level, adjustments can take place quickly, as in the case of insects and bacteria that have rapidly evolved resistance to pesticides and antibiotics, even a slightly increased evolutionary potential for some forms would be significant in terms of the human life span. Virulence in microbes may apparently evolve spontaneously and within only months (see Sect. 7.4.9, 7.6.1).

The recombinant genetic material could in theory cause an immediate increase in fitness, and possible ecological release, by: (1) the specific new capacities that have been engineered (e.g., adding herbivore resistance in plants, increased drought resistance); (2) favorable pleiotropic effects; and (3) heterotic effects, or "hybrid vigor" (Mitton and Grant 1984).

The alternative end of the spectrum of possible results is that the addition of genetic material will *diminish* fitness by: (1) metabolic or biomechanical in-stability, (2) unfavorable pleiotropic effects, and (3) hybrid dysgenesis (see Sect. 7.4.6).

It will take specific information to decide where on this spectrum of enhanced

to diminished fitness a given organism would be located. It is unlikely that all genetically engineered organisms would fall to one end of the spectrum or the other. Only much more research on natural populations and evolutionary processes, and on genetically engineered organisms, can give the perspective needed on these issues.

7.4.4. The Biotically Regulated Species Model

Engineered populations, it is claimed, will be native species (or naturalized domesticated species) that regardless of any genetic modifications will be subject to control by their local competitors, predators, parasites, and pathogens. If the recombinant population were to begin to expand, then there would be a compensatory increase in the populations of predators, parasites, and pathogens that would feedback and return the rogue population to supposed equilibrium and end the expansion.

Can we rely on the basic assumption that populations are necessarily (biotically) regulated at some level? Competitors, predators, parasites, and pathogens certainly are *not* the only factors that prevent the expansion of populations (e.g., Ricklefs 1979). Physical factors as well as biotic factors interact to play a "controlling" role, such as seasonal patterns of temperature and rainfall, the distribution of major and trace resources, availability of mating or breeding sites for animals, and of dispersal agents for plants, soil conditions, drought, salinity, and fire cycles. Moreover, outbreaks of native species do sometimes occur, despite the presence of predators and parasites.

If a genetic modification directly or pleiotropically reduces the role of one or more such restricting factors then ecological release could occur, just as a change in the level of the restricting factor itself can cause a population change. The competitors, predators, parasites, and pathogens might or might not increase their numbers to bring the population back to any supposed equilibrium, depending on their own restricting factors and on how important they had been or could ever become in any population regulation. Hosts do, after all, often have defenses against parasites and pathogens.

Lastly, population explosion is not required to cause environmental damage. Changes in habits also can be important, such as, for example, gray squirrels eating bark and killing trees when introduced to Great Britain.

7.4.5. The Untuned Engine Model

Many of the arguments that recombinant organisms will necessarily be safe involve various versions of a belief that genomes and nature itself have been optimized over a long evolutionary history (the next five models, Sect. 7.6). The simple form of the argument is that any alteration of the genome will harm rather than help the individual or population that is left untended in nature. It is as though the organism is a precision, high-performance machine and any tinkering with it will untune it; selection will eliminate an untended recombinant organism. In effect, nature will cleanse itself of anything artificial.

Such beliefs in optimalization are not at present widely held among spe-

cialists, yet they persist generally because they have common-sensical appeal, and because as idealizations they do have uses in mathematical models (Beatty 1980). Notions of optimalization and balance historically have, nevertheless, created firm beliefs about reality that in turn have instructed the interpretation of opposing facts and logic. One must look closely at these contumacious notions. (See also Regal 1985.)

A recent well-known argument by B. Davis (1984) illustrates the untuned engine model:

> "... in the last analysis the judgments are made, and must be made, largely on the basis of fundamental Darwinian principles. These are the same for the spread of any bacteria—in the human population, on plants, or in the soil. The key principle is that the organisms found in nature have been selected over millions of years, from a virtually limitless supply of variation, for their adaptation to some ecological niche; and any genetic modification introduced in the laboratory is infinitely more likely to impair rather than to improve the adaptation, unless the environment is also changed. (For example, widespread use of antibiotics selects for resistant strains.)"

(See reply by Simberloff and Colwell 1985.)

Paleoecologists have found that community composition is constantly being rearranged in response to climatic and other changes on the scale of hundreds or thousands of years (Davis 1983). Species change their biotic associations over thousands of years, which does not support the idea that millions of years of evolution have optimally crafted every species for a particular role or niche in a biological community.

Whittaker's careful and important studies of the distributions of species along ecological gradients showed that each tends to be distributed independently of the others. This is not what one would predict if species were usually tightly coevolved and invariably interdependent (Whittaker 1970). This is not to say that a community is a random assemblage. A species must be well-constructed and it must be able to fit in with the species that are present.

There is abundant evidence that organisms are only *adequately* adapted for survival and they are *not optimally or perfectly* adapted. Careful biomechanical analysis and comparative studies show that there is usually room for improvement. Phylogeny provides many examples of improved systems replacing fore-runners (see also Williams 1966; Lewontin 1978; Gans 1983). It is often astonishing how well-adapted species are, such as orchids that mimic the females of their male insect pollinators, but it is extremely common to exaggerate a spectacular example and say that the organism is perfectly or ideally adapted. The human eye is a remarkable biological organ, but it is nevertheless a considerably flawed optical device.

7.4.6. The Pregnant Pole-Vaulter Model

It is commonly argued that engineered species will be so burdened with novel combinations of metabolic functions that they will not be able to survive in

nature without the assistance of humans. Surely this could often be the case, and one imagines the proverbial pregnant pole-vaulter. This is the stereotype of a human so burdened with a biological function that it cannot compete in demanding situations, but requires special nurturing.

Real organisms do bear many burdens in nature, however, and still survive. No categorical statement can be made that an expensive feature will reduce fitness, for this is an economic issue pivoting on individual circumstances. Any judgment that an organism bearing a new load of metabolic functions would not flourish in nature would have to be made on a case-by-case basis. If the benefit-to-cost ratio is favorable, then the burden is not too great.

Three examples illustrate this general point. Mast flowering in trees, where enormous numbers of flowers bloom and are shed in a few days, may seem extravagantly expensive when one considers that the average individual in a stable population replaces only itself. The cost to moose of replacing 45 pounds of shed antlers each year seems too high when one first thinks of it. In both cases competitive success in breeding benefits in excess of metabolic and material costs. Endothermy is an enormous metabolic expense, but it is nevertheless paid by birds, mammals, and other creatures. If a genetic modification directly or through pleiotropy benefits an organism sufficiently, then the organism may in theory spread in the ecosystem even if the costs are also high. If the new genes were to have a heterotic effect, or were to increase tolerance to climate, increase resistance to disease, predators, or herbivores, expand the food resource base, increase reproductive rate, give a competitive edge, and so on, then it is possible that the benefits could outweigh the costs. There is no reason to assume that all engineered organisms, or even all of the more bizarre ones, will conform to the stereotype of the proverbial pregnant pole-vaulter, though many doubtless will.

New genetic additions, in theory, could reduce fitness by costs other than economic/energetic. They could have adverse pleiotropic effects or they could theoretically add degrees of error to biochemical control systems by contributing "noise," or cybernetic equivocation (Regal 1977, Shapiro 1983).

7.4.7. The Hopeless Monster Model

The hopeless monster model is based on an analogy to the kinds of mutations that result from radiation damage where a functional DNA sequence is disrupted. The idea is that such mutations are much more likely to harm than to benefit an organism, and they will not spread in nature. So if engineered genetic modifications are the functional equivalent of this type of mutation then they will always be producing organisms that will be essentially crippled in nature.

It is hard to see that the addition of *functional* genetic material from species A into species B is really the same as the random *disruption* of a DNA sequence. Moreover, those organisms that are expected to be vigorous in nature would not be crippled. The analogy seems extremely weak at best.

It may be true that a random genetic mutation is more likely to harm an individual than to benefit it, but surely not "infinitely" more likely. Mutations

induced by x-rays and other radiations are reported to have been used to breed "superior" crop plants in as many as 123 released varieties (Dietz 1978, citing Sigurbörjnsson and Micke 1974). Dobzhansky et al. (1977) review evidence from experiments with *Drosophila:* "although the average effect of radiation-induced mutations is deleterious, a minority proved to be favorable and permitted the populations to exploit more effectively the experimental environment."

Promises are commonly made of various sorts of "genetic leashes," based on limited laboratory models. These include *building in* crippling effects on released organisms. But technical problems aside, it is quite uncertain how long such genes would be maintained in populations under natural selection.

7.4.8. The Bankrupt Venture Model

It is commonly argued that over hundreds of millions of years of evolution surely all combinations of traits have been tried. Unfit forms were eliminated by natural selection. Genetic engineering is unlikely to create anything new. The same forces of natural selection will eliminate engineered organisms that are left untended. In effect, "that business can't make it in this town on its own. It has (must have) been tried."

This idea is astonishingly simplistic and theoretically unsound. Take only *the first step* in a calculation: one human being alone, heterozygous at 6700 structural gene loci, can produce 10^{2017} different kinds of gametes. All the theoretical possibilites will never be realized even for the alleles of this restricted set of genes. The number of atoms in the known universe has been estimated at only 10^{70} (Ayala and Valentine 1979).

The idea also ignores the known causes and patterns of extinction, the time scale involved, and biogeography. Over 99% of all species have become extinct and yet many now-extinct lineages thrived for millions of years. Alexander (1985) points out that if a given species had been eliminated by a competitor that subsequently became extinct itself then the mechanism of elimination is no longer present today.

7.4.9. The Baroque Pest Model

Pests and pathogens sometimes show a variety of complicated traits in their life cycles. A great many genes may be involved. So, it is argued, one would *necessarily* have to add a great number of rather specific genes to a benign organism to transform it into a pest or pathogen.

However, that is true only if one is starting out with pure cytoplasm. Comparative biology, though, shows that adaptations do not evolve from nothing. Many traits allowing the vigor of the pathogen or pest may have already been present in its benign ancestor.

Nearly all organisms have parasites, but most are benign and only a few cause disease. The step from benign parasite or commensal to pathogen may involve simply a shift to a new host and minimal modifications. Avian influenza viruses are known to become virulent very suddenly, and also suddenly to

infect and kill mammals such as seals (Hinshaw et al. 1984; Bean et al. 1985). Soil amoebas sometimes shift from free-living habits to become pathogens of humans. Killer bees are not so different from benign honeybees. Flight, stinger, venom, etc. are not modifications for virulence to humans, though attack requires each. Starlings, benign in Europe and pests in America, in either case are not so different from other song birds.

7.4.10. The Introduced, or Exotic, Species Model

At present, those who have systematically studied the issue of the possible effects of genetically engineered organisms favor the hypothesis (Sharples 1982) that introduced species may be the best model that we have for thinking about the addition of noncrippled recombinant organisms to an environment.

In brief, the model says that most attempts at introduction fail. Some now-troublesome introductions even required repeated initial attempts. Most successful introductions are relatively harmless. A small fraction of intentional and accidental introductions have been disasters—the gypsy moth, fire ant, chestnut blight, killer bees, starlings, some human diseases, and so on. Therefore one must study and think about genetically engineered species on a case-by-case basis to try to prevent any major disaster, but the risks are statistically so low that there is no need to call a generic moratorium on all research and development. I am in qualified agreement with the introduced species model. An organism with new biological properties will be an exotic element in nature whether its origins are the laboratory or a distant continent.

Ecologists may not be able to make predictions about genetically engineered species *in abstract terms;* however, on a case-by-case basis, *when specific details are available,* ecologists with broad field experience working together with molecular biologists should be able to give advice that can improve the predictions considerably. Modern, carefully planned introductions for biological control have been relatively safe (but see Howarth 1983).

7.5. Alien Species Issues Are Important to Genetic Engineering

Reference to "introduced," "alien," or "immigrant" species conjures up a number of images, because a number of ecological issues are involved. The following issues concerning alien species are likely to be important in thinking out the possible effects of recombinant populations that may become free-living by release or by accident.

1. Why have many attempts to introduce species failed?

Implications: One argument that genetically engineered organisms will be safe if they escape or are released is that most attempts to introduce new species to areas have failed. On this basis it has been projected that escaped organisms will not establish themselves.

Comments: Ecologists have a great deal of scattered information about the failure of attempts at introduction. The time of year may have been wrong, the

population size was too small, and so on. It is not scientifically satisfactory to make a prediction about the risk or safety of a particular instance based only on a broad statistical observation. This ecological issue demands that science attempt to improve its understanding of the mechanisms that explain why certain classes of attempted introductions of species have succeeded under given conditions while others have failed.

2. When alien species have become pests, what have been the mechanisms of "ecological release"?

Implications: It is commonly argued in defense of the safety of releasing genetically engineered species that in the past alien species have become pests only because they were free from their native competitors, predators, and pathogens. So, it is argued that the engineering of any native species will be safe.

Comments: The specific dynamics of past outbreaks, both of introduced and of native species, clearly show that the above generalization is inadequate (Sect. 7.4).

3. Under what conditions do benign invasions of natural communities take place? Do invasions into natural communities vary as a function of the community's complexity or successional stage?

Implications: This question has theoretical implications and implications for environmental conservation. It is of theoretical interest to know to what extent natural communities are invasible, and when and why a new species cannot establish itself unless it displaces a native species or is otherwise destructive. Paleoecological studies (Davis 1983) argue against species "saturation" of even mature temperate communities. In terms of conservation, can we safely assume that certain natural communities will be relatively "immune" to invasion by engineered species that resemble or are derived from native species? Tropical communities, for example, are often said to be highly resistant to invasion. Yet we do see some major ecological disturbances, such as those caused by imported cichlid fishes and the aquatic floating weed, *Salvinia,* in the relatively undisturbed Sepik River system of New Guinea.

Comments: Much or most of the earth's land surface is in fact under cultivation, disturbed by previous introductions, grazing, fire prevention programs, industrial chemicals, lumbering, etc. So whether or not certain natural communities will reject new species will often be an irrelevant issue since much land is obviously disturbed and much that looks natural is not really natural. The answers may have more limited specific applicability than it might seem, but it is nevertheless important to place ecological theory on as firm a basis as possible.

7.6. The Vectors: Implications

Gene-splicing is commonly done with bits of DNA that function as carriers. It is often claimed that these act precisely and will cleanly implant the genes with surgical precision. But in fact these genetic elements can have other effects that one should be aware of.

Insertion sequences are small replicating pieces of DNA that can naturally "move" about between chromosomes under certain conditions. Insertion sequences will pick up genes, carry them along, and insert these at chemically specific sites on chromosomes. Because of this last trait they are particularly useful in genetic engineering. Transposons, insertion sequences with their cargos, can sometimes move naturally between cells and individuals when attached to plasmids and viruses. Plasmids are circular extrachromosomal DNA found in bacteria and some higher organisms and will pick up transposons. Smaller nonconjugative plasmids may naturally move between cells if they attach to larger conjugative plasmids that have the mechanisms for such movement. Genomes are thus a great deal more fluid than was once thought.

Laboratory gene-splicing collects and uses such natural vectors. Splicing might involve, for example, cutting up DNA with restriction enzymes, linking select fragments to insertion sequences, linking this composite unit in turn to a plasmid or virus, cloning, and inserting the composite of cargo genes and vectors into a new host. The typical recombinant organism then would have not only the desired new genes, but *new* sets of insertion sequences and plasmids or viruses (Watson et al. 1983).

Since Barbara McClintock discovered and characterized moveable genetic elements in maize in the early 1940s and 1950s, it has been confirmed that they can *turn* active and stimulate major, if temporary, bouts of mutation and chromosomal rearrangement (Nevers and Saedler 1977; Campbell 1981a; Engels 1984; Engels and Preston 1984; McClintock 1984). There has also been speculation on the possible evolutionary significance of this phenomenon. This is a large subject and I will discuss it in more detail elsewhere. But some points should be briefly raised here.

7.6.1. Rapid Evolution?

Bouts of mutation and chromosomal rearrangement caused by vector elements can sometimes so alter a laboratory population of some higher taxa that in effect it becomes reproductively isolated from other populations. Therefore the recombinant population in nature may have an unusual evolutionary potential because of (1) its new engineered traits; (2) a propensity of the vectors to generate added genetic variability; and (3) possibly, an unusual capacity for reproductive isolation, which could promote population differentiation (Ish-Horowicz 1982). Sympatric speciation may not occur in nature without difficulty (Futuyma and Mayer 1980), but genetically engineered organisms might well have a head start toward *any* sort of new adaptation, and even speciation. It is widely agreed that *transportation* of genes by plasmids and viruses between bacteria has had major adaptive evolutionary implications for them at the *population* level, and implications for human health. There is little direct evidence, though, that the *mutations* caused in bacteria by viruses and transposons have been an important constructive force in evolution at *higher* levels (Campbell 1981b; Datta and Hughes 1983).

Why are some diseases such as polio and measles relatively stable and easy

to control while others are evolving actively and thus are difficult to control? The new strains of influenza that sweep through every so often are familiar examples of evolutionarily active pathogens (Palese and Young 1982, Couch and Kasel 1983). These are the result of "an intrinsic property of genetic variability that, when expressed, gives rise to variants with altered antigenic determinants" in *only some* of the known types of influenza viruses. In part this is due to high rates of recombination between strains. *Neisseria gonorrhoeae, Salmonella,* and trypanosomes are also difficult to control and strains are known to have changing antigenic properties due to intrachromosomal rearrangements involving active transposons (discussed in Saunders 1982). One should not draw premature conclusions, but there are uniquely detailed data on disease organisms. I suggest that they are important to consider as potential models of genetically *precipitated* evolution, in contrast to the initiation of change exclusively in response to a shifting environment.

Since the 1950s evolutionary specialists have had various notions of genetic inertia, or genetic homeostasis, a unity of the genome, that retards the genomes of higher organisms from responding to arrays of local potential selection pressures and adaptive options. Waddington, Lerner, Darlington, McClintock, Lewis, Wright, and Mayr have all written on this (references in Mayr 1970). Although there has been evidence for slow speciation, there has also been evidence for rapid speciation (Lewis 1966). The question has been, What factors might tend to counteract genetic inertia, to disrupt the cohesiveness of the genome, and promote rapid adaptation, and even speciation? Mayr gave hybridization as an example of a disrupting mechanism that can force a new equilibrium to be established by natural selection.

Findings in molecular genetics suggest that instabilities in repetitive DNA can be involved in responses to genome shock and the temporary production of variability (Marx 1984). It seems fair to ask whether the vectors either alone or with their cargos of transported genes could trigger genomic instabilities that would lead to microevolutionary volatility, and possibly even speciation, in higher organisms. Will genetically engineered populations become evolutionarily volatile and sometimes even produce outbreaks of pests and pathogens?

Many uncertainties exist that could be resolved by appropriate research. Even in the cases of the pathogenic microbes noted above it is theoretically possible that the harmful evolutionary activity was not initiated by spontaneous internal genetic features, but was environmentally initiated. But since potential mechanisms seem to exist for unwanted change, the range of possibilities must be discussed and investigated before science can claim to understand what the ecological fate of engineered species will be.

7.6.2. Enhanced Lateral Transfer of Genetic Material?

Transposons when linked to conjugative plasmids or viruses are known to be involved in the transfer of genetic material laterally—between species. This is very important in bacterial adaptation and microevolution. The extent of such lateral transfer in higher organisms is unknown but is believed to be quite minor (references in Britten 1982; Sharples 1982; Cold Spring Harbor Symp Quant

Biol Vol 45). A small body of publications instead argues for major evolutionary significance, on the basis of largely circumstantial evidence (discussions in Greenwood and Atkinson 1977; Appleby et al. 1983; Reeves and O'Brien 1984; and especially Went 1971; Syvanen 1984, 1985; Lewin 1985).

Transposons with enhancers that could promote lateral transfer in conjunction with intercellular vectors are being constructed and it has been suggested that these should be very carefully contained until their potentials are completely understood. Could they, if introduced into free-living populations, even facilitate penetration of mammalian tissues (Engels, personal communication; Scavarda and Hartl 1984)?

Again, we need to learn much more before we begin either to dispair or to relax. The host range of at least some vectors is known to be quite limited, for example. But the following considerations range from possible to marginally possible and could prove to be of importance. A science of molecular evolution is only in its infancy. Scientists should be aware: (1) that genes foreign to a taxonomic group may spread from an engineered species A to native species B by hybridization or by molecular vectors; (2) that a new capacity such as herbivore or pathogen resistance, enhanced growth on poor soils, etc. may thus turn B into a pest; (3) that any transfer of evolutionary volatility may promote the rapid evolutionary assimilation of the new trait, or may even promote a rapid adaptive radiation of undesirable organisms.

More research is needed on the ecology of the vector elements. Much exciting work is in progress, but many of the questions raised here will require more detailed knowledge of the ecology of moveable bits of DNA in nature.

7.7. Conclusions

Public discussion on the subject of the safety of free-living engineered organisms may have been misfocused and clouded by sensationalism and emotional debate. Even the calmer and serious discussion on this subject has been seriously flawed by incorrect and oversimplified common-sensical notions about evolutionary ecology. Common sense has, though, proved to be no more a reliable substitute for critical empiricism in ecology or evolutionary biology than it has been in astronomy or physics. Appreciation of this is of utmost importance because *the arguments that recombinant populations will always be safe are now based not on empirical studies but almost entirely on outdated, common-sensical beliefs about ecology and evolution.*

On reanalysis, there is no basis to conclude that all genetically engineered organisms will be safe in nature. Enough possible dangers are associated with *any* exotic organism so that each potential release should be looked at very carefully. The sorts of damage that populations out of control can cause are well known and need not be itemized in a volume such as this. The evaluation procedures for engineered creatures will present special methodological problems since one cannot know what their interactions have been in some native habitat, as one can in the case of a traditional proposed introduction.

There is an obvious need for more basic research in evolutionary ecology and related disciplines so that science can build a firm conceptual basis for understanding and dealing with the coming generations of new organisms, as well as with a new scale of attempts to engineer changes in nature. To paraphrase Voltaire: as long as men believe absurdities they will commit atrocities. With the vast power of genetic engineering, society can be expected to try to manipulate natural systems in ways that can scarcely be dreamed of. The issues of safety and runaway organisms aside, scientists will need much more basic information on the organization and functioning of natural biotic populations and communities if society is not, simply through misjudgment, to multiply the ecological misadventures and disasters of the past, in its future attempts to control nature. The issues are too complex and technical to expect decision makers to be able to deal well with them without good advice from scientists in a range of disciplines.

There is clearly an immediate need to find methods to screen for potentially dangerous recombinant forms, but this is a separate issue from the equally immediate need for more basic research on natural populations and communities and on the ecological and evolutionary properties of recombinant organisms.

The mainstream of research in ecology has already been useful if it corrects the simplistic ideas of the past, on which a false sense of security and a carelessness with gene-splicing might well have become established. Continued research within the mainstream, with an eye on the issues raised here, is to be encouraged. A main point of the present chapter is to suggest the importance of more study on exotic species.

Most research on introduced species has been done from the mission-oriented point of view of controlling them, but more *basic* research might well be done with an aim toward improving ecological and evolutionary theory. For example, too often a rogue species is studied only in its new habitat and not in its native habitat, so that there is little context in which to interpret its ecological expansion. Expeditions to its native habitat may concentrate too much on simply looking for a potential biological control agent and may not gain important ecological information (Howarth 1983 is of some interest here). The genetics of introduced species are usually not studied and compared in both new and native habitats, but in the absence of careful comparative studies it is difficult to decide finally if, for example, hybridization was critical in the explosive expansions of populations of fire ants, *Lantana,* and killer bees.

It would be especially timely and important to encourage greater interaction between microbial ecologists and the main body of researchers in ecology because many of the recombinant organisms that may first become established in nature are microbes (Stotzky and Babich 1984).

The basic biology and especially the population and ecological genetics of genetically engineered species should be studied in the context of the issues raised here. *As soon as organisms are released* scientists should be prepared to study both success and failure, population genetics, and any long-term changes in ecology or genetics in such a way that not only a practical but also

a basic theoretical understanding of the new class of organisms eventually can be gained.

7.8. Acknowledgments

I thank Ernst Mayr and Peter Raven for their encouragement to undertake this general project and for conversations. A very large number of colleagues are also due great thanks for critical help and discussions. Among them, I particularly thank those who made comments on various drafts of the manuscript: P. Abrams, T. Allison, D. Andow, J. Brokow, R. Colwell, K. Corbin, E. Cushing, M. Davis, M. Kottler, M. Levin, H. Mooney, P. Morrow, G. Orians, D. Pimentel, S. Risch, F. Sharples, D. Simberloff, and H. Tordoff. The Cold Spring Harbor workshop in 1984 was an invaluable opportunity for each participant to take an important step toward clarification of information and perspectives on this difficult and complex issue. It also underscored the need for the present attempt at a point-by-point organization and analysis of the issues. I very much would like to thank those from government, industry, and academia who contributed, particularly Jack Fowle, whose good judgment and organizational talents were so critical to its quality.

7.9. References

Alexander M (1985) Ecological consequences: reducing the uncertainties. Issues Sci Technol 1(3):57–68

Appleby CA, Tjepkema JD, Trinick MJ (1983) Hemoglobin in a nonleguminous plant, *Parasponia:* possible genetic origin and function in nitrogen fixation. Science 220:951–953

Ayala FJ, Valentine JW (1979) Evolving: The Theory and Process of Organic Evolution. Benjamin/Cummings, Menlo Park, California

Bean WJ, Kawaoka Y, Wood JM, Pearson JE, Webster RG (1985) Characterization of virulent and avirulent A/Chicken/Pennsylvania/83 Influenza A viruses: potential role of defective interfering RNAs in nature. J Virol 54:151–160

Beatty J (1980) Optimal-design models and the strategy of model building in evolutionary biology. Philos Sci 47:532–561

Brill WJ (1985) Safety concerns and genetic engineering in agriculture. Science 227:381–384

Britten RJ (1982) Genomic alterations in evolution. In: Bonner JT (ed), Evolution and Development. Springer-Verlag, Berlin/Heidelberg/New York, pp 41–64

Brown J, Colwell RC, Lenski RE, Levin BR, Lloyd M, Regal PJ, Simberloff D (1984) Report on workshop on possible ecological and evolutionary impacts of bioengineered organisms released into the environment. Bull Ecol Soc Am 65:436–438

Campbell A (1981a) Some general questions about moveable elements and their implications. Cold Spring Harbor Symp Quant Biol 45:1–9

Campbell A (1981b) Evolutionary significance of accessory DNA elements in bacteria. Annu Rev Microbiol 35–83

Couch RB, Kasel JA (1983) Immunity to influenza in man. Annu Rev Microbiol 37:529–549

Datta N, Hughes VM (1983) Plasmids of the same Inc groups in Enterobacteria before and after the medical use of antibiotics. Nature 306:616–617

Davis B (1984) Science, fanaticism, and the law. Genet Engin News 4(5):4

Davis M (1983) Quaternary history of deciduous forests of eastern North America and Europe. Ann Missouri Bot Gard 70:550–563

Dietz A (1978) The use of ionizing radiation to develop a blight resistant American chestnut, *Castanea dentata*, through induced mutations. In: Proceedings of the American Chestnut Symposium, Morganstown, West Virginia, pp 17–20

Dobzhansky T, Ayala FJ, Stebbins GL, Valentine JW (1977) Evolution. WH Freeman, New York, 572 p

Eisner T (1983) Chemicals, genes, and the loss of species. Nat Conserv News 33(6):23–24

Elton CS (1958) The Ecology of Invasions by Animals and Plants. Methuen, London

Engels WR (1984) The P family of transposable elements in Drosophila. Annu Rev Genet 17:315–344

Engels WR (ms) Precautions for handling P factors. (Manuscript in preparation.)

Engels WR, Preston CR (1984) Formation of chromosome rearrangements by P factors in *Drosophila*. Genetics 107:657–678

Futuyma DJ, Mayer GC (1980) Non-allopatric speciation in animals. Syst Zool 29(3):254–271

Gans C (1983) On the fallacy of perfection. In: Fay RR, Gourevitch G (eds), Perspectives on Modern Auditory Research: Papers in Honor of EG Wever. Amphora Press, Groton, Connecticut, pp 101–112

Gore A Jr (1984) The environmental implications of genetic engineering. Report of the Subcommittee on Investigations and Oversight to the Committee on Science and Technology, US House of Representatives, 98th Congress, 2nd session, Serial V, US Government Printing Office

Greenwood RM, Atkinson IAE (1977) Evolution of divaricating plants in New Zealand in relation to Moa Browsing. Proc N Zeal Ecol Soc 24:21–33

Hardy RWF, Glass DJ (1985) Our investment: what is at stake? Issues Sci Technol 1(3):69–82

Hinshaw VS, Bean WJ, Webster RG, Rehg JE, Fiorelli P, Early G, Geraci JR, St Aubin DJ (1984) Are seals frequently infected with avian influenza viruses? J Virol 51:863–865

Howarth FG (1983) Classical biocontrol: panacea or Pandora's box. Proc Hawaiian Entomol Soc 24:239–244

Ish-Horowicz D (1982) Transposable elements, hybrid incompatibility and speciation. Nature 299:676–677

Krimsky S (1982) Genetic Alchemy: The Social History of the Recombinant DNA Controversy. MIT Press, Cambridge, Massachusetts

Lewin R (1985) Fish to bacterium gene transfer. Science 227:1020

Lewis H (1966) Speciation in flowering plants. Science 152:167–172

Lewontin RC (1978) Adaptation. Sci Am 239(3):213–230

Marx JL (1984) Instability in plants and the ghost of Lamarck. Science 224:1415–1416

Mayr E (1980) Populations, Species, and Evolution. Harvard University Press, Cambridge, 453 p

McClintock B (1984) The significance of responses of the genome to challenge. Science 226:792–801

Mitton JB, Grant MC (1984) Associations among protein heterozygosity, growth rate, and developmental homeostasis. Annu Rev Ecol Syst 15:479–499

Nevers P, Saedler H (1977) Transposable genetic elements as agents of gene instability and chromosomal rearrangements. Nature 268:109–115

Palese P, Young JF (1982) Variation of influenza A, B, and C viruses. Science 215:1468–1474

Reeves RH, O'Brien SJ (1984) Molecular genetic characterization of the RD-114 gene family of endogenous feline retroviral sequences. J Virol 52:164–171

Regal PJ (1977) Evolutionary loss of useless features: is it molecular noise suppression? Am Natur 111(977):123–133

Regal PJ (1985) The ecology of evolution: implications of the individualistic paradigm. In: Halvorson HO, Pramer D, Rogul M (eds.), Engineered Organisms in the Environment: Scientific Issues. American Society for Microbiology, Washington, DC pp 11–19

Ricklefs RE (1979) Ecology. Chiron, New York

Saunders JR (1982) Chromosomal rearrangements in gonococcal pathogenicity. Nature 299:781–782

Scavarda NJ, Hartl DL (1984) Interspecific DNA transformation in *Drosophila*. Proceedings of the National Academy of Sciences, USA 81:7515–7519

Shapiro BL (1983) Down syndrome—a disruption of homeostasis. Am J Med Genet 14:241–269

Sharples FE (1982) Spread of organisms with novel genotypes: thoughts from an ecological perspective. ORNL/TM-8473, Oak Ridge National Laboratory Environmental Sciences Division Publication No 2040, 50 p (reprinted in Recomb DNA Tech Bull 6:43–56)

Sigurbjornsson B, Micke A (1974) Philosophy and accomplishments of mutation breeding, polyploidy and induced mutations in plant breeding. International Atomic Energy Agency, Vienna, pp 37–39

Simberloff D (1981) Community effects of introduced species. In: Nitecki MH (ed), Biotic Crises in Ecological and Evolutionary Time. Academic Press, New York

Simberloff D, Colwell R (1985) Release of engineered orgainisms: a call for ecological and evolutionary assessment of risks. Genet Engin News 4(8):4

Stotzky G, Babich H (1984) Fate of genetically-engineered microbes in natural environments. Recomb DNA Tech Bull 7:163–188

Syvanen M (1984) The evolutionary implications of mobile genetic elements. Annu Rev Genet 18:271–293

Syvanen M (1985) Cross-species gene transfer: implications for a new theory of evolution. J Theor Biol 112:333–343

Watson JD, Tooze J (1981) The DNA Story: A Documentary History of Gene Cloning. WH Freeman, San Francisco

Watson JD, Tooze J, Kurtz DT (1983) Recombinant DNA: A Short Course. WH Freeman, New York

Went FW (1971) Parallel evolution. Taxon 20:197–226

Wet JMJ de (1971) Polyploidy and evolution in plants. Taxon 20:29–35

Whittaker RH (1970) Communities and Ecosystems. Macmillan, London

Williams GC (1966) Adaptation and Natural Selection. Princeton University Press, Princeton, New Jersey

3. Site Characteristics Promoting Invasions and System Impact of Invaders

8. Site Characteristics Favoring Invasions

G.H. Orians

8.1. Introduction

Recent increases in intercontinental invasion rates by organisms of many taxa, brought about primarily by human activity, create both important ecological problems for the recipient lands and opportunities to understand better those factors that favor success as a colonizer and the environmental conditions that favor successful invasions. Elton's (1958) extensive review of invasions by plants and animals concentrated on the ecological conditions in areas where alien individuals were arriving. From his survey, Elton concluded that invaders were more likely to establish viable populations in cultivated or otherwise disturbed and, usually, simplified communities. He also noted that natural habitats on small islands were much more vulnerable to invading species than those on large continents, a point clearly anticipated by Darwin in *The Origin of Species*.

Much of the empirical evidence gathered since Elton wrote his book has supported his general claims. A botanist is likely to recognize roadside plants in most parts of the world, but identification of plants in the less disturbed vegetation beyond demands knowledge of the local flora. Gardeners in the Pacific Northwest wage continual war against a variety of slugs introduced from Europe, whereas the closed forests of the surrounding hills, even those in various stages of succession following clearcutting earlier in the century, are populated almost entirely by native slugs. Urban areas in America support dense populations of starlings *(Sturnus vulgaris)*, house sparrows *(Passer domesticus)*, and

rock doves *(Columba livia)*, introduced species that are largely restricted to agricultural areas outside cities.

Even though Elton's conclusions appear to be withstanding the tests of time, not much progress has been made since 1958 in understanding why disturbed and island communities are more readily invaded than relatively undisturbed mainland communities. Nor do we understand why disturbed communities are invaded by both native and foreign species or why natural ecological comunities are invaded by *some* introduced species. In part, the relative lack of progress is due to the fact that the problems are very complex. Ecologists have great difficulty in identifying the factors regulating distributions and abundances of even well-studied individual species. It is not surprising, then, that we find it difficult to understand problems involving interactions among many species.

In part, however, the lack of progress has been caused by a shift of attention away from the receiving environment to the nature of the colonizing species themselves. Much thought has been given to traits believed to enhance the probability that a species becomes a successful invader in areas into which it is introduced. Among the traits generally believed to favor success are high population growth rates, short life cycles, high levels of allocations of resources to reproduction, good dispersal, and flexible utilization of a variety of environmental resources (Baker and Stebbins 1965; Bazzaz, this volume). Nonetheless, many species possessing most of these traits have not become established when introduced whereas other species quite deficient in these traits have. Demographic traits of individual species may influence probabilities of success, but they do not guarantee it. We need to develop an approach that combines information on the invading organisms and the environments into which they are being introduced. This chapter is a contribution to that effort.

My approach is first to point to the importance of considering the ways different kinds of organisms relate to their environments. Next I will examine the notion of disturbance and show, by subdividing disturbance into various types that affect interactions among organisms differently, that it is then easier to suggest possible causes for success or failure in different taxa. Third, I will examine some aspects of interactions among organisms in undisturbed communities that may influence how newly arriving species are likely to fare when they are inserted into systems of many species. Finally, I will point to the importance of different time frames in thinking about problems of invasions.

When we refer to invasions we usually mean those organisms moved around by human activity during the past few centuries, particularly within the last 150 years. However, in the context of a longer time frame, most organisms are invaders into the communities where they now live. At high latitudes, successive advances and retreats of glaciers produced great shifts in distributions of living organisms, and adjustments in ranges related to climatic amelioration are still occurring. Many present-day British weeds were common during the early part of the postglacial period, also a time of great disturbance. Modern times are ones of reinvasion by species that had declined during the early centuries following the retreat of the glaciers (Godwin 1956). At still longer time frames,

we can learn a great deal from paleontological studies of great invasions such as occurred when the Panamanian Land Bridge was established several million years ago (Simpson 1980).

8.2. Ecosystems Are in the Mind of the Beholder

When we speak of disturbed ecosystems and environments we do so from the perspective of a large mammal. This perspective of environmental grain (Levins 1968) may serve us well for thinking about and analyzing invasions of large terrestrial vertebrates but it is less suitable for thinking about plants, even large ones, and even poorer for thinking about smaller animals, especially insects. A disturbed environment to an herbivorous insect may simply be one in which its host plant is under some physical stress and, hence, has reduced levels of chemical defenses. Or it may be one in which there are slight changes in soil surfaces that increase the probability of pupal desiccation. Therefore, even though Elton's generalization may hold, the underlying mechanisms are likely to differ among different groups of organisms.

The beholder's view of an ecosystem also depends on the nature and extent of its coevolved relationships. An insectivorous bird, for example, should find a wide range of insects in a new continent suitable as food even if none of them are familiar. Moreover, the bird's nest site requirements and its predator escape responses should also function well in many environments. Birds are rarely involved in coevolved mutualisms that create requirements for the presence of other species with particular ecological traits. A shrub or tree with animal-pollinated flowers and animal-dispersed fruits, however, requires,in addition to suitable conditions for germination and growth, suitable pollinators and seed dispersers if it is to establish itself in a new area. Often these interactive requirements can be met by other species in the new areas (many pollinators and frugivores are generalists; Wheelwright and Orians 1982), but the existence of obligate mutualistic associations adds to the requisites of successful invasions.

Evidence that coevolved mutualisms are important for the establishment of vascular plants in new areas does not appear to have been evaluated systematically. The majority of weedy plants that have invaded new continents are wind-pollinated and most do not produce fleshy fruits that are animal-dispersed. Therefore, they lack obligate mutualistic requirements. The two broad-leaved trees most successful in invading disturbed forests in eastern North America, the tree of heaven (*Ailanthus altissima;* Simaroubaceae) and the empress tree (*Paulownia tomentosa;* Scrophulariaceae), both have insect-pollinated flowers but wind- dispersed seeds. On the other hand, almost no conifers introduced into Europe or into North America from elsewhere have established feral populations even though many of them have been extensively planted and most do not require the services of animals for either pollen or seed dispersal (Mitchell 1978).

8.3. Disturbance Is Also in the Mind of the Beholder

Disturbance comes in many forms and it affects different organisms in varied ways. One organism's pollutants are another's nutrients. Disturbance may affect soil surfaces, thereby changing germination and pupation sites, soil surface microclimates, and the food resources available to organisms living in the soil. Disturbance may influence competitive relationships among organisms by imposing heavy mortality on competitors or by enhancing resources through fertilization, eutrophication, and introduction of novel foods. Disturbance may influence predator–prey relationships by selective removal of some predators, introduction of others, or by imposing important changes in the temporal pattern and severity of predation. Disturbance may influence mutualistic relationships by altering abundances of species tied to one another by such relationships, or by adding species that interact in new ways with species already present. Because of the diversity of these effects, the role of disturbance in influencing invasions must be studied by examining the different types of disturbances, specifying their most probable effects, developing ideas about how those changes should influence species with different characteristics and, finally, making predictions or postdictions about the probabilities of success of invaders in the light of those postulates. I limit my analysis of these problems to a few cases picked because they illustrate the approach and because data were available to permit at least preliminary evaluation of the ideas.

8.3.1. Disturbance and Predation Patterns

Human activities greatly alter predation patterns in ecological communities through such processes as extermination of large predators, selective removal of predatory arthropods by extensive use of pesticides to which the predators adapt more slowly than their prey, and by introduction of new grazing and browsing patterns on plant communities. Altered predation patterns may produce varied results. If predators had prevented potential competitors from becoming abundant enough to compete strongly, reduced predation could enhance competition among existing species and, hence, make communities more resistant to invasion. Alternatively, if predators had prevented the establishment of prey with life history traits that made them more vulnerable to those predators, then reduction of predators could allow new species, either aliens or species of other nearby habitats, to exist. In the case of grazing, domestic animals may impose different temporal patterns of grazing that affect which life history patterns of plants are most successful.

One of the most striking invasions into North America has been the establishment and dominance of a number of European and Middle Eastern annual plants into California (Mooney et al., this volume) and the intermountain west (Mack, this volume). The spread of alien plants into the intermountain west, especially cheatgrass (Bromus tectorum), was remarkably rapid (Leopold 1941; Mack 1981). Prior to the invasions of the aliens, much of the intermountain west was covered with open-canopied communities of shrubs, especially of the

families Chenopodiaceae and Compositae, with a rich understory of perennial bunch grasses and cryptogams. These bunch grasses, and some of the smaller shrubs such as winterfat *(Ceratoides lanata)* and bitterbrush *(Purshia tridentata)*, are highly palatable and nutritious to grazing mammals (Kennedy 1903; Hutchings and Stewart 1953; Ellison 1960). Originally, when sagebrush grasslands were burned or heavily grazed, a pattern of succession initiated by snakeweed *(Gutierrezia sarothrae)* and followed by short-lived perennial grasses such as bottlebrush squirreltail *(Sitanion hystrix)* and Sandberg's bluegrass *(Poa sandbergii)* developed. Soon thereafter seedlings of sagebrush and longer-lived perennial grasses appeared. If disturbance was prevented from recurring, revegetation progressed far toward a "climax" community within a decade or so.

This situation changed dramatically around the turn of the century when continual disturbance by fire; heavy grazing; agricultural practices; and construction of railroads, roads, towns, and canals set the stage for the rapid invasion of alien annuals (Young et al. 1972; Yensen 1981). Although cheatgrass has been a very successful invader (Piemeisel 1938; Platt and Jackman 1946) it does not normally invade stands of healthy native vegetation (Young et al. 1972). The rapid decline of the native perennial grasses has led to the widespread notion that these grasses were not adapted for grazing (Mack and Thompson 1982).

An additional component of the success of European annuals in the Great Basin may have been changes in the timing as well as the intensity of grazing. These lands were grazed by medium-sized and large mammals prior to European settlement of the area, but the pattern was primarily one in which the mammals spent the winter in the lowlands grazing on the previous summer's vegetative growth of the perennial grasses, and moved up to montane pastures during the summer drought in the lowlands. If grazing pressures were, indeed, heaviest in the winter, then autumn-germinating annuals such as *Bromus tectorum* would present young and highly palatable tissues. This could result in heavy grazing on, and, presumably mortality of, the seedlings of annuals. Heavy grazing on mature perennials in the winter would remove primarily tissues to be replaced the following spring, and whereas juvenile perennials would also have been subjected to winter grazing, recruitment rates are in any case low among perennials. Modern domestic livestock, however, are pastured such that they produce sustained heavy use of the basin shrub-grasslands in late spring and early summer, when loss of photosynthetic tissues by the perennials is most damaging and when most of the annuals have already set seed prior to the long, dry summer.

Thus, perennial grasses of the intermountain west may have been adapted to at least light to moderate grazing but grazing that was concentrated in the winter. Fall-germinating annuals have difficulty competing with perennials when there is little grazing or grazing primarily in the winter because they are most palatable and most vulnerable then. By contrast, the grasses of the Great Plains were subjected to year-round grazing by bison herds. Substitution of domestic cattle for bison did not change the pattern of grazing substantially and those grasslands have not been extensively invaded by foreign species despite being

heavily utilized. A similar situation should exist in the Middle East and Central Asia, where altitudinal migrations of large mammals probably prevailed prior to the introduction of animal husbandry. Those areas are dominated by caespitose grasses similar to those of the intermountain west (Mack and Thompson 1982) but they may be better adapted to heavy grazing (Caldwell et al. 1981).

8.3.2. Disturbance and Resource Enhancement

One possible consequence of disturbances is to increase the levels of some resources. Many aspects of pollution fall into this category, especially eutrophication of fresh water environments. Some theoretical arguments suggest that the number of species able to coexist in an environment is positively correlated with the amount and variety of resources present above some threshold level of exploitation (MacArthur 1972; MacNally 1983).

The exact processes by which new species are favored rather than increases in the numbers of species already present are unknown but successful invasions of birds into North America are associated with massive resource enhancement. During the past century only three species of birds have established populations in North America by means of natural range extensions. All other foreign invasions involve either escaped cage birds or deliberate introductions for esthetic or harvesting purposes. The three natural invaders are the cattle egret *(Bubulcus ibis)*, which has spread from Africa via South America to become a common breeder over large areas of the southern and central United States, the blackheaded gull *(Larus ridibundus)*, which now breeds sparingly in eastern Canada,

Table 8.1. Escaped cage birds that now breed ferally in North America

Species	Where Established as Feral Populations	Diet
Spotted dove *(Streptopelia chinensis)*	Los Angeles	Seeds
Ringed turtle dove *(S. risoria)*	Los Angeles, Miami Tampa, Houston, Mobile	Seeds
Rose-ringed parakeet *(Psittacula krameri)*	Miami, Los Angeles	Fruits, seeds
Budgeriger *(Melopsittacus undulatus)*	Miami, Southwest Florida	Fruits, seeds
Canary-winged parakeet *(Brotogeris versicolurus)*	Los Angeles, Southeast Florida	Fruits, seeds
Monk parakeet *(Myiopsitta monachus)*	Florida	Seeds, fruits
Red-crowned parrot *(Amazona viridigenalis)*	Southeast Florida, Los Angeles	Fruits, seeds
Yellow-headed parrot *(Amazona oratrix)*	Southeast Florida, Los Angeles	Fruits, seeds

Table 8.2. Escaped waterfowl that have failed to establish feral populations in North America

Species	Native Range
Black swan (Cygnus atratus)	Australia
Red-breasted goose (Branta ruficollis)	U.S.S.R.
Greylag goose (Anser anser)	Europe
Bar-headed goose (Anser cygnoides)	China
Muscovy duck (Cairina moschata)	Central America
Northern shelduck (Tadorna tadorna)	Europe
Ruddy shelduck (Tadorna ferruginea)	Middle East
Mandarin duck (Aix galericulata)	Eastern Asia

and the little gull (Larus minutus), which breeds in a number of areas in the Great Lakes. Gulls are a group in which most of the native North American species have undergone large increases in population sizes, probably as a result of lake eutrophication and their effectiveness in exploiting garbage dumps. Cattle egrets exploit a greatly expanded resource, cattle, as flushers of insects. This resource has replaced a preexisting one provided by bison. Egrets might have been successful had they made the transatlantic crossing prior to the European settlement of North America, but at that time they would have had to cross vast areas in the Neotropics lacking any large grazing mammals to reach the temperate grasslands.

Many of the successful avian escapees from captivity also exploit new food resources found in the rich plantings of exotic trees in urban areas, particularly in the warmer parts of North America (Table 8.1.). Note that all these species are granivores and frugivores, exploiting the resources most conspicuously enhanced in cities. These cases acquire additional significance when contrasted with the list of waterfowl known to have escaped a number of times from captivity but that have failed to establish feral populations (Table 8.2.). Interestingly, waterfowl are a group in which there has been habitat contraction and reductions in numbers of most of the native species. That there is more to the situation, however, is suggested by the success of species of exotic waterfowl introduced into western Europe. The bar-headed goose (Anser indicus) is established as a feral breeder in Sweden. The Canada goose (Branta canadensis) is now widespread in Britain and southern Scandinavia. The Egyptian goose (Alopochen aegyptiacus) now breeds in southeast England, the Mandarin duck (Aix galericulata) breeds in both Britain and Germany, and the ruddy duck (Oxyura jamaicensis) is established in the English Midlands. The success of introduced waterfowl in Europe may relate to the fact that a combination of hunting and

wetland drainage seriously reduced populations of native species during recent centuries. The recent abolition of hunting and creation of many new artificial ponds and lakes may have made available relatively unexploited habitats into which the introduced species have invaded. Some, such as the Mandarin and ruddy duck, are species for which no close ecological equivalent was present in northern Europe.

Among the most extensive invasions of exotic species are freshwater fishes (Courtenay and Stauffer 1984). Most fresh waters have been highly modified by heavy fishing, vegetation changes in their watersheds, fertilization, and introduction of toxicants. Determining the relative influences of these events on the success of introduced species may be difficult, but some of the successful invasions have doubtless been influenced by resource enhancement.

8.4. Biotic Interactions and Invasions

Species in ecological communities are connected through competitive, predator–prey, and mutualistic associations. Some species are more intimately involved with their associates than are others and all ecological communities have subsets of species among which connections are stronger than with other subsets (Paine 1966, 1980). These connections have both dynamic and evolutionary components (MacMahon et al. 1981), and both are likely to influence the fate of invaders. Here I deal with dynamic relationships in ecological time. Later in the chapter I will address the possible importance of evolved and evolving relationships on the temporal pattern of success of invaders.

8.4.1. Chemical Ecology and Invasions

Recent years have witnessed an explosion of information on the importance and complexities of the chemical interactions between trophic levels, especially among plants and herbivores (Brower and Brower 1964; Brower et al. 1968; Caswell et al. 1973; Edmunds and Alstad 1978; Jones 1979; Rhoades 1979; Fox 1981). These interactions have both constitutive and facultative components because many plants respond to herbivore attacks by altering their defensive postures (Haukioja and Niemela 1979). Defensive commitment is usually decreased as a plant responds to physical stress (see Rhoades 1979 for a review). One major theory of the origins of insect outbreaks postulates that they usually get started in areas where the plants are weakened by physical stresses (White 1974, 1978).

If host chemistry is important for the survival of herbivores, two predictions about where herbivores can be expected to invade successfully follow. The first is that invading herbivores should use species of plants that are chemically similar to the ones they attack in their native ranges. The second prediction is that herbivores should establish invasive populations most readily on stressed plants. A possible example of the first prediction is the insects that attack coniferous trees in western North America. Furniss and Carolin (1980) list 17 alien

species of Coleoptera, Diptera, Homoptera, and Lepidoptera that attack western conifers, four of which are considered major pests. All of these attack the same genera in North America as they do in Europe although the balsam woolly aphid *(Adelges picea)* attacks primarily spruce in Europe but firs *(Abies)* in North America, and the larch wooly aphid *(Adelges strobilobius)* attacks both *Larix* and *Picea* in North America, but only the former genus in Europe.

This argument in reverse suggests that insects attacking introduced plants should be drawn from species that normally eat plants of similar chemistry. Much information is being collected on the rate at which introduced plants acquire new herbivore faunas (Southwood 1961; Strong 1974a,b) but for the most part the natural hosts of the invaders are unknown. The extensive planting of trees in new areas offers substantial opportunities to study this important phenomenon and much more effort should be devoted to this end. For example, the relative freedom of *Eucalyptus* species in California from insect attack may stem from the fact that they are members of the Myrtaceae, a family not represented in California. They may be so different from native trees that the "chemical jump" for any herbivore is too great to be easily accomplished. A corollary prediction is that plants distantly related to or chemically distinct from species in the area into which they are introduced should be better able to invade than plants that are taxonomically, and hence chemically, more similar to the flora into which they are introduced.

Success of invaders on stressed plants does not seem to have been investigated specifically. Of the 17 introduced insect pests on western confers recorded by Furniss and Carolin (1980), 13 are more important pests on introduced conifers and individuals grown in plantations than they are on native species growing under more or less natural conditions. The causes for this pattern are not clear, however, because not only are cultivated individuals usually growing under atypical climatic and soil conditions, but they may also be growing in places where natural enemies are rare and where potential predators on the introduced insects are also rare or lacking. Therefore, more study will be needed to determine the relative roles of these and possibly other factors in favoring success of introduced insects on exotic conifers.

8.4.2. Diffuse Competition and Successful Invasions

Altered patterns of availability of resources, changes in levels and types of predation, and stress on native communities may all exert effects through a process commonly referred to as diffuse competition (MacArthur 1969, 1972; Diamond 1975; Pimm 1982; MacNally 1983). Traditionally, following the arguments of Darwin, ecologists have sought to demonstrate the role of competition by comparing closely related species presumed, by virtue of similar morphologies and ecological requirements, to be most likely to compete strongly (Gause 1934; Diver 1940; Pulliam 1975; Connell 1983; Schoener 1983). More recent evidence indicates that competition may be strong among organisms of quite different taxa, such as mammals, ants, and birds exploiting seeds of desert plants (Brown and Davidson 1977; Brown et al. 1979; Davidson et al. 1980).

This, together with the recognition that many species overlap broadly in their use of environmental resources, has led to the perspective that species may be affected by the activities of many other species, each of which exerts only a small effect.

Although the existence of diffuse competition is theoretically and conceptually very attractive, it is, unfortunately, a very difficult process to demonstrate directly. Even in the case of weeds in agricultural crops, a system subjected to intensive study, there are still very few studies on the extent to which weeds depress growth rates and yields of crop plants and vice versa (Mortimer 1984). How disturbance affects diffuse competition and, thereby, influences invasions remains to be studied.

8.5. Time and the Invasion of Ecological Communities

Our views of invasions change with our time frame of reference. Moreover, it is important to think of time not in absolute terms but with reference to the lengths of life cycles of the organisms under consideration. An introduced tree that becomes established after several generations may appear to us to have an unusually long adjustment period compared to an annual plant that becomes established within a decade even though the latter may have failed to expand from the site of its introduction for more generations than the former. The distinction is important because it directs attention to the kinds of adjustments that may have been made within and among species during the time the potential invader was present but not expanding its population.

Whether or not a community can be invaded also depends on our temporal frame of reference. The number of species of organisms found together in an arbitrarily defined community (lake, parasites on a host, woodlot, etc.) is invariably a subset of the pool of potential invaders to that community. Studies of patterns of accumulation of species in such communities suggest that invasions may occur rapidly but that the community reaches an equilibrium in ecological time when invasions and extinctions are more or less in balance (Strong 1974a,b; Lawton and Strong 1981). At the same time, a rich body of evidence suggests that species may continue to accumulate in many communities over long periods of evolutionary time. Equilibrium in ecological time does not mean saturation in evolutionary time.

A prime example of the lack of saturation in evolutionary time is the great variation in the number of species of plants in comparable ecological communities. For example, although plant comunities are very similar in growth form in the five areas of the world having Mediterranean climates (Mooney and Dunn 1970; Mooney et al. 1975; Mooney 1977), they are not similar in species number. African and Australian plant communities are far richer in species than their Californian and Chilean counterparts (Mooney and Dunn 1970; Werger et al. 1972; Parson and Cameron 1974; Parson and Moldenke 1975; Westman 1975; Whittaker 1977). The genus *Erica* alone has over 600 species in the fynbos of South Africa. One possible explanation of the differences

is that Mediterranean climates are of much greater age in Africa and Australia than in Chile and California (Axelrod 1973). Similarly, temperate deciduous forests are much richer in species in eastern North America and eastern Asia than they are in Europe. Again, a historical explanation seems likely (Whittaker 1977). Existing data do not suggest that there is any apparent limit to the number of plant species that can coexist in communities, but it is clear that the accumulation of species is a very slow process. How much of this is due to difficulties of ecological accommodation and how much it reflects rates at which species form is not clear.

The slow rate of accumulation of plant species in ecological communities suggests that immediate success at invading may not be a good predictor of longer term success. In Britain, the probability that a foreign tree has established feral populations is a function of the time since the tree was first introduced (Table 8.3.). These data are not without their problems because very early introductions that failed are likely not to have been recorded. Moreoever, most early introductions involved European plants whereas most later ones came from areas of the world with climates very different from that of present-day Great Britain. Nonetheless, it is difficult to avoid the conclusion that trees may become established if they been grown in an area for many centuries even if they did not spread from the sites of introduction during earlier centuries. Of the 209 species of introduced trees in central California listed by Metcalf (1968), only a few are reported to be established outside the areas where they were planted. Even species of *Eucalyptus* that grow vigorously in California seldom reproduce except directly beneath the parent plant.

Accumulation of insects on native and introduced plants also appears to continue over long time periods (Southwood 1961) but ecological equilibria may occur more rapidly, related in part to the extent of the range of the plant (Strong 1974b) and its structural complexity (Lawton and Strong 1981). Because of these time scale differences, initial judgments about the susceptibility of sites to invasions may differ from those that would be formed over longer time periods.

Table 8.3. Temporal pattern of introduction and estabishment of trees in Britain

	pre–1000	1000–1500	1500–1600	1600–1700	1700–1800	1800–1900	1900–
Gynosperms Time of Introduction							
Feral	0	1	1	0	0	0	0
Not feral	0	1	2	7	18	109	23
Angiosperms							
Feral	5	2	3	2	2	1	0
Not feral	1	3	8	14	41	101	35

Species are entered according to the century of their first introduction and whether or not they have established feral breeding populations.
Source: Mitchell A (1978). In: A Field Guide to the Trees of Britain and Northern Europe. William Collins and Sons, London.

A still longer-term view of the success of introductions is provided by the great faunal exchange across the Panamanian land bridge which began on a large scale about 3 million years ago (Simpson 1980; Marshall et al. 1982). The fossil record for mammals is good enough to provide a general picture of the course of events. In both continents, the number of families of mammals temporarily increased and then subsequently declined. In North America the number of families of land mammals increased from 27 to 34, and in South America from 23 to 36. Today North America has only 23 families of land mammals, fewer than before the interchange, and South America has only 30.

The causes of the eventual decline in species richness of mammals in both continents are believed to be due to a combination of both competitive and predator–prey interactions but the details will doubtless always remain unknown. Interestingly, the South American mammals that still survive in North America include both generalists (opossum, *Didelphis virginiana*) which appear to be rather similar to placental omnivores of roughly the same size, as well as specialists (armadillo, *Dasypus novemcinctus;* porcupine, *Erethizon dorsatum*) which are ecologically very different from any other North American mammals. The porcupine is especially interesting because it is now most abundant in coniferous forests, a vegetation type unrepresented in most of South America and not present in lowland Central America, through which the founding populations must have passed. The spread of pines into Central America is a relatively recent phenomenon, due primarily to cultivation and burning by agricultural peoples (Maldonado-Koerdell 1964; Stevens 1964).

A still longer time frame pertinent to invasions is that provided by the taxon cycle, first described by Wilson (1961) for Melanesian ants. The basic idea of the taxon cycle is that immigrants to islands initially have excellent colonizing abilities and are able to become established in a wide variety of habitats. Often the range of habitats occupied on the island is greater than that occupied by the species on the mainland even though many invaders are relatively common and widespread on the mainland. However, once the immigrant has established itself, selection pressures are changed because it now interacts with a very different set of species, the physical environment is different, and, importantly, wide dispersal is no longer advantageous. As a result, the competitive ability of a species appears to wane with time, resulting in reductions in habitat distributions, local population densities and hence overall population sizes (Ricklefs and Cox 1972). Cases exhibiting this general pattern have been described for birds of the West Indies (Ricklefs and Cox 1972) and the Solomon Islands (Greenslade 1968) and for selected groups of insects (Wilson 1961; Greenslade 1969).

A key aspect of current conceptions of taxon cycles is that they are driven by mutual adjustments among species both of a competitive and predator–prey nature. As a result, the susceptibility of an island ecosystem to invasion may increase with time since the last invasion because subsequent evolution of the most recent colonizer increases the likelihood that a new species with the properties generally associated with good ability to invade (good dispersal, high reproductive rates, r_o, and generalized diet and habitat requirements) is likely to find favorable conditions for its establishment. If so, a cycle of susceptibility

to invasion should result, leading to minor pulses in invasions over short-term evolutionary time. Similar cycles may also occur on mainland areas (Brown 1957; Darlington 1957; Dillon 1966) but more slowly. Also it is more difficult to detect these patterns on the mainland because it is more difficult to age the time of occupancy of invaders and because selection pressures probably do not change as much for invaders of new mainland areas as they generally do for mainland species invading islands. In any case, a general theory of the ecology of invasions will have to incorporate the kinds of concepts developed in taxon cycle theory if we are to understand invasion patterns over time.

8.6. Conclusions

Elton's insights into the importance of disturbance in favoring invasions into ecological communities has generally been upheld by subsequent research. Progress has been made primarily in developing better understanding of how different kinds of disturbances alter competitive, predatory–prey, and physical stresses so that invaders are at a better advantage than they are in the absence of those effects. Much remains to be learned about these factors but already it is evident that autecological differences among organisms in different taxa will figure prominently in our developing theories about invasions.

The new generation of generalizations will, of necessity, be less sweeping than Elton's. As Colwell (1984) points out, this is not to be viewed with alarm. It represents a natural course of events in a science in which the richness of the properties of the interacting units lends uniqueness to the individual cases. It is not, however, impossible that the special cases of our current visions will eventually be accommodated into broader generalizations whose form is not apparent to us today.

8.7. References

Axelrod DI (1973) History of Mediterranean ecosystems in California. In: diCastri F. Mooney HA (eds), Ecological Studies, Vol 7, Mediterranean Type Ecosystems: Origin and Structure. Springer-Verlag, New York, pp 225–277

Baker HG, Stebbins GL (1965) The Genetics of Colonizing Species. Academic Press, New York

Brower LP, Brower JVZ (1964) Birds, butterflies and plant poisons: a study in ecological chemistry. Zoologica 49:137–159

Brower LP, Ryerson WN, Coppinger LL, Glazier SC (1968) Ecological chemistry and the palatability spectrum. Science 161:1349–1351

Brown JH, Davidson DW (1977) Competition between seed-eating rodents and ants in desert ecosystems. Science 196:880–882

Brown JH, Davidson DW, Reichman OJ (1979) An experimental study of competition between seed-eating desert rodents and ants. Am Zool 19:1129–1143

Brown WL Jr (1957) Centrifugal speciation. Q Rev Biol 32:247–277

Caldwell MM, Richards JH, Johnson DA, Nowak RS, Dzurek RS (1981) Coping with herbivory: photosynthetic capacity and resource allocation in two semiarid *Agropyron* bunchgrasses. Oecologia (Berl) 50:14–24

Caswell H, Reed F, Stephenson SN, Werner PA (1973) Photosynthetic pathways and selective herbivory, a hypothesis. Am Natur 107:465–480

Colwell R (1984) What's new? Community ecology discovers biology. In: Price PW, Slobodchikoff CN, Gaud WS (eds), A New Ecology: Novel Approaches to Interacting Systems. John Wiley, New York, pp 387–396

Connell JH (1983) On the prevalence and relative importance of interspecific competition: evidence from field experiments. Am Natur 122:661–696

Courtenay WR Jr, Stauffer JR Jr (1984) Distribution, Biology and Management of Exotic Fishes. Johns Hopkins University Press, Baltimore

Darlington PJ Jr (1957) Biogeography—An Ecological Perspective. Ronald Press, New York

Davidson DW, Brown JH, Inouye RS (1980) Competition and the structure of granivore communities. Bioscience 30:233–238

Diamond JM (1975) Assembly of species communities. In: Cody ML, Diamond JM (eds), Ecology and Evolution of Communities. Harvard University Press, Cambridge, Massachusetts, pp 342–344

Dillon LS (1966) The life cycle of the species: an extension of current concepts. Syst Zool 15:112–126

Diver C (1940) The problem of closely related species living in the same area. In: Huxley J (ed), The New Systematics. Clarendon Press, Oxford

Edmunds GF Jr, Alstad DN (1978) Coevolution in insect herbivores and conifers. Science 199:941–945

Ellison L (1960) Influence of grazing on plant succession of rangelands. Bot Rev 26:1–66

Elton CS (1958) The Ecology of Invasions by Animals and Plants. Methuen, London, 181 p

Fox LR (1981) Defense and dynamics in plant-herbivore systems. Am Zool 21:853–864

Furniss RL, Carolin VM (1980) Western Forest Insects. U.S. Department of Agriculture, Forest Service, Miscellaneous Publication No 1339

Gause GF (1934) The Struggle for Existence. Williams & Wilkins, Baltimore

Godwin H (1956) The History of the British Flora: A Factual Basis for Phytogeography. Cambridge University Press, Cambridge

Greenslade PJM (1968) Island patterns in the Solomon Islands bird fauna. Evolution 22:751–761

Greenslade PJM (1969) Land fauna: insect distribution patterns in the Solomon Islands. Philos Trans R Soc 255:271–284

Haukioja E, Niemela P (1979) Birch leaves as a resource for herbivores: seasonal occurrence of increased resistance in foliage after mechanical damage of adjacent leaves. Oecologia (Berl) 39: 151–159

Hutchings SS, Stewart G (1953) Increasing forage yields and sheep production on intermountain winter ranges. USDA Forest Service Circular No 925, 63 p

Jones DA (1979) Chemical defense: primary or secondary function? Am Natur 113:445–451

Kennedy PB (1903) Summer ranges of eastern Nevada sheep. Nevada State University Agricultural Experiment Station Bulletin No 55, Reno, 55 p

Lawton JH, Strong DR Jr (1981) Community patterns and competition in folivorous insects. Am Natur 118:317–338

Leopold A (1941) Cheat takes over. Land 1:310–313

Levins R (1968) Evolution in Changing Environments. Princeton University Press, Princeton, New Jersey

MacArthur RH (1969) Species packing, and what interspecies competition minimizes. Proc Natl Acad Sci USA 64:1639–1671

MacArthur RH (1972) Geographical Ecology. Harper & Row, New York

Mack RN (1981) Invasion of *Bromus tectorum* L. into western North America: an ecological chronicle. Agro-Ecosystems 7:145–165

Mack RN, Thompson JN (1982) Evolution in steppe with few large, hooved animals. Am Natur 119:757–773

MacMahon JA, Schimpf DJ, Anderson DC, Smith KG, Bayh RL (1981) An organism-centered approach to some community and ecosystem concepts. J Theor Biol 88:287–307

MacNally RC (1983) On assessing the significance of interspecific competition to guild structure. Ecology 64:1646–1652

Maldonado-Koerdell M (1964) Geohistory and paleogeography of Middle America. In: West RC (ed), Handbook of Middle American Indians, Vol 1. Natural Environments and Early Cultures. University of Texas Press, Austin, Texas, pp 3–32

Marshall LG, Webb SD, Sepkoski JJ Jr, Raup DM (1982) Mammalian evolution and the great American interchange. Science 215:1351–1357

Metcalf (1968) Introduced Trees of Central California. University of California Press, Berkeley and Los Angeles, 159 p

Mitchell A (1978) A Field Guide to the Trees of Britain and Northern Europe. William Collins and Sons, London

Mooney HA (ed) (1977) Convergent Evolution in Chile and California. Dowden, Hutchinson & Ross, Stroudsberg, Pennsylvania

Mooney HA, Dunn EL (1970) Convergent evolution of Mediterranean-climate evergreen sclerophyll shrubs. Evolution 24:292–303

Mooney HA, Harrison AT, Morrow PA (1975) Environmental limitations of photosynthesis on a California evergreen shrub. Oecologia (Berl) 19:293–301

Mortimer AM (1984) Population ecology and weed science. In: Dirzo R, Sarukhan J (eds), Perspectives on Plant Population Ecology. Sinauer, Sunderland, Massachusetts, pp 363–388

Paine RT (1966) Food web complexity and species diversity. Am Natur 100:65–75

Paine RT (1980) Food webs: linkage, interaction strength and community infrastructure. J Anim Ecol 49:667–685

Parson DJ, Moldenke AR (1975) Convergence in vegetation structure along analogous climatic gradients in California and Chile. Ecology 56:590–597

Parsons RF, Cameron DG (1974) Maximum plant species diversity in terrestrial communities. Biotropica 6:202–203

Piemeisel RL (1938) Changes in weedy plant cover on cleared sagebrush land and their probable causes. US Department of Agriculture Technical Bulletin 654

Pimm SL (1982) Food Webs. Chapman & Hall, London

Platt K, Jackman ER (1946) The cheatgrass problem in Oregon. Oregon St Fed Coop Ext Serv Bull 668

Pulliam HR (1975) Coexistence of sparrows: a test of community theory. Science 189:474–476

Rhoades DF (1979) Evolution of plant chemical defense against herbivores. In: Rosenthal GA, Janzen DH (eds), Herbivores: Their Interactions with Secondary Plant Metabolites. Academic Press, New York

Ricklefs RE, Cox CW (1972) Taxon cycles in the West Indian avifauna. Am Natur 106:175–219

Schoener TW (1983) Field experiments on interspecific competition. Am Natur 122:240–285

Simpson GG (1980) Splendid Isolation: The Curious History of South American Mammals. Yale University Press, New Haven

Southwood TRE (1961) The numbers of species of insects associated with various trees. J Anim Ecol 30:1–8

Stevens RL (1964) The soils of Middle America and their relations to Indian peoples and cultures. In: West RC (ed), Handbook of Middle American Indians, Vol 1. Natural Environment and Early Cultures. University of Texas Press, Austin, Texas

Strong DR (1974a) The insects of British trees: community equilibration in ecological time. Ann Missouri Bot Gard 61:692–701

Strong DR (1974b) Rapid asymptotic species accumulation in phytophagous insects: the pests of cacao. Science 185:1064–1066

Werger MJA, Kruger FJ, Taylor HC (1972) A phytosociological study of the Cape Fynbos and other vegetation at Jonkershoek, Stellenbosch. Bothalia 10:599–614

Westman WE (1975) Edaphic climax pattern of the pigmy forest region of California. Ecol Monogr 45:109–135

Wheelwright NT, Orians GH (1982) Seed dispersal by animals: contrasts with pollen dispersal, problems of terminology, and constraints on coevolution. Am Natur 119:402–413

White TCR (1974) A hypothesis to explain outbreaks of looper caterpillars, with special reference to populations of *Selidosema suavis* in a plantation of *Pinus radiata* in New Zealand. Oecologia (Berl) 16:279–302

White, TCR (1978) The importance of a relative shortage of food in animal ecology. Oecologia (Berl) 33:71–86

Whittaker RH (1977) Evolution of species diversity in land plant communities. Evol Biol 10:1–66

Wilson EO (1961) The nature of the taxon cycle in the Melanesian ant fauna. Am Natur 95:169–193

Yensen DL (1981) The 1900 invasion of alien plants into southern Idaho. Great Basin Natur 41:176–183

Young JA, Evans RA, Major J (1972) Alien plants in the Great Basin. J Range Manage 25:194–201

9. Biological Invasions of Plants and Animals in Agriculture and Forestry

D. Pimentel

9.1. Introduction

Over the years, many insect pests, plant pathogens, and weed species have invaded agricultural and forest ecosystems in the United States. Once established, they have significantly reduced crop and forest productivity, despite treatments with pesticides and the presence of natural enemies (Pimentel 1983).

The basic ecological structure of both agricultural and forest ecosystems influences the types of interactions that occur between the invading species in these managed ecosystems. In the U.S., most agricultural crop plants have been introduced from many different regions of the world. The majority of annual grain, vegetable, and fruit crops are from the tropics. In contrast, most trees in commercial forests in the United States are managed native species.

All of the introduced crops are attacked by a combination of introduced and native insect, plant pathogen, and weed species. Some of these pest invaders are from the native habitat of the crops whereas others have been brought from habitats foreign to the crops. In addition, many of the pest insects and plant pathogen invaders have moved from their native U.S. hosts and have become serious pests of the introduced annual tropical crops (Pimentel 1977). This has happened when an introduced crop plant has never interacted or coevolved with these pests and has no resistance to the attack from the native pests (Pimentel 1977). The lack of evolution through genetic feedback to establish a balance in these parasite–host systems often has resulted in severe outbreaks of the pests on the various imported crops (Pimentel 1961).

Most U.S forests have native tree species that have evolved genetic balance
with their pests; this is one factor that reduces the severity of insect herbivore
and plant pathogen attacks compared with those of crops. With pest damages
low compared to those in crops, and the economic value of wood products
relatively low, pesticides are seldom used to control insect and pathogen outbreaks in forests (Pimentel and Levitan 1986). However, when insects such as
the gypsy moth and plant pathogens such as the chestnut blight have been
introduced from foreign regions into U.S. forests, severe outbreaks have followed. In both cases, the associations between the introduced species and native
tree hosts were new, no balance had evolved, and severe damage was inflicted
on the native tree species. In general, the ecosystems of both agriculture and
forestry are favorable environments for invading biological species. Both systems are rich in biomass resources for insects and microorganisms and have
ample nutrients and moisture for weeds. This chapter analyzes the ecological
origin of the major plant and animal invaders of agricultural and forestry ecosystems of the United States. Further, the various ecological factors that have
contributed to the success of these invaders is assessed.

9.2. Weed Invaders

In agriculture all plants other than the crop plants in the managed ecosystem
are considered weeds or biological invaders. Some of the weed invaders are
native to the U.S., whereas many are introduced from other regions of the
world. About 73% of the 80 major U.S. weed species originally come from
regions outside of the U.S. (Table 9.1.). About 67% of these came from Europe
and Eurasia, no doubt because of long-term selection and the intensive early
trade between the U.S. and Europe and Eurasia. Ships that arrived in America
were relatively empty of goods, but were filled with soil ballast. To make room
for the American products, the ballast, containing weed seeds, insects, and

Table 9.1. Major weed species in U.S. cultivated crops and the number of invaders
and their origin

Total Number of Major Weed Species	Number of Biological Invaders	Number of Native Weed Species
80	58	22

Origins of Weed Invaders	Number of Species
Europe	22
Eurasia	17
Africa	1
Asia	4
Tropical America	6
Mediterranean	1
Multiple regions	7
Total	58

Source: USDA 1971.

Table 9.2. Major weed species in U.S. pastures and the number of invaders and their origins

Total Number of Major Weed Species	Number of Biological Invaders	Number of Native Weed Species
110	45	65

Origins of Weed Invaders	Number of Species
Europe	17
Eurasia	18
Africa	0
Asia	4
Tropical America	3
Mediterranean	2
Multiple Regions	1
Total	45

Source: USDA 1971.

other small organisms, was emptied on the U.S. shore (Elton 1958; Sailer 1983). In time some of these organisms found their way onto agricultural lands.

The most important means of introducing weed seeds in crops was through mixing with the crop seeds that were brought into the U.S. for planting. Today, effective means exist for removing most of the weed seeds from crop seeds, but before 1900 the production of pure crop seed was very difficult.

The pattern of biological invasions of weeds in U.S. pastures is quite different from that associated with crops. That is, only 41% of the pasture weed invaders came from regions outside the U.S., whereas 73% of the crop weed invaders were introduced species (Tables 9.1. and 9.2.). Similar to cultivated croplands most (78%) of the weed species that were introduced into the U.S. and invaded pastures and hayfields also came from Europe and Eurasia (Tables 9.1. and 9.2.). In addition to being carried in soil and ballast, some weeds probably were transported in straw and hay used in packaging. In contrast to croplands, fewer introduced weeds became established in managed pastures and hayfields because they are more stable perennial-plant systems and are not plowed and disturbed each growing season as occurs in the case of annual-crop ecosystems.

No data exist on the precise number of weed species that were introduced and the number that actually became established. Overall, however, fewer species of weeds appear to have been introduced and become established in natural ecosystems than in the managed agricultural ecosystems. Harper (1965) reported that "it is extremely questionable whether there is any single proven case of the extinction or even the decline in abundance of a native species which can be directly attributed to aggression by an alien species that was introduced." This appears to be especially true for plant species in U.S. forests.

The primary biological characteristics that make weeds highly successful in managed ecosystems, especially in agriculture are as follows:

1. Weed types usually have rapid growth and early flowering (Baker 1965; Mulligan 1965).

2. The plants set seed under a wide range of temperature conditions and day lengths (Baker 1965; Mulligan 1965).
3. The species have a high degree of "phenotypic plasticity" under a wide range of nutritional and photoperiod conditions (Mulligan 1965).
4. The plants are highly productive in the number of seeds produced and dispersed (Baker 1965; Mulligan 1965).
5. Weedy plants may have different growth forms depending on the ecosystem (Mulligan 1965).
6. Weeds often have germination "polymorphisms" that allow some seeds to germinate immediately but others to germinate several years later (Harper 1965). Further information on the type of biological characteristics that make a plant a weedy or nonweedy type is listed in table 9.3.

The type of weed species that became established in crop and pasture/hayfield ecosystems has depended on the kind of plants in the managed agricultural ecosystems. Row crops have mostly annual weeds, whereas pastures and hayfields have mostly perennial and biennial weeds (Mulligan 1965). Further, growth characteristics of the weeds differ. In both row crops and pasture/hayfield ecosystems weeds were erect, whereas in lawns and overgrazed pastures, weed growth was prostrate (Mulligan 1965).

The growth and ecological characteristics of certain plant groups make them particularly suitable as weed invaders. For example, in California the largest number of introduced weed species are annuals and belong to the families of Compositae, Gramineae, and Cruciferae (Stebbins 1965). Those families all contain large numbers of species that are adaptable annual and perennial weed types.

Most of the tree-weed species, such as white poplar, black locust, and hawthorns, found in forests are native species that do not have characteristics suitable for their use as saw timber and pulp wood. No major species of weed tree is introduced.

Table 9.3. Characteristics of weedy and nonweedy genotypes of *Eupatorium microstemon*

Weed Type	Nonweedy Type
Plastic	Not very plastic
Annual	Perennial
Quick flowering	Slow to flower
Photoperiodically neutral	Short-day requirement
Self-compatible	Self-incompatible
Economical of pollen	Plentiful pollen produced
Chromosomes	
n = 4	n = 20

After Baker 1965.

Table 9.4. Major insect pest species in U.S. crops and the number of invaders and their origins

Crop	Total Number of Major Insect Pest Species	Number of Biological Invaders	Number of Native Insect Species
Corn	16	3	13
Rice	2	2	0
Other small grains	11	6	5
Cotton	15	4	11
Sugar cane and Sugar beets	4	0	4
Tobacco	6	1	5
Vegetables[a]	41	15	26
Fruits[b]	40	16	24
Alfalfa/clover, and other legumes	13	10	3
Total	148	57	91

	Number of Species
Europe	25
Africa	1
Asia	4
Tropical America	12
Oceania	1
Multiple regions	14
Total	57

[a] Asparagus, beans, potato, cabbage, onion, peas, cowpeas, cucurbits and carrots.
[b] Apples, citrus, and pears.
Sources: Metcalf et al. 1962; USDA 1974; Davidson and Lyon 1979.

9.3. Insect Invaders

Only about 40% of the insect species that are pests in U.S. agricultural eco-systems are introduced species (Table 9.4.). As with introduced weed invaders, many (50%) of the introduced insect invaders came from Europe (Sailer 1983). These insects probably were transported either in ballast or along with some forage or food crop material.

Several factors in agricultural ecosystems have affected the ability of insect invaders to become established. First, most of the annual grain, vegetable, and fruit crops in the U.S. are tropical and subtropical (Pimentel 1984), with the result that few invaders could survive in the cold winter climate of North America. Indeed, most tropical insect species cannot hibernate or enter diapause to survive the freezing conditions typical of northern U.S.

In addition, some introduced insect species may have found it difficult to survive the attack and feeding pressure that was exerted by native parasites and predators. Introduced insect species are more susceptible to native parasite and predator attack because of the lack of a long evolutionary association be-

tween introduced species and the new enemies. In fact, various native parasites and predator species have been effective in providing effective biological control of introduced insects (Pimentel 1963; Hokkanen and Pimentel 1984). The Japanese beetle, for example, has been kept in control in many regions of the U.S. by a milky disease caused by *Bacillus popillae*. This organism is believed to be native to North America and appeared to have moved from the native scarab beetle to include the Japanese beetle as another host.

Just as the insect invaders are attacked by native parasites and predators, so are the introduced tropical annual crop plants readily attacked by native herbivores. As mentioned, the majority or 60% of the major insect pests on U.S. crops are native pests (Table 9.4.). The introduced crop plants lack genetic feedback evolved-resistance to native insect species because they did not coevolve. Thus, large numbers of native insect herbivores have added introduced crop plants to their supply of food hosts.

This ecological interaction of insect invaders and crops appears to be typical throughout the world. For example, in most regions where sugarcane was introduced and established as a crop, a unique set of pest species has become associated with it. Of 1645 known pest species of sugarcane, "959 occur only in a single region and 156 only in two regions. Only 18 pest species occur in more than 10 regions" (Strong et al. 1977). These data suggest that sugarcane pest groups develop primarily from native arthropods that have moved from local plants and not from arthropods introduced into the region from the native region of sugarcane.

Another example of the movement of native herbivores onto a crop is the invasion on the potato *(Solanum tuberosum)*. When the potato, which originated in Bolivia and Peru (Hawkes 1944), was introduced into southwestern U.S., it acquired its most serious insect pest, the Colorado potato beetle. Native to the U.S. the Colorado potato beetle originally had coevolved with and fed on wild sand bur *(Solanum rostratum)* (Elton 1958). But, when the potato was introduced, the beetle spread onto potato plants, which had never been exposed to this beetle. Because the plant lacked natural resistance to the beetle, the beetle has become the most serious insect pest of potatoes here and in many parts of the world including Europe and Eurasia.

Sometimes, however, both the insect pest and associated crop are newly associated. For example, both corn and the European corn borer were introduced into the U.S. Corn was brought from tropical America whereas the corn borer came from Europe, where it had fed on wild plants. Thus, corn had not evolved any resistance to the corn borer and was easily damaged by the borers.

Another classic example of a new association between imported pest and host plants is that of the Hessian fly from Europe, and wheat that came from the Middle East (Allan et al. 1959). The fly is a serious pest of wheat in Kansas and other wheat-growing regions in the U.S. This again is a new association between an insect and host plant in a different ecosystem.

Although about 60% of the pest insect species in U.S. crops are native, slightly more or 73% in forests are native (Table 9.5.). However, some of the

Table 9.5. Major insect pest species in U.S. forests and the number of invaders and their origin

Origin of Major Pests	Number of Species
Native	51
Europe	15
Other Introduced	4
Total	70

Sources: Metcalf et al. 1962; Davidson and Prentice 1967; Davidson and Lyon 1979.

most serious insect pests in forests are introduced species (Elton 1958). Exactly which species are considered serious forest pests depends as much on the extent of economic damage caused by the pest as the size of the pest population. For example, the white pine weevil, which is relatively rare in white pine forests, is one of the most serious pests of pine because it attacks and thereby kills the growing tip of the pine tree. When this happens another branch becomes the growing tip, altering the shape of the tree and diminishing its value as saw lumber.

For livestock, cattle, horses, sheep, goats, hogs, and poultry, there are a total of 28 major arthropod pest species (Table 9.6.). About 64% of these pests have a worldwide distribution. Most of these pests were distributed with the introductions of these livestock. Only two insect species, the horn fly and face fly of cattle, were introduced independently of the cattle introduction.

Several ecological characteristics of both the ecosystem and the insect invaders help invaders become established (Lattin and Oman 1983). First, the invaders must find suitable and ample food resources in the system. Then too, the life history of the invaders must be synchronized with the climate of the ecosystem; for example, the insect species must have a resting stage to survive the harsh winter temperatures. Further, the invaders must be able to resist successfully the attacking native natural enemies and be able to reproduce at a relatively high rate. Therefore, the chances of an introduced organism becoming established in a new region is relatively small when characteristics of both the new ecosystem and the invading species are taken into account.

Table 9.6. Major arthropod pests of livestock and the number of invaders and their origins

Total Number of Major Arthropod Pest Species	Number of Biological Invaders	Number of Native Arthropod Pest Species	Number of Worldwide Arthropod Pest Species
28	2[a]	8	18

[a] Horn fly and face fly of cattle introduced from Europe.
Sources: Metcalf et al. 1962; Davidson and Prentice 1969; Davidson and Lyon 1979.

Table 9.7. Major plant pathogens of U.S. vegetable crops and the number of invaders and their origins

Total Number of Major Plant Pathogen Species	Number of Biological Invaders	Number of Native Pathogen Species	Number of Worldwide Pathogen Species
155	48	47	60

Origins of Pathogen Invaders	Number of Species
Europe	19
Tropical America	13
Asia	1
Multiple Regions	18
Total	51

Sources: Chupp and Sherf 1960; Purseglove 1972; 1977; CMI 1983.

9.4. Plant Pathogen Invaders

The patterns of plant pathogen invasions in vegetable and grain crops are distinctly different from those of insect and weed invaders. Almost 40% of the 155 major plant pathogen species in vegetables that were studied had a worldwide distribution (Table 9.7.). Frequently, the pathogenic microorganisms were spread where the crop host was introduced into a new region. Although these could be considered introductions, in this analysis they were classified as having worldwide distribution because their associated host plants had such a wide distribution.

The number of plant pathogens introduced into U.S. vegetable systems, other than those having a worldwide distribution, was only 31%, compared to 73% for weeds and 9% for insects (Table 9.7.). Also in contrast to insects and weeds, most of the introduced plant pathogens came from tropical America, and secondly from Europe.

The number of native pathogens that have moved from native vegetation to attack introduced crops is substantial. A total of about 30% of all major plant pathogen species are native to the U.S. (Table 9.7.). Thus, native pests make up a large proportion of the pathogen pests in agroecosystems and this agrees with trends for both insects and weeds of crops.

All except one of the plant pathogens associated with grain crops have a

Table 9.8. Major plant pathogens of U.S. grain crops and the number of invaders and their origins

Total Number of Major Plant Pathogen Species	Number of Biological Invaders	Number of Native Pathogen Species	Number of Worldwide Pathogen Species
74	0	1	73

Sources: Leonard and Martin, 1963; Purseglove 1972; CMI 1983.

worldwide distribution (Table 9.8.). Only one pathogen species associated with corn is native. Most of the grain pathogens were probably spread with the transport of grains worldwide.

9.5. Introduced Biological Control Organisms

At least 39 pest insect and weed species have been controlled in the U.S. by introduced parasites and predators (Hokkanen and Pimentel 1984). On average, seven natural enemies have been introduced before one biocontrol agent provided at least partial control of the target pest. Sometimes, as with the corn borer, as many as 26 natural enemies have been introduced and only limited control has been achieved (Baker et al. 1949; Whitman 1975).

For the gypsy moth, 40 natural enemies were introduced into the U.S. from Europe and Japan, but only 10 have become established (Nichols 1961). However, none of the established 10, except for a virus, is providing substantial control of the gypsy moth. In addition, about 90 native natural enemies attack the gypsy moth (Campbell 1975). Although all introduced and native natural enemies of the gypsy moth provide some control during outbreaks, the group is not sufficiently effective to prevent the gypsy moth from continuing to be a serious pest in northeastern forests.

An analysis of biological control organisms shows that most introduced natural enemies have come from the native habitat of the pest (Pimentel 1963; Hokkanen and Pimentel 1984). On the basis of the number of documented introductions, 86% of the biological control agents were obtained from the native habitat of the pest (Table 9.7.). The assumption has been that pest outbreaks occurred in the new habitat because of the absence of the natural enemies of the introduced pest that have controlled it in its native habitat (Sweetman 1958; DeBach 1964).

Clearly this approach, with only a one in seven success rate, has not been as successful as anticipated. One important reason for this is that when a pest and its natural enemy are brought to a new ecosystem, the genetic feedback balance already achieved between parasite and host continues to operate in the new ecosystem (Pimentel 1961). This was true for the natural enemies of the corn borer and gypsy moth.

Thus an alternative approach to biological control would be to use imported parasites that are unrelated to the pest host. This would also expand the availability of potential biocontrol agents. With this procedure the association between parasite and host would be new and the disadvantages of coevolved balance, which is typical of long-associated parasites and hosts, would be eliminated.

In the past, only a few natural enemies from alternate hosts have been used for biocontrol. Usually these trials have followed several unsuccessful control attempts that were formerly associated parasites (Hokkanen and Pimentel 1984). But when the new associations of parasite and host have been tried, they have proven exceptionally effective.

One example of this is the control of the weedy cacti, *Opuntia* spp., which when introduced from Mexico and other Gulf of Mexico regions developed into a pest weed in the grazing lands of Australia (Dodd 1940). Several herbivorous insects that previously had been associated with the cacti in their native habitat were imported, but failed to give control. When the moth, *Cactoblastis cactorum,* an insect associated with another group of cacti species found only in the southern lower part of South America, was introduced it provided outstanding control of the cacti in Australia (Dodd 1940; Pimentel 1963).

Another example of a new association of a parasite and host, which provided effective control of a pest, is the control of the citrus mealy bug *(Cryptolaemus montrovzier)* by a wasp parasite (Clausen 1978). The pest is native to China, whereas the natural enemy originated from eastern Australia (Hokkanen and Pimentel 1984) so the chance for coevolved balance was eliminated.

9.6. Crop, Livestock, and Forest Losses from Biological Pest Invaders

The damage to crops, livestock, and forests inflicted by pests affects the potential yields from our U.S. agricultural and forest systems. This represents an important loss of food and fiber goods in the marketplace and diminishes the economic return needed by the growers. Earlier it was mentioned that (from the agricultural viewpoint) all pests are biological invaders, because the only desired species in the managed ecosystem is the crop plant or animal. Therefore, in this analysis of losses, the damage caused by pest invaders from foreign areas and those that are from U.S. ecosystems are not separated.

In the U.S., current crop losses attributed to all pests (insects, plant pathogens, and weeds) are estimated to be 37% (Pimentel 1981). Of this, 13% is due to insects, 12% to plant pathogens, and 12% to weeds. These losses have occurred despite the annual use of nearly 500,000 kg of pesticides plus various biological and other nonchemical controls.

According to survey data collected from about 1945, insects caused only an estimated 7% loss (Pimentel et al. 1978). From 1945 to 1984, losses due to insects have nearly doubled, despite a 10-fold increase in insecticide use. The substantial increase in crop losses can be explained, in part, by some of the major changes that have taken place in agriculture since the early 1940s. These include: planting some crop varieties that are more susceptible to insects; destruction of natural enemies with pesticides; increase in pesticide resistance; reduced crop rotations; increased "cosmetic standards" for fruits and vegetables; reduced sanitation in crops; reduced tillage; planting of some crops in climatic regions where pests are more abundant; and some pesticides, such as herbicides, making crops more susceptible to insect attack.

In contrast to crops, losses attributed to plant pathogens have increased only slightly from 10.5% to 12% (Pimentel et al. 1978), whereas losses caused by weeds have decreased slightly from 13.8% to 12%.

In forests, current losses due to all pests are estimated to be about 33%. Of this 21% is due to plant pathogens, 9% to insects, and 3% to other animals

(USDA 1955). Interestingly, only about 10% of the total pesticide used in the U.S. is dispersed in forests (Pimentel and Levitan 1986). In part, this is because the economic return on investment per hectare in forest production is relatively low and is spread over 20 to 100 years. This makes treatments with pesticides expensive and thus rarely used in forest production.

Annual livestock losses due to pests are considered substantial and are estimated to be about 13% of total potential production (Drummond et al. 1977). These losses are caused by both disease and insects. Often insects are vectors of some of the serious livestock diseases including various viral rickettsial and protozoan diseases of livestock.

9.7. Economic Losses Due to Pest Invaders in Agriculture and Forestry

Translating the crop, forest, and livestock losses into dollars gives a better appreciation of what losses all pests cause each year in the U.S. For crops, the loss of potential production is calculated to be $51 billion; for forests, $2 billion; and for livestock, $13 billion (Pimentel 1985). This totals $66 billion, a substantial loss in economic terms.

To give a more realistic picture of the total cost of losses due to pests, about $3 billion should be added as cost of pesticide materials and application each year (Pimentel 1985). When environmental and social costs, calculated to range from $1 to $3 billion (Pimentel et al. 1980), are also added the economic cost total increases dramatically.

This emphasizes the need for more effective pest control techniques. Although pesticide controls still provide a return of about $4 per dollar invested, they have not lived up to their promise. The same can be said of biocontrols. Given the extent of the losses, enlarging the scope of biocontrol agents holds great promise for improved pest controls and decreased losses.

9.8. Conclusions

In the managed ecosystems of agriculture and forestry any organism other than the established crop and forest plant is considered a biological invader. The invading species can be either introduced from outside the U.S. or a native species that invaded the managed ecosystems.

Many more weed species were foreign invaders of croplands than either insects or pathogens, in part because most crops are tropical annuals and the ecosystems are disturbed each year. Indeed the manipulated, simplified crop ecosystem of U.S. agriculture provides an abundance of food resources and moisture and is an ideal environment for many biological invaders.

In forests, however, most weedy tree species are native, as are the insect and pathogen pests. Only a few insects, pathogens, and tree species have been introduced and subsequently developed into major forest pests.

Most insect and pathogen species appear to be able to invade and attack

related host types with which they had never interacted and evolved genetic feedback balance. This principle applies to insects and pathogens introduced from outside of the U.S. as well as to native insect and pathogen species. Many species associated with native hosts have invaded and attacked crops or animals that were introduced into the U.S.

For weedy plants, genetic feedback evolution does not play a role in their ability to invade crops, pastures, and forests. In this case, the invasions appear to be limited by competition and shortages of nutrients (N, P, K, Ca, etc.), water, sunlight, and space.

In addition to pest species, many organisms are imported as biological control agents of pests, which may either be native or introduced into the U.S. Using biocontrol agents lessens our reliance on pesticides and helps minimize the adverse environmental effect of some chemicals. On average seven natural enemies have been introduced before some control has been achieved with one biocontrol agent. Most of the biocontrol agents used have been from the native habitat of the pest where they had the opportunity to evolve a balance with their hosts. Thus, the genetic feedback evolution is transferred to the new habitat and is the reason for the relatively low success rate. It is recommended that biocontrol agents from alternate hosts be sought for biocontrol. This new approach should increase the chances for biocontrol successes and provide an opportunity to control native pest species.

The losses in agriculture and forestry due to biological pest invaders are estimated to be $66 billion annually. Many opportunities exist to reduce these losses.

9.9. References

Allan ER, Heyne EG, Jones ET, Johnson CO (1959) Genetic analysis of ten sources of Hessian fly resistance, their interrelationships and association with leaf rust reaction in wheat. Bull Kans Agric Exp Stn 104

Baker HG (1965) Characteristics and modes of origin of weeds. In: Baker HG, Stebbins GL (eds), The Genetics of Colonizing Species. Academic Press, New York, pp 147–172

Baker WA, Bradley WG, Clark CA (1949) Biological control of the European corn borer in the United States. US Department of Agriculture Technical Bulletin 983

Campbell RW (1975) The gypsy moth and its natural enemies. Agriculture Information Bulletin 381, USDA

Chupp C, Sherf AF (1960) Vegetable Diseases and Their Control. Ronald Press, New York

Clausen CP (1978) Introduced Parasites and Predators of Arthropod Pests and Weeds: A World Review. Agricultural Research Service, USDA Agriculture Handbook No 480

CMI (1983) Distribution maps of plant diseases. Commonwealth Mycological Institute, Maps 1–554, and Indexes

Davidson AG, Prentice RM (eds) (1967) Important Forest Insects and Diseases of Mutual Concern to Canada, the United States, and Mexico. Queens Printer and Controller of Stationery, Ottawa, Canada

Davidson RH, Lyon WF (1979) Insect Pests of Farm, Garden and Orchard, 7th edit. John Wiley, New York

DeBach PH (ed) (1964) Biological Control of Insect Pests and Weeds. Reinhold, New York

Dodd AP (1940) The Biological Campaign Against Prickly-pear. Comm Prickly Pear Board, Brisbane

Drummond RO, Bram RA, Konnerup N (1977) Animal pests and world food production. In: Pimentel D (ed), World Food, Pest Losses, and the Environment. Westview Press, Boulder, Colorado, pp 63–93

Elton CS (1958) The Ecology of Invasions by Animals and Plants. Methuen, London

Harper JL (1965) Establishment, aggression, and cohabitation in weedy species. In: Baker HG, Stebbins GL (eds), The Genetics of Colonizing Species. Academic Press, New York, pp 243–268

Hawkes JG (1944) Potato Collecting Expeditions in Mexico and South America. Imperial Bureau of Plant Breeding and Genetics. School of Agriculture, Cambridge, England

Hokkanen H, Pimentel D (1984) New approach for selecting biological control agents. Can Entomol 116:1109–1121

Lattin JD, Oman P (1983) Where are the exotic insect threats? In: Wilson CL, Graham CL (eds), Exotic Plant Pests and North American Agriculture. Academic Press, New York, pp 93–137

Leonard WH, Martin JH (1963) Cereal Crops. Macmillan, New York

Metcalf CL, Flint WP, Metcalf RL (1962) Destructive and Useful Insects, 4th edit. McGraw-Hill, New York

Mulligan GA (1965) Recent colonization by herbaceous plants in Canada. In: Baker HG, Stebbins GL (eds), The Genetics of Colonizing Species. Academic Press, New York, pp 127–146

Nichols JO (1961) The Gypsy Moth in Pennsylvania—its history and eradication. Pa Sept Agric Misc Bull No 4404

Pimentel D (1961) Animal population regulation by the genetic feed-back mechanism. Am Natur 95:65–79

Pimentel D (1963) Introducing parasites and predators to control native pests. Canad Entomol 95:785–792

Pimentel D (1977) Ecological basis of insect pest, pathogen and weed problems. In: Cherrett JM, Sagar GR (eds), The Origins of Pest, Parasite, Disease and Weed Problems. Blackwell, Oxford, pp 3–31

Pimentel, D (ed) (1981) Handbook of Pest Management in Agriculture, Vols. I-III. CRC Press, Boca Raton, Florida. 597 p, 501 p, and 656 p

Pimentel, D (1983) Effects of pesticides on the environment. In: 10th Internatl Congress of Plant Protection, Vol 2. British Crop Protection Council, Croydon, England, pp 685–691

Pimentel D (1984) Perennial grasses on the prairies: new crops for the future? In: 1985 Yearbook of Science and the Future. Encyclopaedia Britannica, Chicago, pp 124–137

Pimentel D (1985) Agroecology and economics. Presented at National Meetings Entomological Society of America, San Antonio, Texas, December 9-13, 1984

Pimentel D, Andow D, Dyson-Hudson R, Gallahan D, Jacobson S, Irish M, Kroop S, Moss A, Schreiner I, Shepard M, Thompson T, Vinzant B (1980) Environmental and social costs of pesticides: a preliminary assessment. Oikos 34:127–140

Pimentel D, Krummel J, Gallahan D, Hough G, Merrill A, Schreiner I, Vittum P, Koziol F, Back E, Yen D, Fiance S (1978) Benefits and costs of pesticide use in US food production. Bioscience 28:772, 778–784

Pimentel D and Levitan L (1986) Pesticide pollution of water, soil, air, and biota. Bioscience 36:86–91

Purseglove JW (1972) Tropical Crops: Monocotyledons. Halsted Press, John Wiley, New York

Purseglove JW (1977) Tropical Crops: Dicotyledons. Longman, London

Sailer RI (1983) History of insect introductions. In: Wilson CL, Graham CL (eds), Exotic
    Plant Pests and North American Agriculture. Academic Press, New York, pp 15–38
Stebbins GL (1965) Colonizing species of the native California flora. In: Baker HG,
    Stebbins GL (eds), The Genetics of Colonizing Species. Academic Press, New York,
    pp 173–195
Strong DR, McCoy EP, Rey JR (1977) Time and the number of herbivore species. The
    pests of sugarcane. Ecology 58:167–175
Sweetman HL (1958) The Principles of Biological Control. William C. Brown, Dubuque,
    Iowa
USDA (1955) Timber resource review, chapter 1: Timber resources for America's future.
    US Department of Agriculture Forest Service, 129p
USDA (1971) Common weeds of the United States. Agriculture Research Service, Dover
    Publications, New York
USDA (1974) Guidelines for the use of insecticides. US Agricultural Handbook No 542
Whitman RJ (1975) Natural control of the European corn borer. *Ostrinia nubilalis* (Hub-
    ner) in NY. Unpublished PhD thesis, Cornell University, Ithaca, New York

10. Biological Invasions and Ecosystem Properties: Can Species Make a Difference?

P.M. Vitousek

10.1. Introduction

How important are individual species in controlling ecosystem properties? Many studies of ecosystem-level dynamics are conducted with little reference to the influences of the individual species within ecosystems. In some cases this disregard is explicit, and it is argued that whole-system functional properties are better indicators of ecosystem status than is any aspect of species biology (O'-Neill et al. 1977). This point of view has been deplored but not unambiguously disproved (cf. Foin and Jain 1977; McIntosh 1980).

A number of studies have evaluated the functional importance of particular species in controlling community structure and ecosystem-level fluxes of energy, water, and nutrients. Groups studied include intertidal and benthic consumers (Paine 1966, Lubchenco and Menge 1978, Dayton et al. 1984), phytophagous insects (Mattson and Addy 1975), understory trees (Thomas 1969), early successional species (Marks and Bormann 1972; Marks 1974; Foster et al. 1980; Boring et al. 1981), and spring ephemerals (Muller and Bormann 1976; Blank et al. 1980; Peterson and Rolfe 1982).

Most studies of the importance of plant species in controlling ecosystem processes demonstrate that a particular species takes up energy, nutrients, or water out of proportion to its biomass or apparent abundance in the community; most to date have failed to show that ecosystem properties would be different in the absence of that species. For example, the studies of spring ephemerals

cited above demonstrated that the plants took up as much or more nitrogen and potassium as was lost from the system as a whole during their growth period; no study showed that the nitrogen and potassium taken up would otherwise have been lost. Similarly, the studies that examined the importance of early successional species in retaining nutrients on-site after disturbance rarely obtained direct evidence that nutrient losses would have been different in the absence of those species. In fact, Foster et al. (1980) showed that although removal of all successional vegetation significantly increased pool sizes of available soil nitrogen in the first year, removal of only the dominant species had little effect. Harcombe (1977) observed that the presence of vegetation but not its composition was important in controlling soil nutrient pools early in tropical forest succession.

Studies of the importance of animal species in controlling ecosystem properties have often yielded more decisive results. Experimental studies of the influence of predators in the intertidal zone have demonstrated that the presence or absence of the top predator in a system can profoundly alter the cover, diversity, and presumably productivity of that system despite the relatively small fraction of total ecosystem energy that flows through that predator (Paine 1966). This work was central to the formulation of the concept of "keystone species," and candidate keystone species have been identified in many areas. The importance of grazers in controlling community and ecosystem properties in intertidal and benthic ecosytems has also been verified experimentally (Lubchenco and Menge 1978; Dayton et al. 1984).

An evaluation of biological invasions by exotic species can provide unambiguous evidence for or against the importance of individual species in altering ecosystem properties or processes. Biological invasions can have overwhelmingly negative consequences for native species and communities (Elton 1958; Jarvis 1979; Clark et al. 1984), but it is these very consequences that offer an opportunity for rigorous tests of the importance of particular species in ecosystem functioning. If ecosystem properties are substantially altered by biological invasions, then the biology of individual species is clearly important on the ecosystem level. Further, if any such alterations in ecosystem properties affect humans by altering water yield, soil fertility, erosion, air quality, or any other of the myriad "ecosystem services" that benefit humans (cf. Ehrlich and Mooney 1983), it would provide another, perhaps more successful, argument for care where biological invasions are concerned.

10.2. Ecosystem Properties

There is wide variance in definitions of the term "ecosystem properties," and wide differences of opinion concerning whether ecosystems are holistic systems characterized by "emergent properties" or associations of organisms and substrate with merely "collective properties" (cf. Engelberg and Boyarsky 1979; Salt 1979). I will view productivity, decomposition, and nutrient cycling as collective properties of ecosystems. Nevertheless, I believe that the consequences

of these collective properties can alter the growth and interactions of organisms in ways that represent true ecosystem-level feedbacks. An example of such a feedback is efficient internal nutrient cycling within nutrient-poor plants (a physiological process) which slows decomposition and increases nutrient immobilization by soil microorganisms (both collective properties). It appears that this process can lead to reduced nutrient availability and to still more efficient internal nutrient cycling within plants; if so, the whole cycle represents an ecosystem-level feedback (cf. Vitousek 1982, 1983; Shaver and Melillo 1984). I will evaluate the effects of biological invasions on both collective properties and ecosystem-level feedbacks.

10.3. Alterations in Collective Properties

The major collective properties I will consider are productivity (gross and net primary productivity and net ecosystem production), consumption, decomposition, water fluxes (especially the balance between evapotranspiration and runoff), nutrient cycling and loss, soil fertility, erosion, and disturbance frequency. If an invader can be shown to have altered one or more of these characteristics *on an areal basis,* then I will conclude that that species can affect ecosystem properties.

Many exotic species require large-scale, often anthropogenic, disturbances (i.e., agricultural land clearing) to become established. Consequently, it can be difficult to separate the ecosystem-level effects of exotic species from those of the massive, often novel disturbance that allows the species to establish. I will attempt to focus on situations where exotics can invade intact or naturally disturbed ecosystems, on situations where invaders of disturbed sites alter succession by altering soil characteristics more or less permanently, and on situations where the activities of the exotic organisms themselves represent a disturbance. I will start by examining invasions of primary producers, and then consider invasions by consumers.

10.3.1. Invasions by Producers

All primary producers require carbon dioxide, water, and the same suite of nutrient elements. It can be argued that the primary producers in most natural ecoystems fully exploit the more limiting resources (Mooney and Gulmon 1983); evidence for this view can be found in the substantial extent to which primary productivity can be predicted from a knowledge of climate (cf. Rosenzweig 1968).

Given the similarity in requirements among primary producers, it would seem that any invader that alters the collective properties of the natural ecosystem that it invades must be able either (1) to gain access to resources at times that the natives are inactive or from locations that the natives cannot reach; or (2) to utilize those resources more efficiently than the natives. In natural succession, increases in productivity often occur with the addition of new life forms (Odum

1960; Woodwell 1974); this can occur because later successional plants may be able both to obtain resources that short-lived pioneers cannot and to utilize those resources more efficiently. Late-successional life forms (perennials, trees) can obtain more resources by maintaining a perennial root system that goes deeper into the soil, by maintaining evergreen leaves that are active year-round, or by regenerating deciduous leaves rapidly from storage rather than *de novo* from current photosynthate. More efficient resource use can result from the maintenance of internal storage pools of nutrients (Mooney and Gulmon 1983; Fife and Nambiar 1984).

It follows that biological invasions should be able to alter ecosystem properties when they add a life form that is not well-represented in the native flora. Some excellent examples of this sort of change exist; the consequences of invasions of the riparian zone in the arid southwestern United States by salt-cedar (*Tamarix* spp.) (Robinson 1969) and Russian-olive *(Eleagnus angustifolia)* (Knopf and Olson 1984) are among the best-described. Salt-cedar in particular is deeply rooted and has a rapid rate of transpiration. It requires wet areas for germination and early growth, but it is able to maintain itself on water from deep in the soil once it is established. Its transpiration represents a significant pathway of water loss from reservoirs in the arid and semi-arid southwest (Horton 1977), and it also invades natural springs and water courses in desert regions. For example, Eagle Borax Spring in Death Valley was invaded by salt-cedar in 1930s or 1940s. By the late 1960s, water use by salt-cedar had caused the complete disappearence of surface water from what had been a large marsh (Neill 1983; P. G. Sanchez, personal communication). Most interestingly, when managers removed salt-cedar from the site, surface water reappeared and Eagle Borax Spring and its associated flora and fauna recovered (Neill 1983). It would be hard to picture a more convincing illustration of the potential for invading plant species to alter ecosystem properties.

Floating aquatic weeds represent another life form that can greatly alter ecosystem properties. The water fern *Salvinia molesta* is a morphologically diverse floating weed that has been introduced widely in tropical river systems; its effects have been thoroughly studied in Africa (Mitchell 1970 in Mitchell et al. 1980), India (cf. Thomas 1981), and Papua New Guinea (Mitchell et al. 1980). *Salvinia molesta* was introduced to the Sepik River in Papua New Guinea in the early 1970s. Within ten years, it had measurably altered productivity in flood plain lakes, yield from fisheries, and surface water chemistry (Mitchell et al. 1980). In Kerala, India, floating islands of *Salvinia* 1 to 3 m thick were reported (Thomas 1981); paddy cultivation was underway on one of them!

A number of additional examples of the consequences of invasions by new life forms have been documented. Among these are invasions of the Everglades by *Schinus* and *Melaleuca* (Ewel et al. 1981; Ewel, this volume), which can convert wetland to forest, and the ability of North American treeline species to grow well above local treeline in New Zealand (Wardle 1971). It is logical to expect that wherever biological invasions add a major new life form to an area, ecosystem-level changes can be anticipated.

Biological invaders that differ from the natives in subtler ways can also have

ecosystem-level effects. A large number of studies have evaluated the effects of forest plantations, often composed of exotic species, on productivity, nutrient cycling, and soil properties (Nihlgard 1972; Egunjobi and Onweluzo 1979; Perala and Alban 1982; Feller 1983/84). The establishment of forest plantations generally requires substantial human intervention, so studies in plantations cannot fairly represent the influence of an exotic species under natural conditions. Nonetheless, large effects related to the biology of the planted species are often observed. Pine plantations generally yield less water to runoff than the broadleaf forests they replace (Swank and Douglass 1974; Feller 1981); this effect is due more to increased interception and evaporation in the pine canopy than it is to a longer season for transpiration. Additionally, increased litter production and decreased litter decomposition can lead to massive accumulation of leaf litter on the soil surface (Egunjobi and Onweluzo 1979).

I do not know if such effects can be documented outside of plantation conditions. Monterey pine is spreading from plantation areas into adjacent dry sclerophyll forests in Australia, where it is outgrowing the native eucalypts by a wide margin (Burdon and Chilvers 1977; Chilvers and Burdon 1983). It seems likely that this invasion will alter a number of collective properties in those ecosystems.

Can invading plants that do not differ in life form from natives alter ecosystem properties? If an invader differed sufficiently in its effect on soil properties, resource requirements, photosynthetic pathway, or phenology, then it could alter collective properties in the ecosystems it invaded.

One major way in which plants differ in resource requirements is the presence or absence of nitrogen-fixing symbioses. An invader that has the capacity to fix atmospheric nitrogen could clearly be at a competitive advantage when invading a nitrogen-limited community. More importantly, nitrogen fixed by the invader would ultimately (often rapidly) be released in the soil and become available to other organisms. Production in the majority of temperate and boreal ecosystems is limited by nitrogen availability; additions of nitrogen can therefore increase soil fertility, the amount of nitrogen in circulation, and net primary productivity.

Nitrogen fixing invaders are widely distributed, including *Ulex* and *Leucana* on many Pacific islands (Allan 1936; Egler 1942), Scotch broom *(Cytisus scoparius)* in the Pacific Northwest, and *Casuarina* in Florida. Most often, relatively little is known of the nitrogen status of the area invaded. An exception is the recent invasion by the Atlantic shrub *Myrica faya* in young volcanic regions of Hawaii Volcanoes National Park on the island of Hawaii. The sites invaded are extremely nitrogen-deficient, especially on young lava flows and ash deposits (Vitousek et al. 1983; Balakrishnan and Mueller-Dombois 1983), and they contain no native symbiotic nitrogen fixers. *Myrica faya* can form nearly monospecific stands, and it actively fixes nitrogen in these areas (Vitousek and Matson, unpublished). Its invasion could alter productivity, nutrient cycling, and even the rate and direction of primary succession in this region. Moreover, biological invasions by a wide variety of exotic species in Hawaii are most successful on more fertile sites (cf. Gerrish and Mueller-Dombois 1980), so the invasion of

Myrica may increase the probability of further invasions. A similar relationship between soil fertility and the success of exotic species is also observed in western Australia (Bridgewater and Backshall 1981).

Another invader that can alter soil characteristics is the ice plant *Mesembryanthemum crystallinum* in California (Vivrette and Muller 1977) and Australia (Kloot 1983). *Mesembryanthemum* usually requires human intervention to become established, as it generally invades degraded pastures. Once established, it concentrates salt from throughout the rooting zone onto the soil surface, thereby altering soil physcial and chemical properties and interfering with the growth of potential competitors.

I do not know of documented examples where invaders with different photosynthetic capacities or pathways have altered ecosystem-level net primary productivity in the areas they invade. Certainly widespread invasions by C_4 grasses into areas previously dominated by C_3 plants make such effects likely. CAM species, especially *Opuntia,* are also widely successful invaders of ecosystems in which different photosynthetic pathways predominate.

The effects of phenological differences between invaders and natives are somewhat better documented. Mueller-Dombois (1973) described a situation in which an exotic *Andropogon* in Hawaii maintained its life cycle responses to photoperiodic cues from its home range in the southeastern United States. Consequently, in Hawaii it grew during the dry season and was inactive during the rainy season. Transpiration was much reduced under *Andropogon* relative to native rain forest, and boggy conditions developed (Mueller-Dombois 1973).

Invading plant species could also alter ecosystem properties if they differed from natives in their position in succession (cf. Pickett 1976). Individual species are most often argued to be important on the ecosystem level when they invade immediately after disturbance, grow rapidly, and take up nutrients that might otherwise be lost from the site as a whole (but see Introduction). The European exotic *Senecio sylvaticus* apparently plays this role in unburned clearcuts in the Oregon Cascades (Gholz et al. 1985) taking up a substantial fraction of the nitrogen and phosphorus cycled by the intact 450-year-old Douglas-fir forest within 2 to 3 years following clearcutting. It is tempting to speculate that *Senecio* has entered a hitherto unfilled niche in these forest ecosystems, but it must be remembered that it colonizes following a novel disturbance. The functional equivalent of unburned clearcuts (large-scale disturbances that remove the canopy over a wide area but leave the forest floor intact) were not common prior to widespread logging. It is interesting that a native species, *Epilobium angustifolium,* colonizes after both wildfires and slash burning in the Oregon Cascades (Gholz et al. 1985).

In general, I believe that the best-documented illustrations of the importance of invading plants in altering ecosystem properties involve effects on the hydrologic cycle, while effects on productivity or nutrient cycling are usually more speculative. Information on water yield from natural ecosystems has been collected much more precisely and over a longer period of time than information on most other ecosystem properties. Perhaps if information of a similar quality was available for other collective properties of ecosystems, we could detect many more subtle but pervasive effects of biological invaders.

10.3.2. Invasions by Consumers

Invasions of consumer or disease organisms might be expected to have greater short-term effects on ecosytem properties than invasions by producers; consumers are unlikely to have the basically similar resource acquisition/resource utilization requirements that limit the ecosystem-level impacts of most invasions by plants. Moreover, exotic consumers can themselves represent a disturbance (*sensu* Grime 1979) in that they destroy plant biomass, and the ecosystem-level effects of such disturbance are well described and more or less independent of the cause of disturbance (cf. Vitousek et al. 1979; Swank et al. 1981; Matson and Boone 1984). Finally, invading plants and animals may increase rapidly in a new habitat because they have escaped from their predators; invading animals may additionally be favored by escaping from structurally or chemically defended food plants.

An invading animal or disease organism could alter ecosystem characteristics by removing a particular dominant or functionally important plant species, but it is no easier to document the ecosystem-level consequences of removing a plant species than it is to demonstrate the effect of adding a species. Invaders that effectively remove a major element of the flora or fauna of an area are well-described; among the best-known examples are chestnut blight (McCormick and Platt 1980) and Dutch elm disease (Karnosky 1979; Huenneke 1983) in eastern North America. The large-scale dieback of a single dominant species that these invasions caused must have affected productivity, hydrologic cycles, and nutrient losses in the short term. However, the removal of *Castanea dentata* from southern Appalachian forests apparently had little long-term effect on ecosystem structure (McCormick and Platt 1980). Can we find examples where biological invaders appear to have made a long-term difference in the functioning of natural ecosystems?

One of the best documented set of examples involves invasions of feral pigs *(Sus scrofa)*. European boars invaded the Great Smoky Mountains National Park of North Carolina and Tennessee in the 1940s. They now occupy a large portion, but not all, of the suitable habitats within the park. Their "rooting" is concentrated in higher-elevation deciduous forests during the summer months, where they have significant effects on the composition and structure of forest understory communities in the area (Bratton 1975). More importantly (for this chapter),comparisons of soil characteristics in areas with and without rooting (including a fortuitous before-and-after study) showed that heavily rooted areas have much thinner forest floors, mixed organic and mineral soil horizons, and a much greater proportion of bare ground (Singer et al. 1984). These structural changes were accompanied by significantly greater concentrations of nitrogen and potassium in soil solution, even at 100 cm depth, and greater concentrations of nitrate-nitrogen in streams (Table 10.1.) (Singer et al. 1984).

Exotic pigs have been established in Hawaii for a longer period of time than in the Smoky Mountains; they were originally brought in by Polynesians, and larger European varieties were released in the early 1800s. Comparisons of pig-affected and unaffected areas are therefore difficult, but the use of exclosures allows an appraisal of the consequences of pig removal. Sampling in a 14-year-

Table 10.1. Effects of rooting by European wild boars *(Sus scrofa)* on soil characteristics of high elevation deciduous forests in the Great Smoky Mountains National Park

	Without Pigs	With Pigs
Bare ground (%)	0	88
Litter mass (kg/ha)	3095	1830
Soil nitrate-N (μg/g)	19	29
Soil calcium (μg/g)	90	56
Soil phosphorus (μg/g)	58	32
Leachate nitrate-N (mg/liter)	3.8	6.6
Stream nitrate-N (mg/liter)	0.7	1.5

Source: Singer FJ et al. (1984) J Wildlife Manage 48:464–473. Reprinted with permission.

old pig exclosure revealed that the forest floor within the exclosure was significantly thicker than outside, and that net nitrogen mineralization was greater in the heavily rooted areas outside (Table 10.2.).

These results are consistent with those in the Smokies; both demonstrate that exotic pigs can alter the collective properties of natural ecosystems. Exotic pigs are now established in 11 states, including California, and their range is expanding (Wood and Barrett 1979).

Invasions by rabbits or rabbit diseases have also been demonstrated to alter ecosystem-level characteristics. Rabbits apparently invaded Great Britain in the 1100s (Jarvis 1979); of course no studies document the consequences of *that* invasion. Recent reductions in rabbit populations, however, allow an evaluation of their importance. The *Myxoma* virus entered Britain in 1953 and spread rapidly. An immediate consequence of the near-removal of rabbits was a burst of growth and reproduction in the diverse and interesting flora that occupied heavily grazed areas near warrens (Ranwell 1960). This growth was followed by increased shrub invasion, which in turn is now being followed by exotic pines (Hodgkin 1984). Shrub invasion was associated with an increase in soil moisture, organic matter content, and pH. Most importantly, the nutrient supplying capacity of the soil (as demonstrated by bioassays) increased significantly in shrub-dominated areas (Table 10.3.). This increase in soil fertility may lead to a forest dominated by exotic conifers rather than the diverse native flora of the dunes (Hodgkin 1984).

Table 10.2. Effects of rooting by feral pigs on soil structure and nitrogen availability in a Hawaiian montane rain forest

	14-Year-Old Exclosure	Rooted Area
Forest floor depth (cm)	10.1	7.3
Forest floor mass (Mg/ha)	450	370
Nitrogen mineralized (μg/g)	-2.3	11.9

Source: Vitousek PM and Matson PA (unpublished).

Table 10.3. The effect of shrub invasion (following rabbit removal) on soil
characteristics and on soil fertility

	Soil from Shrub Sites	Soil from Shrub-Free Sites
pH	6.2	5.8
Water content (%)	13	7
Loss-on-ignition (%)	7	4
Festuca growth (g/pot)	100	73

Source: Hodgkin SE (1984) Biol Conserv 29:99–119. Reprinted with permission.
Soil fertility was estimated by determining the growth of *Festuca* in soil from shrub-covered and shrub-free plots.

There are numerous additional examples of the effects of animal invaders on ecosystem properties. Among the best-documented are the effects of lamprey-alewife invasion of the St. Lawrence Great Lakes on secondary production (Avon and Smith 1971), the effects of the Australian possum on forest dieback in New Zealand (Batcheler 1983, but see Stewart and Veblen 1983), and the effects of feral goats and sheep in Hawaii on treeline, soil erosion, and community structure (Egler 1942; Mueller-Dombois and Spatz 1975; Mueller-Dombois 1981; Mueller-Dombois et al. 1981).

I believe that it can be concluded unequivocally that invasions by consumers can alter ecosystem properties in otherwise intact ecosystems, and can arrest succession in disturbed ecosystems. Effects are most striking in isolated island groups where the consumer fauna is sparse; for example, large effects are observed in Hawaii, where the only native mammal is a small bat. The effects of an introduced consumer can nonetheless be clearly documented in the most diverse forest area in North America (Singer et al. 1984).

Why does it seem to be easier to document ecosystem-level effects of particular consumer species than of particular plant species? The observation (that it is easier) appears valid both for biological invaders and for studies of the importance of native consumers in natural ecosystems. Certainly keystone species (top predators) alter ecosystem properties in a way that is disproportionate to their abundance (Paine 1966). Perhaps generalist herbivores alter ecosystems because they affect a large fraction of the primary producers, whereas invading plants usually similar enough to natives in resource requirements that they simply replace natives rather than altering the productive capacity of an ecosystem. And perhaps my view of ecosystem properties is one biased towards those properties that are dominated by vegetation (as a whole)—and a view biased towards properties dominated by consumers would lead to a different conclusion.

10.4. Ecosystem-Level Feedbacks

In this section, I examine the possibility that an alteration in ecosystem characteristics caused by a biological invader could propagate through other ecosystem components to increase the scale of the alteration. One such positive

feedback involves alterations of nutrient-use efficiency as a function of changes in nutrient availability (Vitousek 1982; Shaver and Melillo 1984), as discussed in the Introduction. Plantations of exotic pines may initiate this positive feedback loop by producing large amounts of litter that does not easily decompose, leading to nutrient deficiencies and ultimately to further decreases in the quality of the litter produced. Invading species could even alter the type of humus produced (mull vs. mor). Decreased rates of decomposition under invading species have been reported in a number of studies (cf. Cornforth 1970; Florence and Lamb 1975; Egunjobi and Onweluzo 1979). This feedback can also operate the other way—fertilization can lead to the production of nutrient-rich litter which decomposes more rapidly and maintains higher levels of nutrient availability (cf Miller et al. 1976; Weetman et al. 1980). If the nitrogen fixed by *Myrica faya* in Hawaii (above) has this effect, the long-term changes in productivity and nutrient cycling in that primary successional sequence could be profound.

Another potentially important ecosystem-level feedback involves changes in disturbance frequency. Smith (1985) identified 8 species of alien weeds in Hawaii as "fire-enhancers"—species that increased the areal extent of fires in the areas where they are established. The most spectacular example involved two species of *Andropogon* which formed dense, nearly complete canopies in a seasonally dry region once feral goats were removed. The areal extent of fires increased 100-fold after goats were removed, and the exotic *Andropogon* were able to reestablish themselves on the burned sites better than natives. One major native shrub *(Styphelia tameiameiae)* that successfully withstood centuries of feral goat grazing was rapidly eliminated by the fires. Species that alter hydrologic cycles (i.e., the *Andropogon* species described by Mueller-Dombois 1973) can also set ecosystem-level feedback cycles in motion.

A final ecosystem-level feedback that could exacerbate the consequences of biological invasions is positive interactions between exotic plants and animals. Caldwell et al. (1981) showed that *Agropyron desertorum,* an introduced bunchgrass in the intermountain west, produces three to five times more tillers following clipping or defoliation than the otherwise very similar native bunchgrass *Agropyron spicatum.* Grazing pressure from cattle give *Agropyron desertorum* a competitive advantage. In Hawaii, Smith (1985) showed that there is a strong positive relationship between feral pigs and a number of important weedy plants, including strawberry guava *(Psidium cattleianum)* and a number of *Passiflora* species. The pigs consume and disseminate the seeds of the invaders, and they also disturb areas of soil and create excellent conditions for seed germination and seedling growth. Both soil characteristics and a community structure distinctly different from those in native forest are maintained by this interaction.

10.5. Conclusions

The ecosystem-level consequences of biological invasions can be used as an explicit test of the importance of species properties in controlling ecosystem

function. There is excellent evidence that animal invaders can alter a variety of collective properties of ecosystems, including productivity, soil structure, and nutrient cycling. Invading plants can also clearly alter ecosystem characteristics when they differ in life form from natives. The ecosystem-level consequences of invasions by plants that do not differ in life form from natives are less clear, but a number of circumstances under which such invasions are likely to have such consequences have been identified. Finally, invading plants and animals can establish positive feedbacks that accentuate the ecosystem-level consequences of biological invasions.

10.6. Acknowledgments

I thank P.G. Sanchez and C.W. Smith for generously sharing unpublished information on biological invasions. S.P. Hamburg, L.F. Huenneke, and P.A. Matson commented critically on an earlier draft of this manuscript.

10.7. References

Allan HH (1936) Indigene versus alien in the New Zealand plant world. Ecology 17:187–193
Avon WI, Smith SH (1971) Ship canals and aquatic ecosystems. Science 174:13–20
Balakrishnan N, Mueller-Dombois D (1983) Nutrient studies in relation to habitat types and canopy dieback in the montane rain forest ecosystem, island of Hawaii. Pacif Sci 37:339–359
Batcheler CL (1983) The possum and rata-kamahi dieback in New Zealand: a review. Pacif Sci 37:426
Blank JL, Olson RK, Vitousek PM (1980) Nutrient uptake by a diverse spring ephemeral community. Oecologia (Berl) 47:96–98
Boring LR, Monk CD, Swank WT (1981) Early regeneration of a clearcut southern Appalachian forest. Ecology 62:1244–1253
Bratton SP (1975) The effect of the European wild boar, *Sus scrofa*, on gray beech forest in the Great Smoky Mountains. Ecology 56:1356–1366
Bridgewater PB, Backshall DJ (1981) Dynamics of some Western Australian ligneous formations with special reference to the invasion of exotic species. Vegetatio 46:141–148
Burdon JJ, Chilvers GA (1977) Preliminary studies on a native Australian eucalypt forest invaded by exotic pines. Oecologia (Berl) 31:1–12
Caldwell MM, Richards JH, Johnson DA, Nowack RS, Dzurec RS (1981) Coping with herbivory: photosynthetic capacity and resource allocation in two semiarid *Agropyron* bunchgrasses. Oecologia (Berl) 50:14–24
Chilvers GA, Burdon JJ (1983) Further studies on a native Australian eucalypt forest invaded by exotic pines. Oecologia (Berl) 59:239–245
Clark B, Murray J, Johnson MS (1984) The extinction of endemic species by a program of biological control. Pacif Sci 38:97–104
Cornforth IS (1970) Reafforestation and nutrient reserves in the humid tropics. J Appl Ecol 7:609–615
Dayton PK, Currie V, Gerrodette T, Keller BD, Rosenthal R, Ven Tresca D (1984) Patch dynamics and stability of some California kelp communities. Ecol Monogr 54:253–289

Egler FE (1942) Indigene versus alien in the development of arid Hawaiian vegetation. Ecology 23:14–23

Egunjobi JK, Onweluzo BS (1979) Litterfall, mineral turnover, and litter accumulation in *Pinus caribea* L. stands at Ibadan, Nigeria. Biotropica 11:251–255

Ehrlich PR, Mooney HA (1983) Extinction, substitution, and ecosystem services. Bioscience 33:248–253

Elton CS (1958) The Ecology of Invasions by Animals and Plants. Methuen, London, 181 pp

Engelberg J, Boyarsky LL (1979) The noncybernetic nature of ecosystems. Am Natur 114:317–324

Ewel J, Ojima D, Debusk W (1981) Ecology of a successful exotic tree in the Everglades. In: Proceedings of Second Conference on Scientific Research in National Parks, 8:419-422

Feller MC (1981) Water balances in *Eucalyptus regnans, Eucalyptus obliqua,* and *Pinus radiata* ecosystems in Victoria. Aust For Res 44:153–161

Feller MC (1983/84) Effects of an exotic conifer *(Pinus radiata)* plantation on forest nutrient cycling in southeastern Australia. For Ecol Manag 7:77–102

Fife DN, Nambiar EKS (1984) Movement of nutrients in radiata pine needles in relation to the growth of shoots. Ann Bot 54:303–314

Florence RG, Lamb D (1975) Ecosystem processes and the management of radiata pine forests on sand dunes in South Australia. Proc Ecol Soc Aust 9:34–48

Foin TC, Jain SK (1977) Ecosystem analysis and population biology: lessons for the development of community ecology. Bioscience 27:532–538

Foster MM, Vitousek PM, Randolph PA (1980) The effect of *Ambrosia artemisiifolia* on nutrient cycling in a first-year old-field. Am Midl Natur 103:106–113

Gerrish G, Mueller-Dombois D (1980) Behavior of native and non-native plants in two tropical rainforests on Oahu, Hawaiian Islands. Phytocoenologia 8:237–295

Gholz HL, Hawk GM, Campbell A, Cromack K, Brown AT (1985) Early vegetation recovery and element cycles on a clearcut watershed in western Oregon. Can J For Res 15:400–409

Grime JP (1979) Plant Strategies and Vegetation Processes. John Wiley, New York

Harcombe PA (1977) Nutrient accumulation by vegetation during the first year of recovery of a tropical forest ecosystem. In: Cairns J, Dickson KL, Herricks EE (eds), Recovery and Restoration of Damaged Ecosystems. University Press of Virginia, Charlottesville, pp 347–378

Hodgkin SE (1984) Scrub encroachment and its effects on soil fertility on Newborough Warren, Anglesey, Wales. Biol Conserv 29:99–119

Horton JS (1977) The development and perpetuation of the permanent tamarisk type in the phreatophyte zone of the Southwest. In: Johnson RR, Jones DA (eds), Importance, Preservation, and Management of Riparian Habitat: A Symposium. U.S. Department of Agriculture Forest Service General Technical Report RM-43, pp 124-127

Huenneke LF (1983) Understory response to gaps caused by the death of *Ulmus americana* in Central New York. Bull Torrey Botan Club 110:170–175

Jarvis PJ (1979) The ecology of plant and animal introductions. Prog Phys Geogr 3:187–214

Karnosky DF (1979) Dutch elm disease: a review of the history, environmental implications, control, and research needs. Environ Conserv 6:311–322

Kloot PM (1983) The role of common iceplant *(Mesembryanthemum crystallinum)* in the deterioration of medic pastures. Aust J Ecol 8:301–306

Knopf FL, Olson TE (1984) Naturalization of Russian-olive: implications to Rocky Mountain wildlife. Wildlife Soc Bull 12:289–298

Lubchenco J, Menge BA (1978) Community development and persistence in a low rocky intertidal zone. Ecol Monogr 48:67–94

Marks PL (1974) The role of pin cherry *(Prunus pensylvanica* L.) in the maintenance of stability in northern hardwood ecosystems. Ecol Monogr 44:73–88

Marks PL, Bormann FH (1972) Revegetation following forest cutting: mechanisms for return to steady state nutrient cycling. Science 176:914–915

Matson PA, Boone RD (1984) Natural disturbance and nitrogen mineralization: waveform dieback of mountain hemlock in the Oregon Cascades. Ecology 65:1511–1516

Mattson WJ, Addy ND (1975) Phytophagous insects as regulators of forest primary production. Science 190:515–522

McCormick JF, Platt RB (1980) Recovery of an Appalachian forest following the chestnut blight, or, Catherine Keever—you were right! Am Midld Natur 104:264–273

McIntosh RP (1980) The relationship between succession and the recovery process in ecosystems. In: Cairns J (ed), The Recovery Process in Damaged Ecosystems. Ann Arbor Science pp 11–62

Miller HG, Cooper JM, Miller JD (1976) Effects of nitrogen supply on nutrients in litterfall and crown leaching in a stand of Corsican pine. J Appl Ecol 13:233–248

Mitchell DS (1970) Autecological studies of *Salvinia auriculata* Aubl. Ph.D. Thesis, University of London, 669 p

Mitchell DS, Petr T, Viner AB (1980) The water-fern *Salvinia molesta* in the Sepik River, Papua New Guinea. Environ Conserv 7:115–122

Mooney HA, Gulmon SL (1983) The determinants of plant productivity—natural versus man-modified communities. In: Mooney HA, Godron M (eds), Disturbance and Ecosystems: Components of Response. Springer-Verlag, Berlin, pp 146–158

Mueller-Dombois D (1973) A non-adapted vegetation interferes with water removal in a tropical rain forest area in Hawaii. Trop Ecol 14:1–18

Mueller-Dombois D (1981) Vegetation dynamics in a coastal grassland of Hawaii. Vegetatio 46:131–140

Mueller-Dombois D, Bridges KW, Carson HL (eds) (1981) Island Ecosystems: Biological Organization in Selected Hawaiian Communities. Hutchinson Ross, Stroudsburg, Pennsylvania, 583 p

Mueller-Dombois D, Spatz G (1975) The influence of feral goats on the lowland vegetation in Hawaii Volcanoes National Park. Phytocoenologia 3:1–29

Muller RN, Bormann FH (1976) Role of *Erythronium americanum* Ker. in energy flow and nutrient dynamics of a northern hardwood forest ecosystem. Science 193:1126–1128

Neill WM (1983) The tamarisk invasion of desert riparian areas. Educ Bull 83-4, Desert Protective Council, Spring Valley, California, 4 p

Nihlgard B (1972) Plant biomass, primary production and distribution of chemical elements in a beech and a planted spruce forest in South Sweden. Oikos 23:69–81

Odum EP (1960) Organic production and turnover in old field succession. Ecology 41:34–49

O'Neill RV, Ausmus BS, Jackson DR, Van Hook RI, Van Voris P, Washburne C, Watson AP (1977) Monitoring terrestrial ecosystems by analysis of nutrient export. Water Air Soil Pollut 8:271–277

Paine RT (1966) Food web complexity and species diversity. Am Natur 100:65–75

Perala DA, Alban DH (1982) Biomass, nutrient distribution and litterfall in *Populus, Pinus* and *Picea* stands on two different soils in Minnesota. Plant Soil 64:177–192

Peterson DL, Rolfe GL (1982) Nutrient dynamics of herbaceous vegetation in upland and floodplain forest communities. Am Midld Natur 107:325–339

Pickett STA (1976) Succession: an evolutionary interpretation. Am Natur 110:107–119

Ranwell DS (1960) Newborough Warren, Anglesey, III. Changes in the vegetation on parts of the dune system after the loss of rabbits by myxomatosis. J Ecol 48:385–395

Robinson TW (1969) Introduction, spread, and areal extent of saltcedar *(Tamarix)* in the western states. U.S. Department of the Interior Geological Survey Prof Paper 491-A, 12 p

Rosenzweig ML (1968) Net primary productivity of terrestrial communities: prediction from climatological data. Am Natur 102:67–74

Salt GW (1979) A comment on the use of the term "emergent properties." Am Natur 113:145–148

Shaver GR, Melillo JM (1984) Nutrient uptake, nutrient recovery, and nutrient use by marsh plants: efficiency concepts and relation to availability. Ecology 65:1491–1510

Singer FJ, Swank WT, Clebsch EEC (1984) Effects of wild pig rooting in a deciduous forest. J Wildlife Manag 48:464–473

Smith CW (1985) Impact of alien plants on Hawaii's native biota. In: Stone CP, Scott JM (eds), Hawaii's Terrestrial Ecosystems: Preservation and Management. Cooperative Park Studies Unit. University of Hawaii, Honolulu, pp 180–250.

Stewart GH, Veblen TT (1983) Forest instability and canopy tree mortality in Westland, New Zealand. Pacif Sci 37:427–431

Swank WT, Douglass JE (1974) Streamflow greatly reduced by converting deciduous hardwood stands to pine. Science 185:857–859

Swank WT, Waide JB, Crossley DA Jr, Todd RL (1981) Insect defoliation enhances nitrate export from forested ecosystems. Oecologia (Berl) 51:297–299

Thomas KJ (1981) The role of aquatic weeds in changing the pattern of ecosystems in Kerala. Environ Conserv 8:63–66

Thomas WA (1969) Accumulation and cycling of calcium by dogwood trees. Ecol Monogr 39:101–120

Vitousek PM (1982) Nutrient cycling and nutrient use efficiency. Am Natur 119:553–572

Vitousek PM (1983) Nitrogen turnover in a ragweed-dominated first-year old field in southern Indiana. Am Midld Natur 110:46–53

Vitousek PM, Gosz JR, Grier CC, Melillo JM, Reiners WA, Todd RL (1979) Nitrate losses from disturbed ecosystems. Science 204:469–474

Vitousek PM (1983) Nitrogen turnover in a ragweed-dominated first-year old field in southern Indiana. Am Midl Natur 110:46–53

Vivrette NJ, Muller CH (1977) Mechanism of invasion and dominance of coastal grassland by *Mesembryanthemum crystallinum*. Ecol Monogr 47:301–318

Wardle P (1971) An explanation for alpine timberline. N Zeal J Bot 9:371–402

Weetman GF, Roberge MR, Meng CH (1980) Black spruce: 15-year growth and microbiological responses to thinning and fertilization. Can J For Res 10:502–509

Wood GW, Barrett RH (1979) Status of wild pigs in the United States. Wildlife Soc Bull 7:237–246

Woodwell GM (1974) Success, succession, and Adam Smith. Bioscience 24:81–87

4. Modeling the Invasion Process

11. Predicting Invasions and Rates of Spread

J. Roughgarden

11.1. Introduction

Here I briefly review the extent to which invasions can be predicted, focusing
on prediction using mathematical models of population dynamics. At the outset
it is worth stressing that the best prediction is provided by empirical precedent,
and not by a model at all. For example, at different times three species of
Anolis lizards have been introduced to the island of Bermuda, an island that
had not possessed *Anolis* lizards. The course of these three successive invasions
has been chronicled by Wingate (1965). Also, experimental introductions of
anoles have been made and carefully monitored by Schoener and Schoener
(1983a) to small cays in the Bahamas, and by Roughgarden et al. (1984) to a
small cay in the eastern Caribbean. This information provides an excellent basis
for predicting the course of introductions of *Anolis* lizards to other small islands
in the Caribbean region. Nonetheless, we cannot, in practice, develop empirical
precedents for all instances where we might inquire about the likelihood of an
invasion, and the need for models to aid in transferring knowledge from one
situation to another is certain to continue.

The theoretical literature on invasions treats three issues: (1) the rate of
spread of an invasion that has taken, (2) the condition for an invasion to take,
and (3) the properties of a fauna that is assembled by successive invasion. The
theory for (1) is quite robust, has been empirically tested, and is about as reliable
as theory gets. Theory for (2) and (3) is not general. It is addressed to special

kinds of populations and population interactions; it is also largely, though not entirely, untested. A fourth issue I discuss concerns the theoretical possibility of prediction itself and the related issue of whether the developing and testing of models for biological processes offers the most efficient approach to prediction. This topic has not been investigated with particular reference to predicting the course of invasions, but is nonetheless important in the present context.

The literature of theoretical ecology (and population genetics) is filled with phrases such as "invasion when rare" and "initial increase." A standard approach in the analysis of dynamical models is to determine the pattern of trajectories in the neighborhood of all the fixed points of the system (a fixed point is an equilibrium point). In many models there are fixed points corresponding to the absence of one of the components in the system. For example, a model with n compartments may have n fixed points, each of dimension $(n - 1)$ and corresponding to one of the compartments being empty. The strategy of analysis of the full model often is to splice together the pattern of trajectories around each of the fixed points to determine the global pattern of trajectories. In this context phrases such as "invasion when rare" and "initial increase" refer to a mathematical technique, and not at all to the biological phenomenon of invasion. In this chapter I mention only theoretical studies where the intent of the theory is to relate to the biological phenomenon of invasion, and not studies where the idea of an invasion is merely synonymous with the mathematical analysis of the local stability of a boundary equilibrium.

Finally, it is well to be explicit on what it is that one may wish to predict. The questions one may ask are: (1) Can, or will, a particular species invade a particular habitat? This is exemplified by the question, Will the mediterranean fruit fly invade the crops of central California? (2) Can a particular species invade any habitat? This is usually a species-conservation question, e.g., Can the California condor survive anywhere outside of captivity? (3) Will a particular habitat be invaded by something? And if so, what are the most likely candidates for the future invaders? This is usually a habitat-conservation question, e.g., Will the temperate rain forest on the Olympic peninsula be altered in its basic composition by its restriction to narrow corridors along river banks? (4) If an invasion does take, how fast will it spread? This is important in deciding where to establish buffer zones to prevent the spread of an invading pest, and to predict whether the spread of a species exerting biological control on a pest will be fast enough to be useful. (5) Finally, after an invasion has spread, what are the implications for the residents? Will the invasion cause a cascade of extinctions among the previous residents that would not have otherwise occurred? As an example, the invasion of European grasses in the grasslands of California has led to the restriction of many native herbs to marginal soils such as serpentine outcroppings. Could this have been predicted? The theory that pertains to invasions has never been considered as a whole; and it does not address these questions equally. Nonetheless, it has something to offer about them all, although there is certainly room for much more theoretical study, especially using models more specific to particular systems than the models used so far, as well

as for empirical studies that directly relate to theoretical developments and that are not merely considered in a theoretical context as an afterthought.

11.2. The Spread of an Invasion

The main theoretical finding about the spread of an invading species is that the square root of the area occupied by the species increases linearly with time. This result originates in population-genetic models for the spread of a favorable mutation (Fisher 1937; Kendall 1948) and was brought into an ecological context by Skellam (1951), and independently by Nowak (1971). The model consists of an exponentially growing population on a plane. The population is introduced to a point on the plane. As the organisms disperse from their place of birth the population's range expands. An individual is assumed to disperse in any direction with equal probability, with the dispersal distance being normally distributed, independent of where it was born (the plane is homogeneous) and independent of the population size. Exponential growth continues at every place the population reaches. This picture is appropriate to the early phase of an invasion that has successfully "taken." In the model it is shown that if a circular contour is drawn on the plane through all points having a given population density, then the radius of this circle increases linearly with time. In practice the contour representing the points that have a particular population density is not exactly circular and the square root of the area may be used instead of the area. Skellam (1951) analyzed the data on the spread of the muskrat, *Ondatra zibethica,* in central Europe from about 1909 to 1927 from Ulbrich (1930) and observed that the square root of the area occupied by this population increased approximately as a linear function of time during this period. A figure giving the contours that document the spread of this populations also appears in Elton (1958) based on Ulbrich's data. This result is quite robust. For example, in models, a population spreading in a linear habitat, as along a corridor (river valley) or along a coastline, also expands as a linear function of time. This result thus appears to be the basis for the frequent observation that maps of invading species show the leading edge of the invasion progressing at a roughly constant velocity through suitable habitat (e.g., Rapoport 1983, Thresh 1983). Recent extentions to this theory include relaxing the assumption of normally distributed dispersal distances (Mollison 1977), and allowing for multiple foci of introduction and of jumps in addition to diffusive migration (Mack 1985; Baker, this volume).

11.3. The Initiation of an Invasion

Models for the conditions under which an invasion can begin have focussed on two related subjects, the minimum propagule size and the minimum habitat size.

11.3.1. Minimum Propagule Size

Demographic Stochasticity

The most general lower limit to the size of an invading propagule is set by demographic stochasticity. The term demographic stochasticity was coined by May (1973) for variation in population size resulting from ever-present chance variation in the fertility and survivorship of individuals in a finite population. The term demographic stochasticity is distinguished from environmental stochasticity that refers to variation in fertility and survivorship resulting from fluctuations in environmental conditions. The theory of demographic stochasticity has long been studied in statistics using a class of models called "birth-death processes." This theory was first applied to the question of the minimum propagule size and to the related question of the expected time to the random extinction of a propagule by MacArthur and Wilson (1967). Their analysis was extended by Goel and Richter-Dyn (1974). In most formulations extinction is taken as an absorbing boundary and some upper bound to the population size is taken as a reflecting boundary. After comparing several alternative formulations Goel and Richter-Dyn offer the formula that the minimum size for a propagule is approximately equal to $3/ln(b/d)$ where b is the per-individual stochastic birth rate and d the per-individual stochastic death rate. This quantity evaluates to a number between 10 and 20 in many circumstances. The meaning of the minimum size for a propagule in the context of these models is this: the population is nearly certain to drift to its upper bound before it drifts to extinction provided its size exceeds the minimum size. Thus a population may also "take" if its initial size is less than this minimum size should it happen to drift above the minimum size after its introduction. The minimum size may thus be thought of as the minimum size to guarantee that the population will "take," and not as a minimum size for it to have merely a possibility of taking.

MacArthur and Wilson (1967) cited data (then unpublished) of Crowell (1973) concerning *Peromyscus* introduced in 1962 to islands off the coast of Maine. One pair was introduced to each of three islands, two pairs to another, three pairs to another, and eight pairs to still another. By 1965 all the introductions with more than two pairs has taken, but only one of the single-pair introductions has taken. MacArthur and Wilson suggested that these results were quantitatively in accord with their analysis of demographic stochasticity.

Multiple Equilibria

Another class of models that applies to the question of the minimum propagule size consists of deterministic population models. These models share the property that there are at least two simultaneously stable steady-state abundances, one representing extinction and the other representing a positive population size. Between the extinction and positive stable equilibrium is a point, often called a "break point," that is technically an unstable equilibrium, and that can be thought of as demarcating the border between two domains of attraction.

If the population is initially below this demarcation point, then the population becomes extinct, and if the population's initial abundance is above the demarcation point then it "takes." The first general recognition of this phenomenon in ecology was offered in a review by May (1977) that brought together examples involving grazed pasture and rangeland, exploited fish and whale populations, nearshore marine communities, outbreaks of a psyllid on eucalyptus in Australia, the introduction of a European sawfly in New Brunswick, and the spruce budworm in Canada. Examples also arise in host–parasite systems (Anderson 1979). Recently, a possible example stemming from models of the population dynamics of barnacles (Roughgarden and Iwasa 1985) has been proposed for the invasion of an Australian barnacle into the British Isles and the coast of France after a period of heavy shipping during World War II (Crisp 1958). The underlying cause of multiple equilibria is any of many mechanisms that produce nonlinearity in the population dynamics such that at low abundances the birth rate does not exceed the death rate, whereas at high abundances the birth rate can exceed the death rate.

11.3.2. Minimum Habitat Size

The simplest model for the minimum size of a habitat consists of applying the theory of demographic stochasticity discussed above. If one knows the home range of the individuals in a population then the minimum habitat size is the area necessary to contain enough individuals to exceed the minimum propagule size.

Furthermore, deterministic models of population dynamics together with dispersal in a spatially heterogeneous environment often predict that a favorable interval, or "patch," within a region must exceed a threshold value for a population to colonize the region. This type of result was first introduced to ecological models by Skellam (1951), and extended in Gurney and Neisbiet (1975), McMurtrie (1978), Ludwig et al. (1980), Namba (1980), Okubo (1980), Pacala and Roughgarden (1982), and Shigesada and Roughgarden (1982). Predictions of minimum patch size generally parallel similar predictions from population-genetic theory concerning the minimum spatial scale of geographical variation in natural selection that can be evolutionarily "tracked" by a population (Slatkin 1973; Nagylaki 1975; May et al. 1975). Skellam's (1951) model (No. 3.3, Case iii) is the simplest model that predicts the existence of a minimum habitat size. The population is viewed as growing exponentially in an interval of length $2L$; only the right half is analyzed because, in the model, the left half is symmetric to the right half. If the root mean square dispersal distance is l and the exponential growth rate is r, then Skellam obtained the requirement that the habitat half length, L, be greater than $(\pi\sqrt{l^2/2})/(2\sqrt{r})$. Subsequent analysis reveals that this result is sensitive to how the boundary conditions are stipulated. Skellam stipulated that the population size be zero at the edge of the habitat. This requirement implies that the area surrounding the favorable habitat be exceptionally unfavorable; indeed, mathematically Skellam's condition is obtained

in a two-habitat model in the limit that the length of the unfavorable interval adjacent to the favorable habitat approaches infinity and the intrinsic rate of increase within the unfavorable habitat approaches minus infinity. Thus Skellam's condition is one extreme; at the other extreme, if the unfavorable habitat merely has a zero intrinsic rate of increase in it then there is no deterministic requirement on the minimum size of the favorable habitat for a single species to colonize it. Pacala and Roughgarden (1982) showed that even if there is no minimum size to a favorable habitat for successful colonization by a single species, the presence of a competitor leads to a minimum habitat size that depends on many population-dynamic parameters including the strength of competition, and the r and K values in both habitats. All the results above are derived using diffusion-equation representations of population dynamics with dispersal that involve limited migration relative to the scale of the habitats. Roughgarden and Iwasa (1985) proposed a model for the population dynamics of a marine form with pelagic larval dispersal and where the local habitats are all coupled to one another by sharing a common "larval pool." Criteria for the minimum size of suitable habitat are found in this circumstance as well.

Evidence of minimum habitat size exists, but is not sufficiently quantitative to allow testing any of the models that might account for the cause of a minimum habitat size. For example, Brown (1971) has found that species of alpine small mammals are absent in the smaller ranges of Nevada even though appropriate habitat is present, and similar results have been obtained for alpine plants of the Nevada mountains (Harper et al. 1978). Evidence of a minimum habitat size is often contained in studies pertaining to the relation between species diversity and island area, as conventionally studied in island biogeography. For example, Schoener and Schoener (1983b) have presented exceptionally clear evidence of thresholds relating island size to number of lizard species in the Bahamas. For cays near a source area, there are no species on cays less than 10^4 ft^2, one species on cays between 10^4 and 1.6×10^4 ft^2, and two species on cays greater than 1.6×10^4 ft^2. In island biogeography it is customary to focus on explanations of such diversity-area curves by reference to immigration and extinction rates, but the possibility that it reflects a rank-ordered sequence of species-specific habitat thresholds should not be overlooked.

11.4. Faunal Consequences of Invasion

A successful invasion may affect other species in the region, or it may not bring any noticeable effect. To my knowledge, little theory has been devoted to predicting the consequences of a successful invasion for a variety of species interactions and types of population structure. Most of the theory that does exist about the consequences of invasion pertains to predicting the effects of an invasion on species with which the invader may compete. The early work on this subject (MacArthur and Levins 1967) relied on the concept of limiting similarity, a concept referring to the maximum overlap in the use of limiting resources

that can be tolerated consistent with species coexistence according to the Lotka-Volterra competition equations. MacArthur and Levins (1967) determined the limiting similarity for symmetric competition among two and three species, May and MacArthur (1972) extended the theory to n species assuming further symmetries and a criterion for species coexistence relying on environmental stochasticity, and Roughgarden (1974) detailed the conditions under which one or both of the residents were eliminated as a result of the invasion. In these studies the overall consequence of repeated invasions was not explicitly addressed, and it was widely believed that the theory would ultimately be solved to predict that repeated invasions would result in equally spaced niches separated by a constant amount representing the limiting similarity. This constant amount would then represent a generalization of the Hutchinsonian ratios (Hutchinson 1959) originally described for two species to situations involving more than two species. From a theoretical standpoint, results of this theory are somewhat fragile (Lawton and Hassell 1984) and from a biological standpoint shortcomings are that neither invasion nor the possibility of coevolution among the members of the fauna subsequent to an invasion are actually modeled.

These limitations of limiting-similarity theory together with the need to develop a formulation of competition theory that was tailor-made for the biogeography of West Indian *Anolis* led to the analysis in Roughgarden et al. (1983)and Rummel and Roughgarden (1983, 1985) in which two computer algorithms were presented for the "assembly" of a community of competing species. In one algorithm the community was assembled through successive single invasions. The invasions were assumed to occur at, so to speak, relatively unused locations in niche space. In the second algorithm each invasion was followed by an episode during which the fauna was allowed time for coevolutionary adjustments in niche space before another invader was introduced. The output of these algorithms provide a glimpse at how communities formed only by successive invasion may differ from those formed by invasion combined with coevolutionary adjustments. According to this work most of the common beliefs about "what competition theory says" are inaccurate. In particular, there are no circumstances in which these models predict that communities form with equally spaced niches. Only in a special and highly symmetric case are equally spaced niches even approximately attained. Moreover, although invasions at relatively vacant locations in niche space rarely cause the extinction of a resident, extinctions are the rule during the subsequent coevolution of the invaded fauna. Reinvasions after the extinction lead to cycles reminiscent of the "taxon cycles" reported by Wilson (1961) for the ant fauna of Pacific islands. Rummel and Roughgarden (1985) offer a catalogue of differences expected to distinguish communities assembled only from successive single invasions, from communities assembled by successive single invasions each of which follows a period when the community coevolutionarily readjusts. Although Hutchinsonian spacing is observed in almost no community produced by the algorithms, statistical evidence for an overdispersion of niches was found using the method Schoener (1984) originally applied to the analysis of biogeographical body-size

patterns for hawks; Hutchinsonian spacing was found to be an "ensemble" property, and not the property of any particular community. Roughgarden (1985) reviews biogeographical and fossil evidence that a cycle of invasion and extinction during coevolution is occurring among *Anolis* lizard populations of islands in the northeastern Caribbean, together with evidence pertaining to the suitability of this theory for *Anolis* populations. Clearly, even if the application of this theory to *Anolis* populations in the West Indies is appropriate, the theory is sufficiently specialized that its applicability to other types of communities is limited, and, so far as I know, little more can be said at this time about the faunal consequences of invasions.

11.5. The Predictability of Invasions

A general issue facing any attempt at predicting population phenomena using quantitative theory, including the prediction of when and where invasions will occur, and what will happen after an invasion, is whether developing and testing a biologically based model offers an efficient way to make the prediction. It takes work to formulate a model and to solve it mathematically; and it takes still more to validate it in at least one circumstance. Can one make just as good a prediction, though perhaps restricted to the short term, by using ad hoc methods requiring less work, though of a different kind? The answer is probably yes, if recent activity in fisheries biology offers any guide (Shepard 1984). The models reviewed above were developed with the goal of increasing understanding, not with the goal of making predictions. The ability of predict invasions, if any, is a fortuitous byproduct of these models.

A second general issue is whether the models that potentially offer the ability to predict population phenomena are a special and perhaps nongeneric type of model. May (1976) brought to the attention of ecologists the awareness that deterministic population models may produce trajectories that are similar in important respects to stochastic processes, a finding that had been known from physical models in the theory of turbulence (Lorenz 1963). This finding lingered in neglect after the report by Hassell et al. (1976) that most of the (mainly insect) populations they surveyed had population-dynamic parameters that did not produce deterministic trajectories with stochastic features. Recently, however, Schaffer and Kot (1985) have suggested that the dynamics of measle epidemics in New York and Baltimore are produced by a deterministic process that itself leads to stochastic trajectories, and not simply by a deterministic process such as an oscillation that is embedded in a stochastic environment and subjected to recurrent random shocks. As skill in detecting this possibility increases, further examples may accumulate. All the theory pertaining to invasions developed to date assumes that there is a community with knowable properties to be invaded. If the community that is to be invaded is itself sufficiently variable, then predicting anything about an invasion will assume the status of a weather report.

11.6. Acknowledgements

I thank D. Simberloff, two anonymous reviewers, and the editors for useful comments on the manuscript. I also thank the National Science Foundation and the Department of Energy for their support of my research as reviewed here.

11.7. References

Anderson RM (1979) The influence of parasitic species on host population growth. In: Anderson RM, Turner BD (eds), Population Dynamics. Blackwell, Oxford

Brown JH (1971) Mammals on mountaintops: nonequilibrium insular biogeography. Am Natur 105:467–478

Crisp DJ (1958) The spread of *Eliminius modestus* Darwin in northwest Europe. J Mar Biol Assoc 37:483–520

Crowell KL (1973) Experimental zoogeography: introductions of mice to small islands. Am Natur 107:535–558

Elton CS (1958) The ecology of invasions by animals and plants. Methuen, London

Fisher RA (1937) The wave of advance of advantageous genes. Ann Eugen 7:355–369

Goel NS, Richter-Dyn N (1974) Stochastic Models in Biology. Academic Press, New York

Gurney WS, Nisbet RM (1975) The regulation of inhomogeneous populations. J Theor Biol 52:441–457

Harper KT, Freeman DC, Ostler A, Kikoft LG (1978) The flora of Great Basin mountain ranges: diversity, sources, and dispersal ecology. Great Basin Nat Mem 2:81–103

Hassell MP, Lawton JH, May RM (1976) Patterns of dynamical in single-species populations. J Anim Ecol 45:471–486

Hutchinson G (1959) Homage to Santa Rosalia, or why are there so many kinds of animals? Am Natur 93:145–159

Kendall DG (1948) A form of wave propagation associated with the equations of heat conduction. Proc Cambridge Philos Soc 44:591–594

Lawton JH, Hassell MP (1984) Interspecific competition in insects, chapter 15. In: Huffaker CB, Rabb RL (eds), Ecological Entomology. John Wiley, New York

Lorenz EN (1963) Deterministic nonperiodic flow. J Atm Sci 20:130–141

Ludwig DH, Aronson DG, Weinberger HF (1980) Spatial patterning of the spruce budworm. J Math Biosci 8:217–258

MacArthur RH, Levins R (1967) The limiting similarity, convergence and divergence of coexisting species. Am Natur 101:377–385

MacArthur RH, Wilson EO (1967) The Theory of Island Biogeography. Princeton University Press, Princeton, New Jersey

Mack RN (1985) Invading plants: their potential contribution to population biology. In: White J (ed), Studies on Plant Demography: John L Harper Festschrift. Academic Press, pp 127–142

May RM (1973) Stability and Complexity in Model Ecosystems. Princeton University Press, Princeton, New Jersey

May RM (1976) Simple mathematical models with very complicated dynamics. Nature 261:459–467

May RM (1977) Thresholds and breakpoints in ecosystems with a multiplicity of stable states. Nature 269:471–477

May RM, Endler JA, McMurtrie RE (1975) Gene frequency clines in the presence of selection opposed by gene flow. Am Natur 109:659–676

May RM, MacArthur RH (1972) Niche overlap as a function of environmental variability. Proc Nat'l Acad Sci USA 69:1109–1113

McMurtrie R (1978) Persistence and stability of single-species and predator-prey systems in spatially heterogeneous environments. Math Biosci 39:11–51

Mollison D (1977) Spatial contact nodels for ecological and epidemic spread. J R Stat Soc B 39:283–326

Nagylaki T (1975) Conditions for the existence of clines. Genetics 80:595–615

Namba T (1980) Density-dependent dispersal and spatial distribution of a population. J Theor Biol 86:351–363

Nowak E (1971) The range expansion of animals and its causes. Trans from Polish, 1975. Foreign Scientific Publications Department, US Department of Commerce, Washington, DC

Okubo A (1980) Diffusion and Ecological Problems: Mathematical Models. Springer-Verlag, Berlin/ Heidelberg/ New York

Pacala SW, Roughgarden J (1982) Spatial heterogeneity and interspecific competition. Theor Popul Biol 21:92–113

Rapoport E (1983) Aereography. Pergamon Press, Oxford

Roughgarden J (1974) Species packing and the competition function with illustrations from coral reef fish. Theor Popul Biol 5:163–186

Roughgarden J (1985) A comparison of food-limited and space-limited animal competition communities. In: Diamond J, Case T (eds), Community Ecology. Harper & Row, New York p 492–516

Roughgarden J, Heckel D, Fuentes ER (1983) Coevolutionary theory and the island biogeography of *Anolis*. In: Huey R, Pianka E, Schoener T (eds), Lizard Ecology: Studies on a Model Organism. Harvard Unviersity Press, pp 371–410

Roughgarden J, Iwasa Y (1985) Dynamics of a metapopulation with space-limited sub-populations. Manuscript

Roughgarden J, Pacala S, Rummel J (1984) Strong present-day competition between the *Anolis* lizard populations of St Maarten (Neth Antilles). In: Shorrocks B (ed), Evolutionary Ecology. Blackwell, Oxford, pp 203–220

Rummel J, Roughgarden J (1983) Some differences between invasion-structured and coevolution-structured competition competitive communities. Oikos 41:477–486

Rummel J, Roughgarden J (1985) A theory of faunal buildup for competition communities. Evolution 43:1009–1033

Schaffer WM, Kot M (1985) Nearly one dimensional dynamics in an epidemic. Manuscript

Schoener TW (1984) Size differences among sympatric, bird-eating hawks: a worldwide survey. In: Strong D, Simberloff D, Abele L, Thistle N (eds), Ecological Communities: Conceptual Issues and the Evidence. Princeton University Press, Princeton, NJ, pp 254–281

Schoener TW, Schoener A (1983a) The time to extinction of a colonizing propagule of lizards increases with island area. Nature 302:332–334

Schoener TW, Schoener A (1983b) Distribution of vertebrates on some very small islands. II. Patterns in species counts. J Anim Ecol 52:237–262

Shepard JG (1984) The availability and information content of fisheries data. In: May RM (ed), Exploitation in Marine Ecosystems. Dahlem Konferenzen, Berlin, Springer-Verlag p 95–109

Shigesada N, Roughgarden J (1982) The role of rapid dispersal in the population dynamics of competition. Theor Popul Biol 21:353–372

Skellam JG (1951) Random dispersal in theoretical populations. Biometrika 38:196–218

Slatkin M (1973) Gene flow and selection in a cline. Genetics 75:733–756

Thresh JM (1983) Progress curves of plant virus disease. Adv Appl Biol 8:1–85

Ulbrich J (1930) Die Bisamratte. Heinrich, Dresden

Wilson EO (1961) The nature of the taxon cycle in the Melanesian ant fauna. Am Natur 95:169–173

Wingate DB (1965) Terrestrial herpetofauna of Bermuda. Herpetologica 21:202–218

5. Biogeographic Case Histories

12. Alien Plant Invasion into the Intermountain West: A Case History

R.N. Mack

12.1. Introduction

As with all history, a record of invading plants is a narrative of events. If detailed and chronicled, such compilations may be much more than lists of disparate facts; instead such archives may reveal the causes of rapid, pronounced changes in biota. Documentation of the entry, progress, and morbidity of a disease in a population has been long recognized as an essential procedure in medicine that contributes to disease prevention. Similarly, a case history of invasions documenting the entry, proliferation, and consequences of plant incursions could be used for the prevention, control, or even eradication of these aliens.

In the arid largely treeless intermountain region of western North America (hereafter referred to as the Region) plant invasions have been extensive and have radically transformed the vegetation in the last 100 years (Young et al. 1972; Franklin and Dyrness 1973; Yensen 1981). As in disease epidemiology, there often being no trained observers present during many of these invasions, we must take a largely retrospective approach in assembling these case histories. Some firm evidence is available in the form of herbarium specimens, but we rely mostly on anecdotal observations from which we may derive indirect measures of the time and point of entry, the mode, direction, and the rate of spread, and the consequences for the natives. In a sense, the answer to "What happened?"—massive transformation of a landscape—is both evident and documentable today. Instead my goal is to outline the How?, Why?, How much?,

When? Where? and, even to some extent, the Who? that brought about these plant invasions and subsequent changes in vegetation. The ultimate goal is for such detective work to go beyond being a collection of case histories and to become predictive, i.e., to become science. We then could use these compilations of events to predict the course of future invasions.

These invasions occurred in a geologically diverse area bounded by the Rocky Mountains to the east and by the Cascade-Sierra Nevada Ranges to the west. The Region includes the Columbia Plateau and to the south most of the Great Basin. Grasslands (here termed steppe) extend from the Fraser Valley of southern British Columbia south through much of Washington, Oregon, Idaho, Utah, and the northern two-thirds of Nevada and northeastern California. Where the basins and plateaus abut mountains, the steppe usually forms an ecotone with coniferous forest; to the south it borders desert. The areal extent of steppe in the intermountain west is perhaps best delimited by the distribution of perennial grasses, such as *Agropyron spicatum,* along with shrubs, principally *Artemisia tridentata* or one or more members of the Chenopodiaceous genus *Atriplex* (Daubenmire 1978).

The exceptional susceptibility of this steppe vegetation to plant invasion is best understood when cast in the context of the development of the Region's environment.

12.2. Climate

Climate at the latitudes of the Region (approximately 39° to 51° N) is controlled largely by the Prevailing Westerlies, an air mass containing cyclonic storms that seasonally shifts latitudinal position along the west coast. Although these storms provide most of the Region's moisture, it is the timing of this precipitation (more than 60% falling in autumn and winter), and not its absolute amount, that is the Region's single most important climatic feature. Early spring, particularly in the southern part of the Region, can be a secondary rainy season. In sharp contrast to the Great Plains east of the Rocky Mountains, summers are comparatively dry and receive less than 25% of the annual precipitation. Rainless periods of more than 30 days are common through July and August and are broken only by intense isolated storms (Trewartha 1981).

Since mid-Pliocene the on-shore movement of the Westerlies has been impeded by the Cascade-Sierra Nevada Ranges and the coastal ranges. In a classic example of orographic effects moisture in the Westerlies falls on the windward sides of these mountains, regardless of season. Consequently, the Region occurs mainly in a rain shadow of varying intensity all along its 1500-km western border. The 30-year average precipitation varies from lows of 124 mm at Wendover, Utah to 203 mm at Yakima, Washington to highs of 407 mm at Walla Walla, Washington and 437 mm at Salt Lake City (U.S. Environmental Data and Information Service 1981). Yet in all these locations the combination of low annual precipitation, high evapotranspiration, and dry summers precludes trees on zonal soils. With the marine influence of the Westerlies extending far

inland, the air temperatures in the Region are warmer than comparable latitudes to the east, especially since the Rocky Mountains serve as a northern and eastern barrier to polar air (Rumney 1968).

This climate with its wet cool autumns and winters and dry hot summers resembles the climate in a large area extending from Spain in the western Mediterranean Basin to Turkmenistan in Central Asia (Trewartha 1981). Plants from this extensive arid region where Europe, Asia, and Africa meet evolved under a climatic regimen that was similar to the Holocene climate in the intermountain west. Consequently, these plants were preadapted to the gross features of climate in this Region (Young et al. 1972).

12.3. Disturbance and the Importance of Large Congregating Mammals

Equally important in the eventual displacement of native plants by aliens was the dissimilarity between arid Eurasia and the intermountain west in the mode and level of disturbance. Arid Eurasia is characterized by having had disturbance over several millenia through agriculture and a much longer record of disturbance by large congregating mammals. The lack of prominence of either mode of disturbance until modern times in the intermountain west was to have far-reaching consequences as aliens supplanted natives.

The overall lack of large congregating mammals in this Region may have involved a chain reaction of causes and effects set into motion by the annual movements of the Prevailing Westerlies. In response to the annual distribution of precipitation, most perennial grasses in the intermountain west have the C_3 photosynthetic pathway, complete vegetative growth by early summer, and aestivate into autumn. Such plant phenology may have placed a severe constraint on large herbivores, such as bison, as these mammals require green forage for milk production well into summer after calving. The few grasses in the Region that remain green through much of summer, the C_4 species, such as *Distichlis stricta, Sporobolus cryptandrus,* and *Aristida longiseta,* are confined mainly to valleys.

Both the accounts of the first Europeans and the more verifiable Holocene fossil record indicate few large ungulates in the Region (Schroedl 1973). Lewis and Clark did not sight bison, elk, or deer as they followed the Columbia River (Thwaites 1905), an observation shared by other explorers (Cooper 1860; Simpson 1876) further south in Utah and Nevada. It is probably not a coincidence that all bison remains found on the Columbia Plateau, the northern third of the Region, have been found in river valleys. Herds may have been restricted to such sites for much of the year, thereby limiting their size (Mack and Thompson 1982).

Unlike steppe in the Great Plains few native plants other than cacti have thorns, spines, or prickles. Those plants with zoochores are invariably aliens. Furthermore, the dominant grasses, including *Agropyron spicatum, Festuca idahoensis, Oryzopsis hymenoides,* and *Sitanion hystrix,* are caespitose or

bunchgrasses, as opposed to the rhizomatous growth form of many grasses that dominate the Great Plains. Although the caespitose habit does not preclude tolerance of a grazing environment (Caldwell et al. 1981), leaf area lost in grazing must be replaced, a characteristic that these native bunchgrasses do not display readily. The intolerance of these grasses to trampling is equally important. As a consequence, native plants, such as the caespitose grasses, were poorly adapted to the almost simultaneous introduction of livestock plus the introduction of plants that both tolerated the climate and the interaction with ungulates. Within less than 50 years the steppe environments in this Region were converted to either rangeland or cultivated fields.

12.4. The Role of Humans in Plant Invasions

12.4.1. The Possibility of Pre-European Introductions

Plant invasions require transport, a phenomenon that humans are particularly adept at providing. Such human intervention was especially important for transport to a region that was somewhat insulated from transoceanic immigrations by being in a continental interior and isolated further from interregional transport by climatic barriers and mountains. There remains at least the possibility however that plant introduction in the Region did not begin with European settlers. Commerce, the most likely means by which alien plants are transported, was well established long before European settlers arrived, particularly among the aborigines of the Pacific Northwest. This commerce had been organized to a sophisticated level, in terms of the diversity of goods, the quantity traded annually, and most importantly here, the distances traversed. Trade ultimately connected tribes in the Pacific Northwest with groups residing on the Great Plains to the east (Wood 1972).

Lewis and Clark reported that the Wasco and the Wishram, tribes occupying opposite banks of the Columbia River near what is now The Dalles, Oregon, were virtual middlemen in a high-volume trade that moved smoked salmon, obsidian, and marine mollusc shells east in exchange for goods, including buffalo robes, flour, and *seeds,* coming west (Wood 1972). The opportunity for trade goods to have borne the seeds of aliens that later became established must have been high. Acquisition of the horse by tribes in the Region by ca. 1750 (Haines 1938) would have increased the volume of interregional contact and trade. But these routes had been established much earlier; sites on the Columbia Plateau dated at 600 to 700 B.P. contain pottery diagnostic of tribes in Arizona (F. C. Leonhardy, personal communication).

The ethnography of the aborigines is important to this case history of alien plants in another respect. The native peoples throughout the Region were not horticulturists (Drury 1963; Josephy 1965). Consequently, there was no selection among the native plant colonizers before European settlement by the selectional forces encompassed by agriculture.

12.4.2. The Arrival of Europeans

Whatever the level of trade among the aborigines, here as elsewhere in North America, it was dwarfed by the impact of European exploration, disturbance, and settlement. The region was settled comparatively late, however, both because of its isolation from the burgeoning population centers along the east coast and also because much of it had been erroneously dismissed as worthless by early explorers who mistook the prominent *Artemisia tridentata* as a sign of low soil fertility (e.g., Simpson 1876). Nevertheless, settlements were established before the mid-19th century first by British and American fur trading companies and later by missionaries and settlers, most notably in southeastern Washington and along the Great Salt Lake in Utah. In addition semiferal herds of cattle were maintained locally, a practice that intensified after gold was discovered in the Sierra Nevada in 1859 (Elliott 1973) and in the mountains bordering the Columbia Plateau after the American Civil War (Meinig 1968).

Establishment of these isolated communities provided early opportunities for alien plant entry as all these settlements, including the trading outposts, had to be largely self sufficient. As a result aliens, most likely introduced in the contaminated seeds of crops, were detected early in the century. At the steppe-forest ecotone near Spokane, Washington, Geyer (1846) in 1843 or 1844 found aliens, including *Erodium cicutarium, Hordeum pusillum,* and *Camelina sativa,* in cultivated fields. Although it is not always possible to discern Geyer's locations, he saw *Erodium cicutarium* and *Thlaspi* (probably *T. arvense*) and *Cotula* (possibly *C. coronopifolia*) near one verifiable location, the American Missionary Station (established only 7 years before his visit), at Walla Walla, Washington. Other records of alien entry on the Columbia Plateau before 1885 are very scant, although *Echinochloa crusgalli* had arrived in the Yakima Valley by 1883 (Scribner 1883), *Bromus mollis* at White Salmon, Washington along the Columbia River in 1882 [W.N. Suksdorf (s.n.) WS], and *Bromus brizaeformis* at Spangle, Washington by 1884 [W.N. Suksdorf (s.n.) WS]. Aliens had also arrived in the mainland of British Columbia by 1875, including *Capsella bursa-pastoris, Erysimum cheiranthoides,* and *Rumex acetosella* (Macoun 1877).

The recurring detection of *Erodium cicutarium* early in the 19th century may have foretold its even earlier introduction in the Region. The plant's seeds have barbs on the carpel and coiled persistent styles (Stamp 1984). Seeds could have been introduced while entangled in the tails of the horses that were being acquired by the aborigines from Spanish settlements in the Southwest.

Aliens were also detected soon after settlers arrived at the Great Salt Lake in 1847. An army expedition collected *Agropyron repens* and *Erodium cicutarium* on two islands in the Great Salt Lake in 1850 (Tanner 1940). Certainly the most comprehensive listing of alien plants in the era before extensive settlement was assembled by Watson based on collecting in 1869 in northern Nevada and Utah (Watson 1871). His list includes species found "in among the sagebrush" (noncultivated sites):

Anthemis cotula	*Brassica nigra*
Brassica campestri	*Capsella bursa-pastoris*

Chenopodium album   *Marrubium vulgare*
Chenopodium botrys   *Vaccaria segetalis*

and other species found associated with agriculture or along streams:

Amaranthus cruentus   *Poa annua*
Avena fatua   *Polygonum convolvulus*
Coriandrum sativum   *Polygonum lapathifolium*
Datura stramonium   *Polypogon monspeliensis*
Eragrostis cilianensis *Portulaca oleracea*
Melilotus alba   *Rorippa nasturtium-aquaticum*
Melilotus indica   *Setaria viridis*
Nepeta cataria   *Verbascum thapsus*
Phleum pratense

As will be demonstrated later the species Watson found are common contaminants of seed lots of alfalfa, clover, oats, and wheat.

The rapid development of the railroad system accelerated human immigration to the Region. By 1870 one line of a transcontinental railroad system was in place. Further east–west lines were added and extensive cross-linkages in the system were quickly built in the intermountain west (Meinig 1968). Unlike the 19th century settlement elsewhere in North America, here many settlers arrived by train rather than wagon. In the 30 years after the Civil War, the steppe in British Columbia, Washington, Oregon, Idaho, Utah, and Nevada was transformed from a sparsely settled section of the Far West to an extensively settled region of farms and ranches (Meinig 1968; Elliott 1973).

Aliens Become Prominent

Since the bulk of these settlers aspired to become farmers rather than ranchers, it is not surprising that the circumstantial evidence points to contaminated seeds of crops as being the major means by which alien plants arrived in the Region. Not only were the methods for sieving seeds primitive or unavailable in these remote farming districts, but seed lots were often deliberately altered with the seeds of weed or lesser-valued plants. For example, Red clover (*Trifolium pratense*) was commonly adulterated with *Medicago lupulina* (Hillman 1900).

Impure seeds of crops plagued the western farmer in the 19th century. The leading miller in Walla Walla County in southeastern Washington pointed out in 1863 that there were not "...100 bushels of pure wheat in the county" and the local newspaper editor urged farmers to obtain "...pure white seed" (as cited in Meinig 1968). In this era before any legislation governing seed purity, finding pure seeds was a formidable challenge. Hillman (1900) found 68 aliens as contaminants in the commercial lots of clover seeds available to Nevada farmers. The lots of wheat and alfalfa seed were heavily contaminated, often with parasitic dodder, *Cuscuta epithymum* (Hillman 1900). Extensive contamination of cereals with alien grasses continued well into this century (Washington

Department of Agriculture 1914). By the time Seed Purity Laws were enacted around the turn of the century many of the aliens that were to play major roles in conversion of the steppe had already arrived, including *Bromus tectorum, Cirsium vulgare,* and *Salsola kali.*

The earliest indication that aliens were becoming established (as opposed to simply arriving) in the Region comes from range assessors and plant collectors. Repeatedly in the last quarter of the 19th century "special agents" of the U.S. Department of Agriculture criss-crossed the intermountain west to assess the status of rangelands (Tracy 1888; Shear 1901; Griffiths 1902). Tracy (1888), who restricted his remarks mainly to grasses, found aliens such as *Avena fatua* already "very abundant from California eastward to Central Nevada, and occasionally found in the Salt Lake Valley." Elsewhere in Nevada and Utah he encountered other aliens, including *Bromus mollis, Bromus racemosus,* and *Poa annua. Bromus secalinus* was already "The most common species on low, dry, sandy land" at Ogden, Utah. *Erodium cicutarium,* although not a grass, was nonetheless widely introduced (Tracy 1888). By 1902 introduced bromes *(Bromus secalinus, Bromus rubens,* and *Bromus mollis)* were locally conspicuous in northern Nevada (Griffiths 1902).

To the north in the Columbia Basin of Washington *Verbascum thapsus, Erodium cicutarium, Anthemis cotula,* and *Vaccaria segetalis* were occupying fields and gardens by 1893; alien bromes included *Bromus secalinus* and *Bromus racemosus.* In the Walla Walla area the alien brome, *Bromus mollis* was conspicuous by 1899 (Shear 1901). None of these aliens were as yet considered noxious (Sandberg and Leiberg n.d.); perhaps because much of this area had been brought into cultivation less than 20 years earlier. Consistent with the overall lack of prominent aliens was Cotton's (1904) finding that succession on the rangelands was still dominated by natives, including *Festuca microstachys, Festuca octoflora, Descurainia pinnata, Machaeranthera canescens,* and *Plantago patagonica;* the aliens *Salsola kali* and *Corispermum hyssopifolium* were however locally abundant.

In Nevada in the 1890s, as elsewhere in the Region, a newly established agricultural experiment station accepted its mission to alert the public to the danger of alien weeds. F.H. Hillman prepared several comprehensive bulletins on weeds and weed impurities arriving in contaminated commercial lots of clover seed (Hillman 1900). In his zeal to ensure that farmers would be provided with the means to identify such seed contaminants as *Capsella bursa-pastoris, Verbascum thapsus,* and *Lepidium intermedium,* he not only published bulletins for statewide distribution with descriptions and lithographs of the weeds (Hillman 1893a,b), but also had dried specimens of the weeds *with seeds* affixed to the pages of each copy of the bulletins! I do not know whether he took the precaution to sterilize the seeds of these as yet infrequent aliens in the Nevada flora. But the seeds are mounted with paste and could have been lost with handling or released when the bulletins were discarded. Among the myriad means by which aliens may be dispersed (Ridley 1930) this is the only instance that I am aware in which the very medium alerting the public to impending danger may have become the means by which the danger materialized.

Two Examples of Invasions

Within 25 years after the initial wave of large-scale immigration by settlers the rapid rise in the abundance and distribution of noxious weeds was causing official concern (e.g., British Columbia Department of Agriculture 1893; Coville 1896; Shear 1901). Areal spread was monitored for those aliens known to be noxious, such as *Salsola kali,* which had already invaded wheat-growing areas in the Dakotas (Dewey 1894). The plant had reached the Snake River Plain in southeastern Idaho as early as 1890 as seen in an annotated photograph of a railroad embankment (Yensen 1980). As late as 1894 Craig stated that *S. kali* had not yet been reported for Oregon, although only 3 years later it was occurring locally in the northeastern portion of that state (Dewey 1897). Further rapid spread was reported along the Oregon Short Line, tracks linking Boise with Oregon at Huntington (Henderson 1898). Spread was doubtlessly also facilitated by river traffic; another photograph (ca. 1890) shows *S. kali* at Walters Ferry (Yensen 1980), and Dewey (1897) reports its spread in irrigated lands along the Snake River. Piper (1898a) reported that the species was first found in eastern Wahington in 1897 at seven locations. But by 1898 a comprehensive search revealed that the plant had spread to several dozen new locations via the railroad (Piper 1898b). Systematic survey soon ceased as the plant overwhelmed eradication efforts and became abundant soon thereafter (Cotton 1904; Foster 1906).

Salsola kali is interesting in another context, for its prominence reveals how quickly even our impression of the Region's pristine vegetation has been altered. Each plant forms a large hemispheric mass of overlapping branches. At death this mass usually becomes detached from the roots and may be moved by the wind several km along the ground, spreading seeds along its path. *Salsola kali* (and to a lesser extent the alien *Sisymbrium altissimum*) are commonly termed "tumbleweeds," and their appearance has been popularized as part of the landscape seen by the pioneers in western North America. Ironically, these shrublike herbs are but further signs of a "New West," overrun by aliens, rather than of the "Old West" with which they are erroneously identified.

The invasion and proliferation of *Bromus tectorum* (cheatgrass) is unusual both for the speed with which this annual grass spread throughout the Region and for the wealth of information that can be assembled for documenting the causes and consequences of its spread. Circumstantial evidence suggests that *Bromus tectorum* arrived in the Region in the 1880s in impure seed. The earliest records of the grass are all from inland wheat-growing districts (1889 at Spence's Bridge, British Columbia; 1893 near Ritzville, Washington; 1894 at Provo, Utah). Additional pre-1900 specimens were collected in Washington at Spokane, Pullman, and Pasco and in northeastern Oregon (Mack 1981). The grass also appears to be recorded in a 1898 photograph from Ada County in southern Idaho (Yensen 1980).

In the first decade of the 20th century the grass was frequently encountered, although it was prominent only locally, such as along railroad right-of-ways (Cotton 1904). By 1915 the continuing contamination of seed lots, a caryopsis that readily adheres to fur as well as remains viable in dung, and cereal agri-

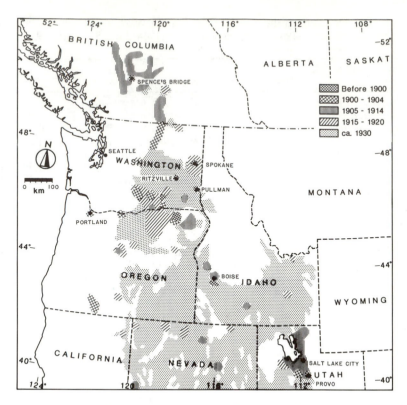

Figure 12.1. The spread of *Bromus tectorum* (cheatgrass) in the intermountain west between 1889, its first recorded occurrence, and 1930, the date at which it reached the maximum extent of its new range. Records from Mack (1981) and Yensen (1980).

culture in which the annual cycle of husbandry matched the alien's phenology, had combined with rapid dissemination by the railroad to cause an explosive increase in the new range of this cleistogamous grass. By 1920 the clamor was growing over a means to cope with this grass in fields of alfalfa and wheat. In rangelands the plant was alternatively deemed a threat or possibly acceptable forage. Nevertheless by 1930 the grass had reached its current distribution in the intermountain west. For the first time reference was made to "cheatgrass *(B. tectorum)* lands," range or fields solely dominated by a grass that had been introduced at most only 50 years earlier (Mack 1981) (Fig. 12.1.).

12.5. The Columbia Plateau: A Microcosm of the Invasions

Local Floras from the Columbia Plateau of southern Washington assembled by 1892, 1901, 1914, and 1928 provide an unusually comprehensive picture of the growing diversity of aliens within part of the Region (Table 12.1.). By 1892

Table l2.1. The 200 aliens in the Columbia Plateau of Washington listed in Suksdorf (1892), Piper and Beattie (1901), Piper and Beattie (1914), and St. John (1937). Nomenclature follows Hitchcock et al. (1955–1969)[a,b]

A. Aliens Listed in All Four Floras	Use	Crop		Use	Crop
Amaranthus retroflexus		l	Plantago lanceolata		l
Anthemis cotula		l	Plantago major		l
Brassica campestris		l	Poa compressa		c
Cerastium vulgatum		l	Poa pratensis	f	
Chenopodium botrys	m	l	Polygonum lapathifolium		l
Cirsium vulgare		l	Portulaca oleracea	o	l
Convolvulus arvensis		c	Rumex acetosella		l
Dactylis glomerata	f		Senecio vulgaris		
Echinochloa crusgalli	f	l	Sisymbrium officinale		l
Erodium cicutarium	f	l	Sonchus asper		g
Festuca pratensis	f		Spergula arvensis		l
Holcus lanatus	f	c	Stellaria media		l
Hordeum murinum			Taraxacum officinale		l,c
Lactuca serriola		l	Tragopogon porrifolius	v	l,c
Lamium amplexicaule		l	Trifolium pratense	f	
Marrubium vulgare	m	l	Vaccaria segetalis		c
Nepeta cataria	m		Verbascum blattaria		l
Phleum pratense	f	l	Verbascum thapsus	o	l

B. Aliens Listed Once	Use	Crop		Use	Crop
(1892)					
Abutilon theophrasti		l	Geranium molle		
Agropyron intermedium	f		Rumex acetosa	v	
Avena sativa	v	c	Sonchus oleraceus		g
Bromus racemosus		c	Trifolium procumbens		
(1901)					
Arabis glabra			Camelina sativa	x	
			Spergularia diandra		
(1914)					
Setaria italica	f		Veronica officinalis		l
(1937)					
Aegilops cylindrica		c	Kochia californica		
Agrostis interrupta		c	Lamium purpureum		l
Ailanthus altissima	o		Lathyrus latifolius	o	
Alopecurus myosuroides			Lepidium perfoliatum		l
Anchusa azurea	o		Linaria dalmatica	o	
Anchusa officinalis	o	l	Lithospermum arvense		c
Arctium minus	m	c	Lynchis alba	o	l
Asperugo procumbens		l	Maclura pomifera	o	
Atriplex rosea			Melilotus officinalis	f	
Berberis vulgaris			Mentha piperita	o	
Bromus erectus			Mollugo verticillata		
Bromus inermis	f		Morus alba	o	
Cardaria pubescens		g	Myosotis micrantha		l
Centaurea jacea			Myosotis scorpioides	o	
Centaurea repens		c	Myriophyllum brasiliense	o	
Centaurea solstitialis		c	Papaver argemone		

Table 12.1 *Continued*

B. Aliens Listed Once	Use	Crop		Use	Crop
Chenopodium virgatum			*Papaver rhoeas*	o	
Chenopodium murale			*Physalis lanceolata*	o	
Chorispora tenella		c	*Poa bulbosa*	f	l
Chrysanthemum balsamita	o		*Potentilla argentea*	o	
Cichorium intybus	v		*Puccinellia rupestris*		
Cnicus benedictus	o		*Rubus laciniatus*	o	
Conium maculatum	m	l	*Rubus patientia*		
Conringia orientalis		l	*Sagina procumbens*	o	
Corispermum nitidum			*Saponaria officinalis*	o	l
Cuscuta epilinum		l	*Sclerochloa dura*		
Digitaria sanguinalis			*Silybum marianum*	o	
Elymus caput-medusae		c	*Sisymbrium loeselii*		
Erigeron annuus			*Tanacetum vulgare*	m	
Erysimum repandum		l	*Taraxacum laevigatum*		l,c
Euphorbia esula			*Tragopogon dubius*		l,c
Glecoma hederacea			*Tragopogon pratensis*		l,c
Gypsophila paniculata	o		*Tribulus terrestris*		
Hesperis matronalis	o		*Trifolium agrarium*		
Holosteum umbellatum		l	*Veronica persica*		l
Hyoscyamus niger	m		*Vicia villosa*	f	
Hypericum perforatum	o	l	*Zizania aquatica*	f	

C. Aliens Listed in Two or Three Flora[c]	Use	Crop		Use	Crop
Agropyron repens[d]	f		*Euphorbia cyparissias*[e]	o	
Agrostemma githago[e]		c	*Festuca myuros*[d]		c
Agrostis alba[e]	f		*Geranium pusillum*[e]		l
Alyssum alyssoides[e]		l	*Hemizonia pungens*[f]		
Amaranthus cruentus[e]	o		*Iva xanthifolia*[d]		
Ambrosia psilostachya[e]			*Linaria vulgaris*[f]	o	
Ambrosia trifida[f]			*Linum usitatissimum*[f]	x	
Anthemis arvensis[d]			*Lolium perenne*[d]	f	
Arenaria serphyllifolia[d]		l	*Lychnis coronaria*[d]	o	
Arrhenatherum elatius[d]	f		*Malva neglecta*[f]		l
Artemisia absinthium[f]	m		*Medicago hispida*[d]	f	
Avena fatua[d]		c	*Medicago lupulina*[d]	f	
Brassica kaber[e]		l	*Medicago sativa*[f]	f	
Brassica nigra[d]	v	l	*Melilotus alba*[e]	f	
Bromus brizaeformis[e]		c	*Mentha spicata*[f]	o	
Bromus commutatus[f]		c	*Phalaris arundinacea*[d]	f	
Bromus mollis[d]		c	*Poa annua*[d]		c
Bromus rigidus[f]		c	*Poa nemoralis*[d]	x	
Bromus secalinus[d]		c	*Polygonum convolvulus*[d]		c
Bromus sterilis[e]		c	*Polypogon monspeliensis*[d]	o	
Bromus tectorum[e]		c	*Prunella vulgaris*[e]	o	l
Camelina microcarpa[f]		l	*Robinia pseudo-acacia*[d]	o	
Capsella bursa-pastoris[d]		l	*Rorippa nasturtium-aquaticum*[d]	v	

Table 12.1 *Continued*

C. Aliens Listed in Two or Three Flora[c]	Use	Crop		Use	Crop
Centaurea cyanus[d]	o	c	*Rumex crispus*[e]		l
Centaurea nigra[e]			*Salsola kali*[e]		c
Cerastium viscosum[d]		l	*Setaria viridis*[d]		l
Chenopodium album[d]		l	*Silene noctiflora*[f]		l
Chenopodium capitatum[d]			*Sisymbrium altissimum*[f]		l
Chenopodium hyridum[d]			*Solanum dulcamara*[e]	o	
Chrysanthemum leucanthemum[d]			*Solanum rostratum*[f]		
Chrysanthemum parthenium[d]	o		*Spergularia rubra*[e]		
Cirsium arvense[e]		l	*Stellaria graminea*[e]		
Convolvulus sepium[d]	o	c	*Thlaspi arvense*[e]		l
Datura stramonium[e]	m		*Trifolium arvense*[e]		
Daucus carota[d]	v		*Trifolium hybridum*[e]	f	
Descurainia sophia[e]		l	*Trifolium repens*[d]	f	
Dipsacus sylvestris[e]			*Valerianella locusta*[d]	v	
Draba verna[d]			*Veronica arvenis*[d]		
Erysimum cheiranthoides[d]					

[a] Synonymities are in accordance with Hitchcock et al. (1955–1969). Identification of species as aliens conformed mainly to descriptions in Hitchcock et al. (1955–1969), with additional verification using Gray's Manual (Fernald 1950) and Munz (1959). Except for Suksdorf's list which makes no distinction between aliens and natives, I have relied extensively on the information in each Flora for whether a plant was considered an alien. Consequently, some aliens (originally thought to have been native) may have been overlooked. I have omitted species such as *Festuca rubra* that were considered aliens in these earlier Floras but for which the actual native distribution is unknown and could include the intermountain west.

[b] If deliberate introduction was possible, its use is given in the first column: f, forage; m, medicinal; o, ornamental; v, victuals; x, some other use (e.g., lawns, oil) (as compiled from Bailey 1949). The crop in which an alien was likely to have been dispersed is given in the second column: c, cereals and other grasses; l, small-seeded legumes; g, occurring in any regional crop.

[c] Flora in which each species was first included: [d]1892; [e]1901; [f]1914.

Suksdorf, a professional plant collector, had catalogued 1185 taxa within approximately a 70 x 70 km area north of the Columbia River at Bingen, Washington (Fig. 12.2.). Among the more than 80 aliens that he reported some had been deliberately introduced (e.g., *Agropyron repens*, *Trifolium repens*, *Poa pratensis*, and *Phleum pratense*); others had been associated with earlier European settlement over much of the continent (e.g., *Plantago lanceolata*, *Plantago major*, *Cirsium vulgare*, *Erodium cicutarium*, and *Rumex acetosella*); and many others are commonly found as seed contaminants (*Draba verna*, *Bromus secalinus*, *Bromus racemosus*, *Avena fatua*, *Stellaria media*, and *Capsella bursa-pastoris*).

The number of aliens was about the same by the turn of the century in a 35-km radius around Pullman, Washington about 200 km east of Suksdorf's collection area (Piper and Beattie 1901). Perhaps because their collecting area was very small, they did not find such widely distributed aliens as *Brassica nigra*,

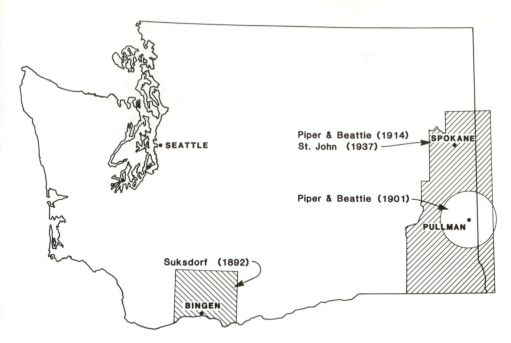

Figure 12.2. The collection areas in eastern Washington for the Floras of Suksdorf (1892), Piper and Beattie (1901), Piper and Beattie (1914), and St. John (1937). Aliens from these Floras are compiled in Table 12.1.

which had been found earlier by Suksdorf. Their Flora did include however such future steppe dominants as *Bromus tectorum* and *Salsola kali* along with *Cirsium arvense* and *Dipsacus sylvestris*. For their 1914 Flora Piper and Beattie expanded their collecting area to include much of eastern Washington and a small part of adjacent Idaho; yet the overall number of aliens changed little and comprises fewer than 100 taxa. Nevertheless new aliens were detected including *Sisymbrium altissimum, Brassica kaber, Bromus rigidus,* and *Artemisia absinthium,* all commonly found today.

The largest increase in the alien flora appears to have taken place in the decades bracketing World War I, a period in which the land in wheat production was maximized (Meinig 1968; Yensen 1980). Seventy-five new aliens were reported out of a total of over 180 (St. John 1937) within the same area in which Piper and Beattie had collected for their 1914 Flora. The magnitude of this increase in the alien flora is all the more surprising since it came after passage of Seed Purity Laws throughout western United States and British Columbia. Included within this list of new arrivals are several species that are common today *(Arctium minus, Centaurea jacea, Centaurea repens, Holosteum umbellatum, Hypericum perforatum, Poa bulbosa)*. Their detection is additional testimony to the speed with which relative abundance can change within a regional flora in which aliens are continually being introduced. *Elymus caput-medusae* also first appears in St. John's Flora, although it had been collected

within eastern Washington by 1901 (McKell et al. 1962). Both *Hypericum perforatum* and *Linaria dalmatica* were to become such weed problems that biological control campaigns were launched against them (Huffaker and Kennett 1959; Andres et al. 1976).

Most of the aliens in these Floras are natives of Eurasia, but some are natives of eastern North America (e.g., *Ambrosia trifida, Datura stramonium, Erigeron annuus, Physalis lanceolata*) and California (*Hemizonia pungens, Kochia californica*) (St. John 1937); commerce from the more established agricultural regions to the east and south was contributing to the deluge of aliens. Thirty-six aliens appear in all four Floras and include cosmopolitan weeds such as *Taraxacum officinale, Cirsium vulgare, Rumex acetosella, Poa pratensis, Brassica campestris*, and *Verbascum thapsus*. Not only did these species arrive early in the settling of the Columbia Plateau, but they have persisted and become common, if not abundant.

These aliens persist today in cultivated fields or rangelands, or both, yet some have even entered sites supporting otherwise undisturbed natural communities. In his comprehensive study of steppe in Washington Daubenmire (1970) restricted his sampling to sites in which there had been no known disturbance by humans. Furthermore, aliens were either absent or were represented by only a few individuals. Despite these restrictions on site selection, 48 aliens appear in the species lists of his 40 plant associations. Seven aliens in his lists were not detected in the four local Floras: *Agropyron triticeum, Alopecurus pratensis, Arabidopsis thaliana, Bassia hyssopifolia, Centaurea diffusa, Cerastium viscosum*, and *Festuca bromoides*. I have seen some of these species establish through naturally occurring disturbances, such as around rodent burrows; others probably require no disturbance.

For a list of aliens in a region that is mostly steppe, the floristic distribution of the species presents few surprises. Among the 32 families represented the families with the largest number of representatives are Poaceae (42), Asteraceae (32), and Brassicaceae (22); other familes with six or more representatives include Boraginaceae, Caryophyllaceae, Chenopodiaceae, Fabaceae, Lamiaceae, and Scrophulariaceae. The importance of these families in areal coverage and prominence is coincidentally in rough agreement with this listing; certainly the most common aliens are grasses with composites second. Perhaps only the relative importance of the Chenopodiaceae (e.g., *Salsola kali* and *Chenopodium* spp.) is not reflected by the nine species included among these Floras.

More unexpected, however, is the number of these aliens that probably were deliberately introduced. Of these 200 species, 83 are cited by either Bailey (1949) or Hitchcock et al. (1955–1969) as having been used either as forage or food for human consumption, or as having medicinal or ornamental use. The herb or vegetable gardens of the European settlers apparently became the foci from which now common aliens spread, including *Arctium minus, Centaurea cyanus, Daucus carota, Nepeta cataria*, and *Verbascum thapsus*.

With the severe depletion of the native caespitose grasses with overgrazing and trampling (Mack 1981), many alien grasses were introduced as possible new sources of forage; some have become locally prominent (e.g., *Bromus*

inermis, Dactylis glomerata, Poa bulbosa, Poa pratensis). Legumes were also introduced for forage, and the prominence of legumes in these Floras provides further insight as to how many other species were accidentally introduced and disseminated. Legumes (*Trifolium* spp. and *Medicago* spp.) were extensively planted by early settlers because of the importance of clovers and alfalfas for livestock as well as being sources of nitrogen in a crop rotation system with cereals. At least 53 other species in Table 12.1. were likely contaminants of the seed lots of legumes because of their similar seed size; even today these species are found as contaminants in commercial lots of those legumes with small seeds (K.M. Steen, personal communication). Altogether at least half the species in Table 12.1. are contaminants today of commercial lots of seeds.

Comparisons drawn from the compilation of local Floras admittedly have limitations. Only the two most recent Floras (1914 and 1937) are based on the same collection area, and Suksdorf's ca. 1892 collection area was west of the other three (Fig. 12.1.). Furthermore, much of the habitat in which Suksdorf collected was low-elevation forest, although I have excluded from consideration here any aliens that he collected only in the heavily forested area west of the Wind River. Whether omissions between succeeding Floras represent the failure of species to persist versus different levels of collecting intensity are unresolvable. But it is difficult to attribute the omission in Piper and Beattie's 1914 Flora of *Draba verna, Capsella bursa-pastoris,* and *Avena fatua,* some of the most common weeds on the Columbia Plateau, to oversight. The lack of a substantial increase in aliens between 1892 and post-1914 is curious. But the lack of new records reflects neither substantially earlier settlement nor more intensive farming in south central versus eastern Washington. Despite these limitations, the overriding impression is of an area in which after the major wave of European settlement in the late 19th century there was only one brief pause early in the 20th century when aliens were not being added to the flora at a rapid rate.

12.6. Twentieth Century Invasions

The case history of *Halogeton glomeratus* (Chenopodiaceae) illustrates that despite the enactment of Seed Purity Laws and the erection of plant quarantine barriers, aliens continue to arrive in the Region. The rapid spread of this shrub-like annual between 1935 and 1960, despite often intensive eradication efforts, further illustrates the difficulty in controlling some aliens, as well as the problems in predicting the new range and eventual role of the invader.

First collected in 1934 near Elko, Nevada (Blackwell et al. 1979) the shrub was rapidly disseminated throughout northeastern Nevada and adjacent Utah over the next 10 years (Holmgren 1942). From an estimated range in 1937 of a 40-km radius around its first known location (Holmgren 1943), it had been transported by 1950 into southern Idaho as well as across the Wasatch Mountains into eastern Utah (Holmgren 1950). By the early 1950s *H. glomeratus* also had been found in southern Oregon (Blackwell et al. 1979). Its range was

estimated as more than 600,000 ha in 1952 (Pierson 1952 as cited in Tisdale and Zappettini 1953), culminating in an estimated range of about 4 million ha by 1958 (Morton et al. 1959). Much of this spread was along highways, although seeds were also disseminated in animal feces and in fleeces (Stoddart et al. 1953).

The spread of *Halogeton* after 1942 was followed more closely than for most immigrants once the shrub was found to be poisonous to livestock. Official concern that it would destroy the Region's sheep ranges even caused congressional action by 1953 for its eradication (Blackwell et al. 1979). Yet much of the initial concern about its potential range has subsided. While the shrub can occur on many soil types, it is found most commonly on the saline sites dominated by *Atriplex nuttalli* or *Eurotia lanata* after intense grazing (Tisdale and Zappettini 1953). On other habitats *Halogeton* has failed to achieve the dominant role once predicted for it.

The initial isolated discovery of *H. glomeratus* in northern Nevada in the absence of earlier records at likely points of entry on the west coast has fueled speculation as to the mode and source of its entry into the Region. The suggestion that *Halogeton* arrived with Karakul sheep, although unconfirmable, is at least plausible since these sheep were repeatedly introduced into northern Nevada from central Asia early in the 20th century. Furthermore, one of the trails across which these animals would have been driven passes through Lassen County, California, the site of the earliest (1946) collection of *H. glomeratus* in northeastern California (Blackwell et al. 1979). Here, as in probably many other examples, the assumed earliest dates of first occurrence are probably underestimates; herbarium collecting is neither uniform nor continuous and a new alien may escape early detection unless it has unusual features.

More recently other aliens appear to be increasing their range. The annual grasses *Aegilops cylindrica* and *Ventenata dubia,* the former present since at least 1917 [FL Pickett (1318), WS] and the latter detected more recently, have become prominent in eastern Washington in the last 10 years. *Isatis tinctoria* has recently expanded its range in northeastern Utah to noticeable proportions (JR Richards, personal communication). Of the recent immigrations with which I am familiar the appearance of *Bryonia alba,* a solanaceous vine, in the last 10 years in southeastern Washington is most intriguing for vines are rare in the Regional flora.

12.7. The Modern Vegetation

After such massive disturbance by livestock and tillage the common dominants of these steppe communities today are generally no longer the original caespitose grasses and shrubs. European settlers were very persistent in reaching even the most isolated locations in the Region. For example, islands in the Columbia River could have served as refugia for native vegetation. But stockmen were particularly attracted to these inaccessible sites because there was no danger that their livestock would stray or be killed by predators (R. Daubenmire, per-

sonal communication). Except along railroad right-of-ways and very steep slopes, the native species have been replaced most often by *Bromus tectorum, Salsola kali, Sisymbrium altissimum,* and *Erodium cicutarium. Poa pratensis* is often prominent on moist sites. In the southern half of the Region *Elymus caput-medusae* may become dominant, whereas *Halogeton glomeratus* is still conspicuous in northern Nevada and adjacent Utah. Not all natives have decreased in importance within these communities; often *Chrysothamnus nauseosus* and *Chrysothamnus vicidiflorus* increase with grazing along with *Plantago patagonica*. In a region characterized by so many plant invasions, the native *Matricaria matricaroides* is unusual. Although this weedy composite is not a dominant in the Region today, it has become established along roadsides and pathways in northern Europe, and other temperate habitats (Moe 1977)— a rare example of a transoceanic emigration and establishment by a native of the Region's flora.

The current incidence and intensity of fire in this Region illustrates how aliens may alter the environment and, in doing so, foster their own persistence. Fire certainly was part of the pre-settlement environment. But the community structure, with its caespitose grasses and shrubs in an open matrix of annuals, produced infrequent, relatively low-temperature fires that probably had little influence on species composition (Daubenmire 1975). Fires deliberately set by the European settlers had the intended effect of eradicating woody plants, such as *Artemisia tridentata,* over large areas. But this practice along with the destruction of the native grasses by livestock created bare areas for the establishment of aliens (Daubenmire 1968). Once established the alien annual grasses radically changed the pattern of fire. For example, the litter production in a stand of *Bromus tectorum* may be several-fold greater than that of the native steppe communities that it replaces (cf. Daubenmire 1970; Rickard et al. 1977); such litter readily fuels fire in summer. In such an environment the native perennials are at a disadvantage. Despite such fires, buried seeds of the aliens survive. In less than 100 years a cycle has become established in which the annual build-up of grass litter predisposes a site to a devastating fire; by causing a release of nutrients, the resulting fire ensures the site will again support swards of annual grasses (Daubenmire 1975).

12.8. The Search for Causes

As proposed at the beginning of this chapter, case histories could provide an empirical base from which we may see trends in the characteristics of invaders and in the seemingly disparate events in invasions. Once chronologically assembled, compared, and contrasted, this evidence allows formation of testable hypotheses, experimentation, and prediction.

What trends emerge? As elsewhere (Elton 1958; Harper 1965) a sudden increase in disturbance played an essential role in facilitating the establishment of aliens. The dominant aliens in the Region today consistently owe their role to the acceleration of disturbance that coincided with European settlement.

The native dominants of the similarly altered grasslands in Australia, New Zealand, and California share with the species in this Region an evolutionary history without frequent disturbance by large congregating mammals or their functional equivalents (Mack and Thompson 1982). Disturbance, whether through the introduction of agriculture as in this Region, or simply a change in agricultural practice, can be sufficient to allow establishment of aliens. No region seems immune; the numbers of aliens and their abundance have increased markedly even in the Middle East in the last 40 years (Dafni and Heller 1980).

The species that became established in this Region have tended to be annuals, most commonly grasses. The seeds of these aliens are either similar in size to the seeds of a crop or have devices for animal disperal, or both. In addition to tolerating or requiring the Region's climatic regimen, these species are generally fecund, produce seeds that have few dormancy requirements, and in turn produce seedlings that can tolerate a wide range of soil nutrient conditions (Hulbert 1955; Mack and Pyke 1983). Almost all are monoecious, but whether cleistogamy has adaptive significance, or is merely common among these species, is unclear (cf. Young et al. 1972). The persistence of the most prominent aliens, such as *Bromus tectorum,* is enhanced with disturbance, but they can also maintain seemingly permanent populations on sites unaltered by human activity. Once entry to the Region was provided for such species, did settlement only quicken spread that would have inevitably occurred?

Such comparisons among the autecologies of several hundred aliens can already reveal common features. Much less can be currently determined however as to whether the rates of spread for these aliens follow a common pattern, and thereby are predictive, or is each invasion unique? In the only example from this Region for which there is enough information, the invasion of *Bromus tectorum,* range expansion displayed logistic growth. Initially (ca. 1900), the populations appear to have been in lag phase, the length of which was probably dependent on the few foci (points of introduction), the distance between the foci, the rapidity of dispersal, and the limitations on population growth included under the intrinsic rate of increase. Sometime after 1910 *B. tectorum* expanded into new areas as if in log phase; the alien's rate of new range occupation was seemingly limited only by its own intrinsic rate of increase. By 1930 the rate of advance was nil (Mack 1981, 1985).

Accounts of the spread of other aliens, such as *Salsola kali* and *Halogeton glomeratus,* seem consistent with the pattern for *B. tectorum,* but detailed information is elusive on how much area was occupied at different times. Whether logistic growth would describe adequately both the area occupied through time as well as the numerical increase in the invader populations would require documentation of many current invasions (e.g., *Aegilops cylindrica, Centaurea* spp., and *Ventenata dubia*).

In summary, prediction of the course of plant invasions whether in this Region or elsewhere remains a challenge. And yet, as the regional case history outlined here illustrates, the vegetation across vast areas can be permanently altered if strict quarantine measures are not implemented. For the steppe in the intermountain west the question is not whether these invasions could be reversed—there is little evidence that even intense reseeding of native grasses produces

lasting results. Instead information on invasions will prove important here and elsewhere in checking possible invasions by species even less desirable than those that have already become naturalized. If this is not done, agriculture in substantial portions of the world may become restricted to those small plots that can be temporarily cleared of cosmopolitan plants only by the continual application of herbicides. Maintenance of such artificial ecosystems poses inherent and well-known ecological dangers.

12.9. Conclusions

The sudden introduction of agriculture in the latter half of the 19th century swiftly changed the environments that had supported steppe in much of the intermountain west of North America. Unlike the steppe in arid Eurasia, the vegetation in the Columbia Plateau and much of the Great Basin had been dominated by caespitose grasses and shrubs, especially *Artemisia tridentata,* in an environment where disturbance by large congregating mammals had been infrequent. The Region's climate, including the annual distribution of precipitation, however, is similar to that of arid Eurasia. Consequently, as species tolerant of disturbance (e.g., *Bromus tectorum, Salsola kali, Elymus caput-medusae*) were introduced from Eurasia with European settlement, they have become permanent occupants of the *new* environments created by tillage and livestock; in some cases they have even supplanted native species on undisturbed sites.

Although some aliens arrived before 1850, many of the aliens most prominent today arrived between 1870 and 1900, the major period of farm establishment. Most aliens probably arrived as contaminants in seed lots of wheat, clover, and alfalfa and were spread rapidly by a comprehensive railroad system. Livestock may have been the initial vector of some aliens, but more often domestic animals facilitated the intraregional spread of aliens. In some cases immigrants were either introduced or spread as a result of deliberate human intervention. New aliens continue to appear in the intermountain west despite plant quarantine barriers, and in some cases their potential new ranges may be very large.

The prominence of aliens today in this Region illustrates the biological consequences when species comprising the vegetation on different continents evolve under similar climates but under strikingly different disturbance regimens. If disturbance suddenly increases where it was once low and the barriers to immigration are relaxed, new range expansion will occur among species that have evolved in the environment with more frequent disturbance. Such events are directly tied to human activity since the most pervasive functions of humans in communities are as agents of dispersal and disturbance.

12.10. Acknowledgments

I thank H.A. Mooney for organizing the excellent meeting at Asilomar, California in October 1984 that led to the writing of the chapters in this book. Additionally, I thank E.A. Kurtz, F.C. Leonhardy, J. Mastrogiuseppe, B.F. Roche,

K.M. Steen, and R.E. Steever for valuable suggestions or assistance and R.A. Black, S.S. Higgins, K.J. Rice, and J.N. Thompson for critical review of the manuscript.

12.11. References

Andres LA, Davis CJ, Harris P, Wapshere AJ (1976) Biological control of weeds In: Huffaker CB, Messenger PS (eds), Theory and Practice of Biological Control. Academic Press, New York, pp 481-499

Bailey LH (1949) Manual of Cultivated Plants, 2nd edit. Macmillan, New York

Blackwell WH, Haacke JD, Hopkins CO (1979) *Halogeton* (Chenopodiaceae) in North America. Sida Contrib Bot 8:157–169

British Columbia Department of Agriculture (1893) Second Report (1892) Victoria BC, p 936

Caldwell MM, Richards JH, Johnson DA, Nowak RS, Dzuree RS (1981) Coping with herbivory: Photosynthetic capacity and resource allocation in two semiarid *Agropyron* bunchgrasses. Oecologia (Berl) 50:14–24

Cooper JG (1860) No. 1 Report on the botany of the route. Part II. Botanical Report. In: Reports of Exploration and Surveys to Ascertain the Most Practicable and Economical Route for a Railroad from the Mississippi River to the Pacific Ocean, Made Under the Direction of the Secretary of War, 1853–55. Vol 12. Washington, DC, pp 13–39

Cotton JS (1904) A report on the range conditions of central Washington. Washington Agricultural Experiment Station Bulletin 60

Coville FV (1896) The sage plains of Oregon. Nat Geogr 7:395–404

Craig M (1894) Five farmers' foes. Oregon Agricultural Experiment Station Bulletin 32

Dafni A, Heller D (1980) The threat posed by alien weeds in Israel. Weed Res 20:277–283

Daubenmire R (1968) Ecology of fire in grasslands. Adv Ecol Res 5:209–266

Daubenmire R (1970) Steppe vegetation of Washington. Washington Agricultural Experiment Station Technical Bulletin 62

Daubenmire R (1975) Plant succession on abandoned fields, and fire influences, in a steppe area in southeastern Washington. Northwest Sci 49:36–48

Daubenmire R (1978) Plant Geography. Academic Press, New York, pp 204–212

Dewey LH (1894) The Russian thistle. US Department of Agriculture, Division of Botany, Bulletin 15

Dewey LH (1897) Migration of weeds. US Department of Agriculture Yearbook, 1896, pp 263–286

Drury CM (ed) (1963) First White Women over the Rockies, Vol 1. AH Clark, Glendale, California, 131 p

Elliott RR (1973) History of Nevada. University of Nebraska Press, Lincoln, 118 p

Elton CS (1958) The Ecology of Invasions by Animals and Plants. Methuen, London, p 63

Fernald ML (1950) Gray's Manual of Botany, 8th edit. American Book, New York

Foster AS (1906) Observations on the vegetation of the Wallula gorge. Plant World 9:287–291

Franklin JF, Dyrness CT (1973) Natural vegetation of Washington and Oregon. USDA Forest Service General Technical Report PNW-8, US Department of Agriculture, Washington DC

Geyer A (1846) Notes on the vegetation and general character of the Missouri and Oregon Territories, made during a botanical journey from the state of Missouri, across the

South-Pass of the Rocky Mountains, to the Pacific, during the years 1843 and 1844. Lond J Bot 5:511–512; 521

Griffiths D (1902) Forage conditions on the northern border of the Great Basin. US Department of Agriculture Bureau of Plant Industry Bulletin 15

Haines F (1938) The northward spread of horses among the Plains Indians. Am Anthropol 40:429–437

Harper JL (1965) Establishment, aggression, and cohabitation in weedy species. In: Baker HG, Stebbins GL (eds), The Genetics of Colonizing Species. Academic Press, New York. pp 245–265

Henderson LF (1898) Twelve of Idaho's worst weeds. Idaho Agricultural Experiment Station Bulletin 14

Hillman FH (1893a) Nevada Weeds I. Nevada Agricultural Experiment Station Bulletin 21

Hillman FH (1893b) Nevada Weeds II. Nevada Agricultural Experiment Station Bulletin 22

Hillman FH (1900) Clover seeds and their impurities. Nevada Agricultural Experiment Station Bulletin 47

Hitchcock CL, Cronquist A, Ownbey M, Thompson JW (1955-1969) Vascular plants of the Pacific Northwest. (1969) Part 1. Vascular Cryptogams, Gymnosperms, and Monocotyledons. (1964) Part 2. Salicaceae to Saxifragaceae. (1961) Part 3. Saxifragaceae to Ericaceae. (1959) Part 4. Ericaceae through Campanulaceae. (1955) Part 5. Compositae. University of Washington Press, Seattle

Holmgren AH (1942) A handbook of the vascular plants of northeastern Nevada. US Department of the Interior Grazing Service and Utah Agricultural Experiment Station, p 42

Holmgren AH (1943) New poisonous weed invades western ranges. Utah Farm Home Sci 4:3, 11

Holmgren AH (1950) Halogeton invading eastern Utah winter ranges. Utah Farm Home Sci 11:87

Huffaker CB, Kennett CE (1959) A ten-year study of vegetational changes associated with biological control of Klamath weed. J Range Manage 12:69–82

Hulbert LC (1955) Ecological studies of *Bromus tectorum* and other annual bromegrasses. Ecol Monogr 25:181–213

Josephy AM (1965) The Nez Perce Indians and the Opening of the Northwest. Yale University Press, New Haven

Mack RN (1981) Invasion of *Bromus tectorum* L. into western North America: An ecological chronicle. Agro-Ecosystems 7:145–165

Mack RN (1985) Invading plants: Their potential contribution to population biology. In: White J (ed), Studies on Plant Demography: John L. Harper Festschrift. Academic Press, London, pp 127–142

Mack RN, Pyke DA (1983) The demography of *Bromus tectorum:* Variation in time and space. J Ecol 71:69–93

Mack RN, Thompson JN (1982) Evolution in steppe with few large, hooved mammals. Am Natur 119:757–773

Macoun J (1877) Report of Professor Macoun, on the botanical features of the country traversed from Vancouver Island to Carleton, on the Saskatchewan. Geological Survey of Canada. Report of Progress, 1875–76, pp 110–232

McKell CM, Robison JP, Major J (1962) Ecotypic variation in medusahead, an introduced annual grass. Ecology 43:686–698

Meinig DW (1968) The Great Columbia Plain. University of Washington Press, Seattle

Moe LM (1977) Ecological and systematic studies on the discoid Matricarias of North America. PhD thesis, University of California, Berkeley

Morton HL, Haas RH, Erikson LC (1959) Halogeton and its control. Idaho Agricultural Experiment Station Bulletin 307

Munz PA (1959) A California flora. University of California Press, Berkeley

Piper CV (1898a) The Russian thistle in Washington. Washington Agricultural Experiment Station Bulletin 34

Piper CV (1898b) The present status of the Russian thistle in Washington. Washington Agricultural Station Bulletin 37

Piper CV, Beattie RK (1901) The Flora of the Palouse Region. Allen Brothers, Pullman, Washington

Piper CV, Beattie RK (1914) Flora of Southeastern Washington. New Era Printing, Lancaster, Pennsylvania

Rickard WH, Uresk DW, Cline JF (1977) Productivity response to precipitation by native and alien plant communities. In: Andrews RD, Carr RL, Gibson F, Lang BZ, Soltero RA, Swedberg KC (eds), Proceedings of the Symposium on Terrestrial and Aquatic Ecological Studies of the Northwest. Eastern Washington State College Press, Cheney, Washington

Ridley HN (1930) The Dispersal of Plants Throughout the World. L. Reeve, Kent, U.K.

Rumney GR (1968) Climatology and the World's Climates. Macmillan, New York, pp 342–362

Sandberg JH, Leiberg JB (n.d.) Report upon a botanical survey of portions of eastern Washington. In: Records of the Bureau of Plant Industry, Soils, and Agricultural Engineering. Division Botany, Record Group 54, US National Archives, Washington, DC, 122 pp

Schroedl GF (1973) The archaeological occurrence of bison in the southern plateau. PhD thesis, Washington State University, Pullman

Scribner FL (1883) A list of grasses from Washington Territory. Bull Torrey Bot Club 10:63–66.

Shear CL (1901) Field work of the Division of Agrostology: A review and summary of the work done since the organization of the Divison, July 1, 1895. US Department of Agriculture Division of Agrostology Bulletin 25

Simpson JH (1876) Report of explorations across the Great Basin of the territory of Utah for a direct wagon-route from Camp Floyd to Genoa, in Carson Valley, in 1859. U.S. Government Printing Office, Washington DC

St. John H (1937) Flora of Southeastern Washington and of Adjacent Idaho. Students Book, Pullman, Washington

Stamp NE (1984) Self-burial behaviour of *Erodium cicutarium* seeds. J Ecol 72:611–620

Stoddart LA, Cook CW, Gomm, BP (1953) Halogeton may be spread by animals. Utah Farm Home Sci 14:37, 46

Suksdorf, WN (1892) Flora Washingtonensis. A catalogue of the Phaenogamia and Pteridophyta of the State of Washington

Tanner VM (1940) A chapter on the natural history of the Great Basin, 1800 to 1855(1). Great Basin Nat 1:33–61

Thwaites RG (ed) (1905) Original journals of the Lewis and Clark expedition, Vol 3. Dodd, Mead, New York

Tisdale EW, Zappettini G (1953) *Halogeton* studies on Idaho ranges. J Range Manage 6:225–236

Tracy SM (1888) Report of an investigation of the grasses of the arid districts of Texas, New Mexico, Arizona, Nevada and Utah in 1887. US Department of Agriculture, Division of Botany Bulletin 6

Trewartha GT (1981) The Earth's Problem Climates. 2nd edit. University of Wisconsin Press, Madison, pp 272–273; 308–309

US Environmental Data and Information Service (1981) Climatological Data: National Summary. Annual Summary 31(13)

Washington Department of Agriculture (1914) First Annual Report 1913-14. Seed Laboratory Report, pp 101–111

Watson S (1871) United States Geological survey of the Fortieth Parallel, Vol 5. Botany. US Government Printing Office, Washington, DC, p xii

Wood WR (1972) Contrastive features of native North American trade systems. In: Voget FW, Stephenson RL (eds), For the Chief: Essays in Honor of Luther S. Cressman. University of Oregon Anthropology Paper 4, pp 153-169

Yensen D (1980) A grazing history of southwestern Idaho with emphasis on the birds of prey study area. US Department of the Interior–Bureau of Land Management, Snake River Birds of Prey Research Project. Boise, Idaho

Yensen D (1981) The 1900 invasion of alien plants into southern Idaho. Great Basin Nat 41:176–183

Young JA, Evans RA, Major J (1972) Alien plants in the Great Basin. J Range Manage 25:194–201

13. Invasibility: Lessons from South Florida

J.J. Ewel

13.1. Introduction

South Florida contains more conspicuous introduced plants and animals than any other region in the continental United States. At the same time the region also encompasses one of the largest contiguous complexes of preserved eco-systems in the eastern U.S. (Fig. 13.1.). Everglades National Park, dedicated in 1947, covers about 2500 (terrestrial) km^2; the Big Cypress National Preserve, established only a decade ago, occupies 2300 km^2; and the Fakahatchee State Preserve, whose acquisition by the State of Florida began in 1974, contains about 200 km^2. An additional 3600 km^2 are included in the three diked basins with modified hydroperiods controlled since 1949 by the South Florida Water Management District. Most of the introduced species that cause concern in South Florida were present before government agencies gained control of these lands.

Naturalized species represent a large fraction of the total number of species in a wide range of taxonomic groups (Table 13.1.). I will focus on invasions by tree species, and I will begin with an introduction to the flora, environment, and ecosystems of South Florida.

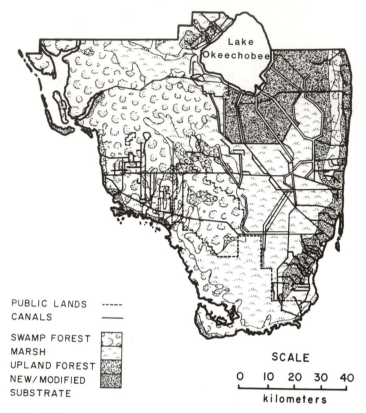

PUBLIC LANDS   -----
CANALS   ———

SWAMP FOREST
MARSH
UPLAND FOREST
NEW/MODIFIED
SUBSTRATE

SCALE

0   10   20   30   40

kilometers

Figure 13.1. South Florida's ecosystems, protected public lands, and major canals and dikes.

13.2. Phytogeography of South Florida

The flora of South Florida, which includes nearly 1650 species, has originated from two main sources (Long and Lakela 1971; Long 1974). About 1000 species (61%) are of tropical origin, and most of these (91%) are "landed immigrants" that reached Florida from the West Indies. These are species—many of which happen to be bird-dispersed pioneers in more tropical settings—that make the region's flora so unique when compared to that of the rest of the mainland.

A second suite of about 642 species, accounting for 39% of the flora, colonized from the temperate north (Long 1974). The animals of South Florida differ strikingly from the plants in this regard. Most of the region's depauperate native fauna consists of temperate-zone taxa: for example, 88% of the breeding land birds are northern species (Robertson and Kushlan 1974).

There are nearly 300 naturalized introduced species in South Florida and they account for approximately 18% of the flora (Ewel and Conde 1979 based on descriptions in Long and Lakela 1971). All of South Florida's naturalized

Table 13.1. Numbers of native and naturalized exotic species in South Florida

Taxonomic Group	Total Number of Species	Number of Exotics	References	Notorious Examples
Plants	1647	±250	Long and Lakela (1971)	*Casuarina* spp. (Australian pines) *Melaleuca quinquenervia* (melaleuca) *Schinus terebinthifolius* (schinus)
Fish[a]	80	13	Courtenay et al. (1984); Loftus and Kushlan (1986)	*Cichlosoma bimaculatum* (twospot cichlid) *Clarius batrachus* (walking catfish) *Pelonesox belizanus* (pike killifish)
Amphibians	18	4	Wilson and Porras (1983); W.F. Loftus (personal communication)	*Bufo marinus* (marine toad) *Hyla septentrionalis* (Cuban tree frog)
Reptiles	52	22	Wilson and Porras (1983)	*Anolis equestris* (knight anole)
Birds	296[b]	±15	Owre (1973, personal communication), Robertson and Kushlan (1974)	*Myiopsitta monachus* (monk parakeet) *Pycnonotus jocosus* (red-whiskered bulbul) *Brotogeris versicolurus* (canary-winged parakeet) *Melopsittacus undulatus* (budgerigars)
Mammals	44	10	Layne (1974)	*Dasypus novemcinctus* (nine-banded armadillo)

[a] Includes freshwater species only.
[b] Of these, 116 species breed in South Florida; an additional 83 species are known from < 10 credible records.

tree species were introduced intentionally, some as ornamentals, some for fruit, and some to afforest the Everglades marshes.

13.3. The South Florida Environment

Few sea-level land areas at latitudes of 25° to 27° receive as much rainfall as South Florida (Fig. 13.2.). Mean annual rainfall averages about 1400 mm, and most of it falls between May and October.

Most rainfall in the region is convectional and accompanied by thunderstorms. The incidence of lightning strikes is among the world's highest, and fires are common, even when the soil is flooded. Fire is important to almost every eco-system in South Florida (Wade et al. 1980) and often influences the success or failure of species' invasions.

South Florida's landscape is one of the geologically youngest on the North American continent; much of it has probably been above sea level for less than 5000 years (Fairbridge 1974; but see Robbin 1984). Its young flora, although species-rich, may not fully occupy all resources. This undersaturation may ex-plain, in part, why so many introduced species have successfully colonized the region.

The southern extremity of peninsular Florida is a limestone platform that barely emerges above sea level. In some places it is covered with a veneer of sand, marl, or peat; elsewhere the limestone itself is the substrate for plant growth. Much of southern Florida floods during the summer rainy season. The water flows slowly southward in sheets over the surface, and well-defined stream channels are almost nonexistent. Therefore its hydrology—and its vegetation—

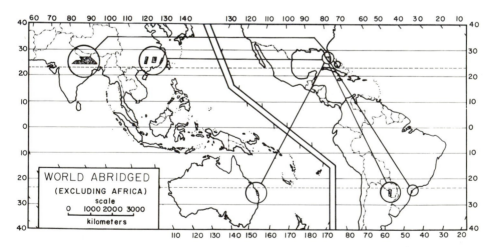

Figure 13.2. World distribution of areas with latitude, elevation, and amounts of rainfall similar to South Florida. Africa has contributed relatively few species (except some important grasses) to South Florida's non-native flora because its climates at comparable latitudes and altitudes are drier than Florida's.

are very susceptible to drainage modifications at distant locations. Ditching and diking were common land-reclamation procedures in South Florida from the 1880s until the late 1940s, dubbed the "Age of Rapacious Drainage" by Gleason (1984).

13.4. The Ecosystems of South Florida

Several detailed treatments of the plant communities of South Florida have been published (e.g., Davis 1943; Egler 1952; Craighead 1971). Four broad groups of communities that reflect major differences in physiognomy, floristics, and environment—upland forests, swamps, marshes, and human-created communities—are delineated in Figure 13.1.

There are two main kinds of upland forest in South Florida, pine forests (dominated by *Pinus elliottii* var. *densa*) and mixed-broadleaved forests (dominated by trees of West Indian origin). Introduced species invade the pinelands readily, especially if they are not burned regularly or if they are severely burned. The mixed broadleaved forests, locally called tropical hammocks, rarely flood and seldom burn. Although they occupy little area in South Florida, their unique and rich species composition has made them important targets for ecological studies and conservation efforts. In urban areas they become havens for many introduced species, but in preserves non-natives usually play a relatively unimportant role.

There are three main types of swamps in South Florida: mangroves, cypress forests, and mixed-species swamp forests. Florida contains most of the mangrove swamps of the U.S., and the Everglades National Park alone contains 1500 km^2, or about two-thirds of the statewide total. Intact mangrove communities are relatively resistant to invasion by introduced species, even though several of Florida's naturalized introduced species can tolerate saltwater. When opened by cutting, lightning, or storms—or if ditched for mosquito control—mangroves are readily invaded. Cypress forests, dominated by *Taxodium ascendens* and *T. distichum,* are especially important west of the Everglades-proper. Human-induced changes in the water regimen of South Florida in recent decades have produced three conspicuous changes in these forests: disappearance of peat due to severe fires, expansion of the distribution of pines and palms *(Sabal palmetto)* into areas formerly dominated by cypress, and invasions by introduced plant species, especially *Melaleuca quinquenervia*. The third, and most floristically complex, type of swamp in the region is the mixed swamp forest. Introduced species invade these stands around their fringes, but the interiors of the mixed swamps seem to be resistant to invasion.

Much of South Florida is covered by seasonally flooded herbaceous vegetation. Like many semiaquatic plants, the dominants in some of these marshes are not mycorrhizal. Sawgrass (actually a sedge, *Cladium jamaicense*) covers vast areas and dominates what may be one of the world's most extensive nonmycorrhizal, monospecific communities. When water regimens change the soil

sometimes becomes better oxygenated. This permits mycorrhizal plants—including several notorious aliens—to invade habitats from which they would normally be excluded.

The sands, marl, and limestone that support most South Florida ecosystems are not especially good substrates for plant growth. When people modify these substrates they unwittingly create new habitats, unlike any that occur naturally in the region. Farming increases soil fertility and aeration. Dredging creates canals (now inhabited by introduced species of fishes and plants) in a land that had few ponds and streams; the spoil banks become dry islands surrounded by marshes. Bulldozing and road construction create topography in a landscape that is otherwise billiard-table flat. These new substrates are not colonized by the same assemblages of plants that originally blanketed South Florida. Instead, new species combinations—frequently dominated by introduced species—grow there. The new substrates themselves occupy thousands of hectares. More importantly perhaps, they harbor dense concentrations of introduced species, so become the staging areas from which these aliens disperse into natural communities. The effectiveness of these new ecosystems as facilitators of introduced-species invasions is enhanced by their shape. Because many of them result from canal and road construction, they are long and narrow, so have a high edge-to-area ratio. Introduced species that colonize such habitats therefore penetrate well inside unmodified ecosystems so their seeds must travel only short distances to reach large areas of native communities.

13.5. Two Successful Introduced Species

Of the dozens of species that have become naturalized in the nonagricultural ecosystems of South Florida, I selected the two best-studied and most conspicuous ones for discussion here: *Melaleuca quinquenervia* (hereafter melaleuca) and *Schinus terebinthifolius* (hereafter schinus). Both species have undergone dramatic range expansions in recent decades and they are still expanding rapidly. Thus, they offer a unique opportunity to study colonization and invasion while they are occurring, rather than after the fact. Both species are evergreen, subtropical trees that were intentionally introduced into Florida within the past century and both have become dominants in landscapes that were formerly treeless, or nearly so. Their northward migration up the Florida peninsula is checked by frost.

There are important differences between melaleuca and schinus, and a comparison of their ecological and life-history traits should prove instructive. They differ with respect to their relationships to fire and water regimes and they tend to invade different communities. They also differ in their degree of dependence on human modification of natural conditions to facilitate invasion. Regardless of each of their impacts on local ecosystems, neither species seems to be a strong competitor of the other. Good invaders that they are, they seem to have divided the spoils fairly equitably.

13.5.1. Melaleuca

Melaleuca quinquenervia (Cav.) Blake (called melaleuca, cajeput, or punk tree) is a myrtaceous tree native to Australia, New Caledonia, and New Guinea. In its native habitat it grows in coastal lowlands where it forms open, nearly monospecific stands that burn regularly. It was introduced into southeastern and southwestern Florida around the turn of the century as a prospect for afforesting the Everglades. It has been the object of many studies in Florida, the most detailed of which were by Meskimen (1962) and Myers (1983, 1984).

Once established, melaleuca tolerates a broad range of site conditions. It becomes established more readily on sand than on marl, but can survive on almost any soil in South Florida. It tolerates extended flooding, moderate drought, and some salinity.

Melaleuca is almost perfectly adapted to fire. It has thick, spongy bark that insulates the cambium. The outer layers of bark are flaky and burn vigorously. This conducts flame into the canopy, igniting the oil-laden foliage. The leaves and small branches are killed, but dormant lateral buds on the trunk germinate within weeks. This prolific resprouting greatly increases the surface area of small branches and therefore the tree's reproductive potential. Furthermore, melaleuca can flower within weeks after a fire.

Each serotinous capsule on a melaleuca contains about 250 tiny seeds (36,000 per g), and these are released after a burn, frost, or any other event that severs the vascular connections to the fruit. A burned melaleuca can release millions of seeds, which are dispersed short distances by wind and water. A melaleuca colony frequently consists of a series of annulae of even-aged trees surrounding a clump of founders. With its within-capsule seed bank, melaleuca can potentially reproduce any time during the year.

13.5.2. Schinus

Schinus terebinthifolius Raddi (schinus, Florida holly, Brazilian pepper, Christmas berry) is a member of the Anacardiaceae native to Brazil, Paraguay, and Argentina. In its natural habitat it is a sparse species and never dominates the landscape as it does in South Florida. It has dark green foliage and produces copious quantities of bright red drupes in late December. Schinus was introduced as an ornamental to South Florida more than 100 years ago, but did not begin to explode across the landscape until the 1950s.

Like melaleuca, schinus grows on a broad range of sites in South Florida, ranging from mangroves to pinelands. It thrives on disturbed soils and in the newly created habitats that result from drainage and farming. It is more exclusively ruderal than melaleuca and tends to prefer better-drained sites.

The relationship of schinus to fire is very different from that of melaleuca. Whereas melaleuca is a tall, slender-crowned tree, schinus is short and squat (Fig. 13.3.). The broad crowns of adjacent schinus trees intertwine. This creates a dense shade and results in almost no herbaceous understory vegetation that might burn. Because schinus produces new leaves throughout the year, and

Figure 13.3. *Melaleuca quinquenervia* (left) and *Schinus terebinthifolius* (right) in South Florida. (Courtesy of J.R. Snyder.)

because its litter decomposes quickly (J.R. Snyder, unpublished) little leaf litter builds up on the forest floor. When schinus does burn (as it frequently does when it colonizes open pinelands), the above-ground parts are killed, but the tree promptly resprouts from the base.

Another important contrast between melaleuca and schinus concerns their reproduction. Schinus flowers synchronously in October and is pollinated by a native syrphid fly, *Palpada vinetorum*. Its fruits ripen in December through February and the pea-sized drupes are dispersed long distances by mammals and birds. Most dispersal is effected by the huge flocks of robins *(Turdus migratorius)* that periodically (but not annually) congregate in South Florida during the winter. Introduced species, such as the red-whiskered bulbul, *Pyononotus jocosus,* also disperse their seeds (Owre 1973). During the late winter months when schinus seeds are dispersed, there is little reproductive activity by native trees. This exploitation of a different time of reproduction may help to explain schinus' success in South Florida.

13.6. Invasibility

Invasibility is a measure of a community's susceptibility to colonization by a particular species. Thus, a given community might have high invasibility with respect to one prospective colonist and low invasibility with respect to another.

It is often argued that most undisturbed communities have low invasibility with respect to almost all newcomers. Before examining the evidence that supports this contention (and an example that refutes it) we should consider the commonly held view that introduced species invasions are determined primarily by species' abilities to gain access to new communities.

13.6.1. Invasions as a Problem of Site Access

The reasoning associated with this view usually goes something like this: A new species is introduced (usually by people) to a region from which it was previously excluded by geographical barriers. It aggressively invades the local community, outcompeting the native species. It is an especially effective competitor because the biological controls that kept it in check at home did not get transferred with it. There are two ways to contain this invader: (1) attack it with all the weapons available, killing as many individuals as possible, and (2) study the species in its native habitat so that the herbivores and diseases that attack it there can be identified, screened, and eventually introduced into its new territory.

By this view, successful invaders are regarded as a disease, rather than a symptom. Sometimes a community has high invasibility because its biota has been disrupted or because the site has been devegetated. Frequently, however, the weakening is more subtle and results from environmental changes that are not readily perceptible to casual observation. Changes in hydrology due to

Figure 13.4. Kudzu *(Pueraria lobata)* blanketing a plantation of sugi cedar *(Cryptomeria japonica)*, both in their native Japan. (Courtesy of K. Hara.)

drainage or dike construction in the northern reaches of the Everglades' marshes is one such example; the slow eutrophication of the Great Lakes is another.

Furthermore, it is not true that naturalized species are invariably kept under control in their native habitats by local competitors, herbivores, and parasites. Witness, for example, the growth of that scourge of the southeastern U.S., kudzu *(Pueraria lobata)*, on a sugi cedar *(Cryptomeria japonica)* stand in its native Japan (Fig. 13.4.).

Admittedly, many introduced species have successfully invaded presumably pristine ecosystems. Examples might include the circumglobal introductions of pigs, goats, and rats by early explorers onto uninhabited islands as well as some of the early bird introductions on the Hawaiian Islands (see Moulton and Pimm 1983 and this volume). Most cases involve species-poor communities, such as those on remote islands (or peninsulas, such as Florida, which jut into a climatic zone different from that of the continent from which they protrude). Some localities commonly cited in the literature on introduced species invasions, e.g., Great Britain and New Zealand, probably reflect both circumstances: a depauperate biota and disturbed native communities. Few, if any, introduced mammals or trees have become naturalized in the mature forests of the Amazon Basin, Zaire, or Borneo.

13.6.2. Tests of Invasibility

Invasibility can be inferred a posteriori by observing communities and the species they contain. One danger of this approach is that it is subject to unintentional bias. Observers tend to be more likely to generalize from a few observations of successful invasion into a community type than from many observations of communities that have not been invaded.

Invasibility can also be tested experimentally, and this is the approach I and my co-workers have taken in studying melaleuca and schinus in South Florida. We conduct these tests by introducing seeds of the potential invader into an array of communities representing various successional stages, degrees of disturbances, hydroperiods, fire regimens, and soil types. Seeds are introduced at intervals throughout the year to test ecosystem invasibility during as many conditions as possible. Myers (1983), for example, introduced at least 2 million melaleuca seeds into each of eight communities over a 2-year period. In the Everglades National Park we introduced more than 20,000 schinus seeds into each of nine communities over a 2-year period (Ewel et al. 1982).

The results of those seed introductions led to similar conclusions regarding both species: introduced seeds yielded more seedlings in disturbed communities than in mature ones (Table 13.2.). In some cases seed germination was higher in a mature community (e.g., the dwarf cypress forest studied by Myers), but in no case did the numbers of surviving seedlings in the mature communities approach those in the disturbed communities. In many communities the seed introductions were complete failures and yielded no surviving seedlings.

Suppose, however, that the timing of the seed-introduction experiments did not coincide with the chance juxtaposition of conditions required to ensure

Table 13.2. Germination of the seeds of two exotic trees introduced into mature and disturbed ecosystems in South Florida

Exotic Species	Total No. Seeds Introduced	Mature Communities		Disturbed Communities	
		Number[a]	Germination (%)	Number[a]	Germination (%)
Schinus terebinthifolius	200,000	5	1.27	14	2.62
Melaleuca quinquenervia	22,000,000	6	0.01	2	0.14

Data for *M. quinquenervia* are summarized from Myers RL (1983) J Appl Ecol 20:645–658.
[a] Number of communities into which introductions were made.

both germination and seedling establishment? To guard against this possibility, both Myers (1983) and Ewel et al. (1982) bypassed the crucial germination and early survival phases and planted seedlings of melaleuca and schinus in the same communities that seeds of these two species had been sown in.

Melaleuca and schinus responded differently (Fig. 13.5.). The outplanted melaleuca seedlings had lower than 50% survivorship in five of the eight communities. Survivors grew best in two disturbed ecosystems (a severely burned pine-cypress ecotone and a drained cypress forest) and one undisturbed community (a dwarf cypress forest).

Schinus, on the other hand, survived everywhere it was planted. The outplanted seedlings grew best in disturbed, open communities, but mortality was low in even the most diverse and dense communities studied. Furthermore, schinus seedlings that looked like they were barely surviving in dense shade proved capable of responding to altered environmental conditions when gaps were formed. A schinus seedling, once established, is a potential canopy tree in almost any forest in South Florida.

13.6.3. Displacement of Cypress by Melaleuca

It is clear that, as postulated earlier, disturbance facilitates invasion. But is it an essential prerequisite? The answer, of course, is "not always," and the work of Myers (1984) on the invasion of malaleuca into the ecotone between pine and cypress forests demonstrates this nicely.

Pond cypress, *Taxodium ascendens,* is one of those native species that colonized South Florida from the north temperate zone. Myers (1984) argues that because there are few trees in South Florida that are well adapted to both fire and flooding, cypress underwent ecological release and occupies habitats from which it might otherwise be competitively excluded. One such habitat lies be-

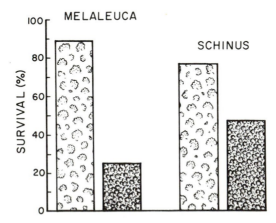

Figure 13.5. Survival of outplanted seedlings of two introduced species of trees, *Melaleuca quinquenervia* and *Schinus terebinthifolius,* in disturbed (open shading) and mature (dense shading) communities in South Florida. [Data on *M. quinquenervia* are from Myers (1983)].

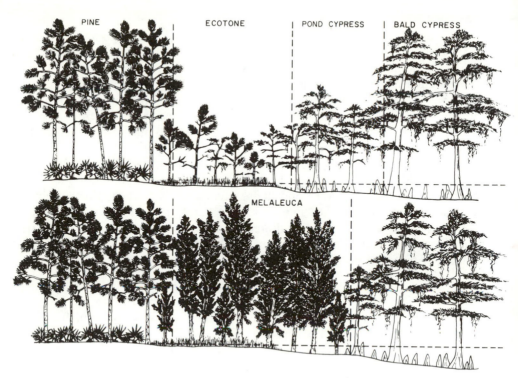

Figure 13.6. Invasion of the ecotone between pine and cypress forests in South Florida by *Melaleuca quinquenervia*. [From Myers RL (1984) In: Ewel KC, Odum HT (eds), Cypress Swamps. Reproduced with permission of University of Florida Press.]

tween pinelands and swamps on soils that are too wet for good growth of pines yet drier than those where cypress grows best (Fig. 13.6.). Cypress stands in these ecotones are short, open-canopied, and subject to frequent fires.

Based on invasibility experiments (Myers 1983), studies of melaleuca distribution and range expansion (Capehart et al. 1977), and a decade of field observations, it seems likely that Myers' (1984) conclusion is correct. Melaleuca is capable of invading the zone between pine and cypress forests in South Florida and successfully displacing cypress therefrom. It is likely, then, that we can anticipate further expansion of melaleuca into relatively undisturbed ecosystems, especially in southwestern South Florida where there is an extensive mosaic of pine and cypress forests.

13.7. Lags in Diffusion

Both schinus and melaleuca were present in Florida long before they became conspicuous elements in the landscape, a phenomenon that is well known with respect to invasions of many other introduced species. What accounts for the

long lag between the time these species were introduced and the time their populations became noticeable? There are at least four possible explanations.

First, schinus and melaleuca may have been introduced during a time when South Florida's ecosystems were more pristine—and therefore more invasion-resistant—than they are today. These two aliens may have exploded across the landscape in response to drainage, farming, and urbanization.

Second, these species may have been undergoing rapid—yet unnoticeable—expansion. An exponentially growing population appears to undergo a long lag phase of initial buildup before its numbers escalate so dramatically that it becomes a conspicuous element in the landscape. The same phenomenon may be occurring today with more recently arrived plants such as downy rosemyrtle *(Rhodomyrtus tomentosus)* and colubrina *(Colubrina asiatica)*.

Two other possible explanations for the long time lag between introduction and population explosion concern new ecosystems as reservoirs of potential invaders. Disturbed areas colonized by introduced species may be staging areas from which invading species shower the surrounding landscape with seeds. A second idea is that populations occupying disturbed habitats eventually produce genetic variants adapted to invasion of undisturbed local communities.

It may have taken several decades for melaleuca and schinus to build up populations large enough to have significant reproductive potential (or "infection pressure," sensu Salisbury 1961). A small population would have dispersed relatively few propagules, and until those seeds hit the right combination of conditions for germination, survival, growth, and reproduction, there would have been no range expansion. The probability of hitting the right combination of environmental conditions increases with time and with the number of propagules dispersed. Colonization of heavily disturbed habitats may have permitted schinus and melaleuca to establish pockets of infestation from which large numbers of seeds were dispersed into healthier communities. Eventually, populations became established there, too. The end result may have been inevitable, but its occurrence may have been hastened by the establishment of island-like populations from which infection could radiate outward.

Baker (1965) suggested an alternative to Salisbury's "infection pressure" explanation of the lag often observed in the colonization pattern of a new invader. He suggested that a new colonist might be confined to restricted habitats until appropriate genotypes become available through recombination or introgression. Although no genetic studies of schinus or melaleuca in Florida have been done, this seems a likely possibility, especially in the case of schinus, which is comprised of four varieties (Barkley 1944). Schinus has many of the characteristics listed by Baker (1965) as attributes of an ideal weed, yet it also has many traits more typical of mature-system species, including high tolerance of shade. Its behavior is analogous to that of a sit-and-wait predator: it becomes established in the understory of dense forests, then captures the site when gaps occur in the canopy. This makes it an unusually formidable species to control. To what extent its physiological characteristics are shared by all genotypes in South Florida is unknown, but it seems likely that there is substantial variation within the population.

13.8. Conclusions

There are few areas in the world with climate–soil combinations similar to those of South Florida. In spite of this paucity of equivalent landscapes likely to act as species donors, naturalized introductions comprise 15 to 25% of most major taxonomic groups in South Florida, giving the region the continent's most exotic-laden biota. Although its geological youth and depauperate fauna may be related to the success of many introductions, its rich flora argues against undersaturation as a major factor in the colonization of thousands of hectares by introduced trees such as melaleuca and schinus.

Species invasions often reflect the condition of the community being invaded rather than uniquely aggressive traits of the invader. Human modification of South Florida's ecosystems has made them especially susceptible to invasions for two reasons. First, some human activities cause changes in community structure that lower the competitiveness of natives and facilitate invasions by introduced species. Land clearing is an obvious example, but more subtle changes induced by aseasonal burning or modest changes in hydroperiods can be equally important. Second, new habitats have been created by drainage, ditching, diking, and farming, and introduced species are often better adapted than native species to these new environments.

On new substrates we can expect to see new communities develop, comprised of combinations of native and introduced species. Naturalized species often develop mutualistic interactions with indigenous species. Schinus, for example, provides food for native organisms that pollinate its flowers and disperse its seeds. It is likely that South Florida's canals, spoil banks, and anthrosols will always support vegetation in which introduced species are important.

This does not mean that resource managers must relinquish intact, native ecosystems to introduced species. However, to be successful a management program must be tailored to fit both the invader and the invadee: i.e., the autecological and life-history attributes of the species as well as the biotic and abiotic attributes of the ecosystem being protected.

For example, schinus is such a well-dispersed invader of ruderal habitats that it is almost impossible to keep it off of well-drained, abandoned farmland in South Florida. The risk of massive invasion of relatively undisturbed forests, however, can probably be reduced substantially by ferreting out individuals that colonize naturally occurring gaps and by burning pine forests frequently.

Melaleuca requires a different approach. It colonizes relatively wet sites, thrives on fire, and is capable of invading some undisturbed communities. It is probably the limited mobility of melaleuca seed that has confined its invasion thus far. Resource managers might be well advised to concentrate on eliminating seed sources nearest the pine-cypress ecotones into which melaleuca is preadapted to spread, rather than expending their resources on pockets of melaleucas near other, less susceptible habitats.

In considering introduced species in South Florida, we must recognize that: (1) some permanent changes in species composition resulting from species introductions are inevitable; (2) invasions often involve intricate—and perhaps

positive—interactions between introduced and native species; and (3) an intact native ecosystem is often the best prophylactic against exotics. To consider only the invading species themselves in developing management programs or in recommending regulatory actions is tantamount to curing symptoms and not disease.

13.9. Acknowledgments

Field work on melaleuca and schinus in South Florida was supported by the U.S. Department of the Interior. Several scientists provided me with information on the status of introduced species in specific taxonomic groups: W.R. Courtenay (fish), J.N. Layne (mammals), W.F. Loftus (fish and herps), and O.T. Owre (birds). I owe special thanks to J.R. Snyder and R.L. Myers for helpful reviews of the manuscript plus many stimulating discussions on species' introductions and South Florida's ecosystems over the past several years.

13.10. References

Baker HG (1965) Characteristics and modes of origin of weeds. In: Baker HG and Stebbins GL (eds), The Genetics of Colonizing Species. Academic Press, New York, pp 147–172

Barkley FG (1944) *Schinus* L. Brittonia 5:160–198

Capehart BL, Ewel JJ, Sedlick BR, Myers RL (1977) Remote sensing survey of *Melaleuca*. Photogra Engin Rem Sens 43:197–206

Courtenay WR Jr, Hensley DA, Taylor JN, McCann JA (1984) Distribution of exotic fishes in the continental United States. In: Courtenay WR Jr, Stauffer JR Jr (eds), Distribution, Biology, and Management of Exotic Fishes. Johns Hopkins University Press, Baltimore, Maryland, pp 41–71

Craighead FC Sr (1971) The Trees of South Florida, Vol 1. The Natural Environments and Their Succession. University of Miami Press, Coral Gables, Florida, 212 p

Davis JH (1943) The natural features of southern Florida, especially the vegetation and the Everglades. Fl Geol Sur Bull 25, 311 p

Egler FE (1952) Southeast saline Everglades vegetation, Florida, and its management. Vegetatio Acta Geobot 3:213–265

Ewel JJ, Conde LF (1979) Seeds in soils of former Everglades farmlands. In: Linn RM (ed), Proceedings of the First Conference on Scientific Research in the National Parks, Vol 1, National Park Service Transactions and Proceedings Series No. 5, US Department of the Interior, Washington DC, pp 225–234

Ewel JJ, Ojima DS, Karl DA, DeBusk WF (1982) Schinus in successional ecosystems of Everglades National Park, USDI, National Park Service, South Florida Research Center Report T-676, 141 p

Fairbridge RW (1974) The Holocene sea-level record in South Florida. In: Gleason PJ (ed), Environments of South Florida: Present and Past. Memoir 2: Miami Geological Society, Miami, Florida, pp 223–232

Gleason PJ (1984) Saving the wild places: a necessity for growth. In: Gleason PJ (ed), Environments of South Florida Present and Past II. Miami Geological Society, Coral Gables Florida, pp viii–xxv

Layne JN (1974) The land mammals of South Florida. In: Gleason PJ (ed), Environments of South Florida: Present and Past. Memoir 2: Miami Geological Society, Miami, Florida, pp 386–413

Loftus WF, Kushlan JA (1986) The freshwater fishes of southern Florida. Bulletin of the Florida State Museum, Biological Sciences Series (in press)

Long RW (1974) Origin of the vascular flora of South Florida. In: Gleason PJ (ed), Environments of South Florida: Present and Past. Memoir 2: Miami Geological Society, Miami, Florida, pp 28–36

Long RW, Lakela O (1971) A Flora of Tropical Florida. University of Miami Press, Coral Gables, Florida, 962 p

Meskimen GF (1962) A silvical study of the *Melaleuca* tree in south Florida. Master's Thesis, University of Florida, Gainesville, Florida, 177 p

Moulton MP, Pimm SL (1983) The introduced Hawaiian avifauna: biogeographic evidence for competition. Am Natur 121:669–690

Myers RL (1983) Site susceptibility to invasion by the exotic tree *Melaleuca quinquenervia* in southern Florida. J Appl Ecol 20:645–658

Myers RL (1984) Ecological compression of *Taxodium distichum* var. *nutans* by *Melaleuca quinquenervia* in southern Florida. In: Ewel KC, Odum HT (eds), Cypress Swamps. University of Florida Press, Gainesville, Florida, pp 358–364

Owre OT (1973) A consideration of the exotic avifauna of southeastern Florida. Wilson Bull 85:491–500

Robbin DM (1984) A new Holocene sea level curve for the Upper Keys and Florida reef tract. In: Gleason PJ (ed), Environments of South Florida: Present and Past II. Miami Geological Society, Coral Gables, Florida, pp 437–458

Robertson WB, JA Kushlan (1974) The South Florida avifauna. In: Gleason PJ (ed), Environments of South Florida: Present and Past. Memoir 2: Miami Geological Society, Miami, Florida, pp 414–452

Salisbury EJ (1961) Weeds and Aliens. Macmillan, New York

Wade D, Ewel J, Hofstetter R (1980) Fire in South Florida ecosystems. USDA Forest Service, General Technical Report SE-17, Southeastern For Exp Sta, Asheville, North Carolina, 125 p

Wilson LD, Porras L (1983) The ecological impact of man on the South Florida herpetofauna. University of Kansas, Museum of Natural History, Special Publication No 9

14. Species Introductions to Hawaii

M.P. Moulton and S.L. Pimm

14.1. Introduction

What determines whether species have the potential to invade new environments? And how should we evaluate community properties to determine
whether a biological system is vulnerable to invasion? These questions are fundamental to a basic understanding of the structure of ecological communities.
The ability of a species to invade, and of a community to resist invasion, are
central features in models designed to look at species diversity, niche overlap,
limiting similarity and community change. Looking at invasions—why they
succeed or fail—may be the most direct way of testing a number of important
theories. Here we address these topics by looking at the patterns of organisms
introduced to the Hawaiian Islands.

   The advantages of looking at invasions (defined in a very broad way) to test
ecological ideas are discussed by Diamond and Case (1986). We shall consider
only species directly introduced by humans. The study of intentional introductions offers several advantages over that of natural invasions. First, among
naturally invading species, those with high dispersability will predominate. These
species are likely to form but a small subset of all possible species, and so our
understanding of properties of successful species and the vulnerability of ecological systems may be more limited. Species introduced by man represent a
broader cross-section of species—some of which might otherwise have been
unable to cross even small stretches of unfavorable habitats.

A second advantage lies in the sheer number of introductions that have been undertaken: In the state of Hawaii, 28% of all the insects (Simberloff, this volume), and about 65% of the plant species, are introduced (St. John 1973); more than 50 species of birds, 20 species of mammals, and 20 species of reptiles and amphibians have been introduced (not all successfully). In fact, throughout most of the lowland areas of the Hawaiian Islands, all the vertebrates and most of the plants and insects encountered are introduced species.

The Hawaiian Islands offer the usual island advantages of isolation, discreteness and replication, and they also possess a variable climate that has fostered a diversity of habitats.

This chapter is divided into four major sections. A brief discussion of the Hawaiian climate and its flora and fauna is followed by a general description of patterns of species introductions and the methods used to document them. In the third and fourth sections, we concentrate on the vertebrates (particularly the passerine birds) and look at what factors predispose a species to be successful (Sect. 3) and what properties of the ecological community influence a species' chance of success (Sect. 4).

14.2. Hawaii

14.2.1. Climate

The Hawaiian Islands are perhaps the most remote islands in the world. They are mountainous, volcanic, and tropical. They may be conveniently divided into two groups: the eight main islands and their associated islets, and the leeward islands. We shall deal only with the main islands: from south to north these are Hawaii, Kahoolawe, Maui, Lanai, Molokai, Oahu, Niihau, and Kauai (Fig. 14.1.). The leeward islands are much smaller; access is strictly limited and is, in any case, difficult.

The Hawaiian climate is oceanic. Temperatures are moderate. The highest temperature recorded in the Islands was 38°C at Pahala on the island of Hawaii (elevation 259 m). The lowest temperature ever recorded was −10°C at the summit of Mauna Kea (elevation 4197 m) also on Hawaii (Ruffner 1978). This represents a temperature range of 48°C across an elevational gradient of nearly 4000 m. In comparison to Hawaii, we can look at similar statistics for a land-locked area that also possesses a gradient in elevation. In Colorado, for example, the highest temperature ever recorded was 48°C (elevation 1672 m) whereas the lowest temperature recorded was −51°C (elevation 2806 m) (Ruffner 1978). This represents a temperature span of about 100°C across an elevational gradient of only about 1150 m.

In contrast to the moderate temperatures, there is a dramatic rainfall gradient. Differences in maximum annual rainfall range from less than 170 mm per year, at Kawaihae on Hawaii, to more than 12,350 mm per year, at Mt. Waialeale on Kauai (Mueller-Dombois 1981).

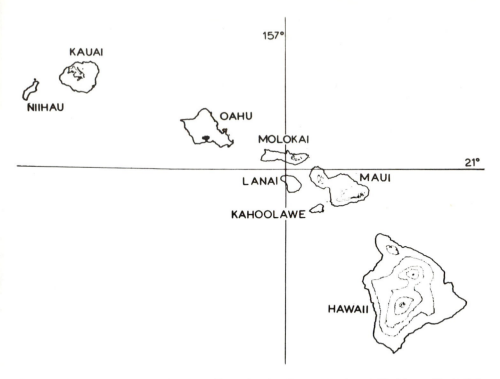

Figure 14.1. Map of the main Hawaiian Islands. (From Moulton MP, Pimm SL (1986) In: Diamond JM, Case TJ (eds), Community Ecology. Reproduced with permission of Harper & Row.)

14.2.2. Biota

Flora

Hawaiian habitats may be classified crudely as either wet or dry. Rock (1913) classified the vegetation emphasizing this contrast. He was of the opinion that the dry leeward forest was much richer in tree species than the wet forest. This dry lowland (as well as leeward) forest was largely destroyed by humans, and Rock's conclusions were based on the study of mere remnants of this forest type. Since the wet forest did not suffer a similar degree of disturbance, he may have underestimated the difference in diversity between these two general forest types.

Additional vegetation classifications have been proposed since Rock published his book on indigenous trees of Hawaii, all of which include a distinction between wet and dry habitats. This difference is also evident in the exotic habitats that have replaced the native ones—as we shall discuss below.

Fauna

The original terrestrial vertebrate fauna of the main Hawaiian Islands included at least 59 passeriform species of birds. Thirty-eight species have survived into

historic times (Pyle 1983). Twenty-one species are known as fossils (Olson and James 1982). These include the endemic honeycreeper subfamily, Drepanidini. There is a single species of bat *(Lasiurus cinereus)*, but there are no native terrestrial species of amphibians or reptiles.

The total number of native land bird species that inhabited the islands when the early Hawaiians arrived is unknown. The native Hawaiians seem likely to have caused many extinctions after they arrived (Olson and James 1982). European man has observed the extinction of 11 species since arriving in the early 19th century (Pyle 1983).

14.2.3. Ecological History

We will probably never know what the Hawaiian habitats were like prior to human contact. Rock (1913) has provided an inkling of what lay before the first Hawaiians, based on his observations of remnants of the dry forest. Evidence now overwhelmingly supports the hypothesis that the first Hawaiians had an enormous impact on these habitats (Kirch 1982a). The first Hawaiians cleared much of the lowland vegetation on all the islands with fire. Kirch (1982b) estimates that approximately 80% of the vegetation below about 460 m was severely altered by the year 1600.

The arrival of the first Europeans brought about further destruction of the native habitats (e.g., Berger 1981). Europeans brought a variety of grazing mammals including pigs, goats, sheep, and cattle (Tomich 1969). Europeans also introduced the Hawaiians to commerce in the form of the sandalwood trade (Schmitt 1977). These two factors undoubtedly led to further destruction of native habitats.

Native species of birds were (and still are) closely linked to native habitats. With the loss of native habitats below about 600 m there was a corresponding loss of native birds. Olson and James (1982) have reported large deposits of bird bones representing many previously undescribed species. These authors argue that destruction of native habitats by the first Hawaiians led to extinction of these species. There was also a loss of native species following the arrival of the Europeans. The reasons for these losses must include more habitat loss due to the introduction of grazing herbivores. But the possibility of extinction through introduced diseases and their various vectors must also be considered (Warner 1968). Finally, there may have been extinctions due directly to predation. Atkinson (1977) has suggested that the roof rat *(Rattus rattus)* might have been responsible for killing off large numbers of native birds.

Introductions of exotic species began with the arrival of the first Hawaiians. These early colonists brought the pig *(Sus scrofa)*, dog *(Canis familiaris)*, jungle fowl *(Gallus gallus)*, rat *(Rattus exulans)*, geckos, and skinks (Kirch 1982a). These reptiles were identified from fossil remains, so species identification is uncertain. The Europeans arrived in the late 1700s and brought a variety of mammals, mostly in the form of domestic livestock. Some of these species established feral populations.

But these introductions represent a small percentage of all the introductions. Most of the introductions came after the mid-1800s. Initially these species were

brought in to further agricultural interests. For example, all the frogs and toads, the mongoose, the bats, and many of the passerine birds were brought in specifically to control insect or mammalian pests (Caum 1933; Oliver and Shaw 1953; Tomich 1969). More recently, there have been numerous introductions due to cage escapes both of birds and reptiles.

In summary, there were probably two major periods of extinction of native species. The first was due to the arrival of the early Hawaiians; whereas the second was due to the arrival of the Europeans. In both cases there was widespread habitat alteration in association with these extinctions.

By the mid to late 1800s, native lowland habitats had been severely altered and many native species of birds had been exterminated. Native lowland habitats were replaced by a variety of new habitats including exotic forest, exotic shrubland, sugarcane, and urban areas. These new habitats apparently did not support populations of native birds. Large numbers of non-native species were then introduced into these replacement habitats.

14.3. The Basic Features of Species Introductions

We have compiled lists of species introductions using the published literature, and have attempted to determine the fate of each introduction. For plants and insects this has not been possible. Our data are for the four groups of introduced terrestrial vertebrates. Even for these groups, the data differ greatly in detail. The published information regarding passerine birds is voluminous, whereas that for the mammals, reptiles, and amphibians is rather limited. The data for nonpasserine birds (principally game birds) are so extensive we have not yet been able to synthesize them. For the most part these differences are due to differences in the behavior of these groups: Passerine birds are likely to be observed by birdwatchers, whereas reptiles, amphibians, and some mammals are not.

The principal problem caused by these differences in detail lies in assessing failed introductions. Observations of passerine birds have been regularly reported, and this makes it a fairly simple matter to determine whether or not a species is still present, but this has not been the case for other groups. We do not want to give the impression that difficulties with passerine data do not exist: We still encounter problems in deciding whether or not an introduction has failed. Moulton (1985) attempted to solve this problem by considering passerines to be extinct after a 2-year period with no reported observations. Observations of a species after a 2-year "absence" might well be due to reintroduction. This may be adequate when there are regularly reported observations, as for the passerine birds. But there are no comparable data for the other major taxa.

14.3.1. Passerine Birds

At least 50 species of passerine birds have been directly introduced 84 times to seven of the main islands (Hawaii, Maui, Molokai, Lanai, Oahu, Kauai, Niihau).

Passerine birds now live on Kahoolawe, too, but it is unknown whether any of these were directly introduced or if they colonized from nearby Maui and Lanai. Species, and the islands to which they were introduced, are listed in the appendix. Many species have successfully colonized other islands from their island of introduction, but it is impossible to determine how many of these natural colonizations have failed. To avoid underestimating failed introductions, we excluded natural colonizations from our analyses. Thus our analyses include only introductions performed directly by man. Of the 84 known introductions there were 31 failures (37%).

This large proportion of failed introductions is not attributable simply to vagaries in the data. Of the 84 species introductions we considered, 63 were *intentional* introductions made by private citizens or agency personnel whose desire was to establish populations of these species. Only 21 introductions could be considered as accidental (i.e., unintentional) and of these only six failed. Thus even if these introductions were truly unintentional, they were still generally successful (15/21 = 71%).

14.3.2. Mammals

Twenty-one species of mammals have been introduced to the eight main Hawaiian Islands a total of 86 times (Tomich 1969). These species are listed in the appendix. In the case of the bats, it is clear that intentional introductions took place as a measure to control insects in sugarcane fields (Tomich 1969). The two bat species listed by Tomich were considered by him to be hypothetical, due to the uncertainty of their specific identification.

In any case, it is clear that the bat introductions, whatever the species, failed. Tomich (1969) suggests that the bats in the shipment that included *Pipistrellus javanicus* might have been in poor condition on arrival and that this might have led to their failure. The native bat is rather rare on the islands (Tomich 1969)— neither of us has ever seen it. This may be due to the low densities of nocturnal insects, which are conspicuous by their absence around street lights in Hawaii's towns and cities.

Of the remaining mammals, the "failed" introductions were generally due to extermination by man. Exceptions to this are *Equus caballus* and *Sus scrofa*. The horse apparently died out on its own on Maui and Hawaii, as did the pig on Kahoolawe and Lanai (Tomich 1969). Of the 86 introductions only six have failed naturally (6/86 = 7%).

14.3.3. Amphibians and Reptiles

We compiled our amphibian and reptile data from four main sources: Stejneger (1899); Oliver and Shaw (1953); Hunsaker and Breese (1967); McKeown (1978). As many as eight species of amphibians have been introduced a total of 19 times. The amphibian species listed in Oliver and Shaw (1953) were, with few exceptions, represented by collected specimens. The exceptions to this are *Bufo boreas, Rana clamitans,* and *Rana nigromaculata*. In the case of *B. boreas*

the species identification is uncertain, but Oliver and Shaw (1953) apparently do not question that a small *Bufo* was present on Oahu at some time, although it no longer exists there. In the case of *R. clamitans* and *R. nigromaculata,* no specimens were observed by Oliver and Shaw (1953). We thus include these species with caution.

The reptile data from the above sources are accepted at face value; however, we point out that there are several differences between the reports of Oliver and Shaw (1953) and McKeown (1978) in the islands listed as being inhabited by different species. The list in our appendix of species and islands inhabited reflects these differences. Of the 58 introductions of 15 species of reptiles there were only five known failures (5/58 = 9%).

14.4. How to Be a Successful Invader

What determines if an introduced species will be successful? Might a successful invader tend to be ecologically generalized, so that it can survive in the new environment? (We might not be able to test this directly, but there is an indirect measure: the size of the species' geographical range.) Is it possible that species from particular regions of the world might generally succeed, whereas species from other regions might generally fail? Are tropical species more or less likely to succeed than temperate ones? Do successful invaders generally have higher per capita growth rates (MacArthur and Wilson 1967) than species that fail? And are there differences in success probabilities between the major groups?

It would be interesting to test these ideas using all groups of introduced species. But, as we have already noted, we do not have records of failed introductions for other than the terrestrial vertebrates. Indeed, we are often unable to answer these questions for groups other than the passerine birds, because of insufficient data. We shall consider how some of these gaps in our knowledge may be filled in our discussion.

14.4.1. Are Successful Invaders More Widespread than Unsuccessful Invaders?

Graves and Gotelli (1983) have observed that widespread bird species were disproportionately common on several Caribbean Islands. Why should this be so? There are two answers. First, species that are widespread on a mainland may have a higher probability of colonizing an island. Second, species with large geographical ranges might generally inhabit more different habitats than species with restricted distributions. In natural situations, high colonization probabilities and habitat generalization might go hand in hand. But with introduced species colonization probability is irrelevant.

To see if widespread species are predisposed to success we compared the geographical range sizes of successfully (i.e. the species survived on at least one island) versus unsuccessfully introduced passerine birds. We estimated range sizes using maps in Long (1981). Range maps were available for all but one species *(Garrulax caerulatus).* We excluded this species because we could not categorize it using the same criteria as for the other species. For each species

Table 14.1. A. Range sizes of species that were introduced but were not successful

Species	Range Size[a]
Melanocorypha mongolica	0.45
Luscinia akahige	0.03
Luscinia komadori	0.03
Copsychus saularis	1.32
Cyanoptila cyanomelana	0.39
Rhipidura leucophrys	1.00
Garrulax albogularis	0.32
Garrulax chinensis	0.24
Parus varius	0.14
Grallina cyanoleuca	1.00
Vidua macroura	2.20
Lagonosticta senegala	2.01
Estrilda troglodytes	0.45
Serinus leucopygius	0.59
Paroaria dominicana	0.18
Passerina cyanea	0.70
Passerina ciris	0.29
Passerina leclancherii	0.05
Pezites militaris	0.69

Table 14.1. B. Range sizes of species that were successfully introduced

Species	Range Size[a]
Alauda arvensis	5.27
Pycnonotus jocosus	0.72
Pycnonotus cafer	0.60
Mimus polyglottos	1.09
Copsychus malabaricus	0.70
Cettia diphone	0.72
Garrulax pectoralis	0.51
Garrulax canorus	0.42
Leiothrix lutea	0.48
Zosterops japonicus	0.61
Acridotheres tristis	0.56
Gracula religiosa	0.98
Passer domesticus	6.77
Uraeginthus bengalus	1.17
Estrilda caerulescens	0.44
Estrilda melpoda	0.77
Estrilda astrild	1.49
Amandava amandava	0.63
Lonchura malabarica	1.37
Lonchura punctulata	1.12
Lonchura malacca	0.77
Lonchura oryzivora	0.10
Serinus mozambicus	1.44
Carpodacus mexicanus	0.89
Sicalis flaveola	1.12
Tiaris olivacea	0.20
Paroaria coronata	0.34
Paroaria capitata	0.28
Cardinalis cardinalis	0.73
Sturnella neglecta 1.35	

[a] Range sizes are scaled to the size of Australia; thus the range size of *Melanocorypha mongolica* is 45% the size of Australia.

we placed a transparent grid of 2-mm squares onto the range map and counted the number of intersections. This was done three times for each map. We then calculated the mean number of intersections. To account for differences in scale of these maps (there were two scales) we calculated the number of intersections for Australia for each of the two scales and divided this into our mean number of intersections. Thus the mean number of points for each species was scaled to the size of Australia. These estimates are listed in Table 14.1. We compared the means of these two groups (failed vs. successful) using a Kruskal-Wallis test, and found a significant difference between these two groups, ($\chi^2 = 4.69$; P <0.03). Thus for passerine birds, size of geographical range appears to be a predictor of success of invasion: Species with larger ranges are apparently more likely to be successful.

14.4.2. Do Species from Certain Zoogeographical Regions Have Higher Probabilities of Success?

Species from different parts of the world have evolved under different conditions. Selective pressure due to competition and predation as well as abiotic factors must vary widely from region to region. If species within certain regions vary consistently then species from some parts of the world might be inherently better suited to invade new systems. If this were the case, we would expect species from certain regions to have higher probabilities of success than species from other regions.

To categorize the species' distribtuions we made plastic overlays depicting the six zoogeographical regions and the Tropics of Cancer and Capricorn. We used the map in Udvardy (1969, p 270) to identify the zoogeographical regions. These overlays were then dropped on the range maps shown in Long (1981), and the species were categorized accordingly. In cases where the distributions were included in more than one zoogeographical region, we assigned the species by virtue of location of the largest portion of the range. We then used a 2 × 6 test of independence to evaluate the null hypothesis that the ratio of successful introductions to total introductions was equal across zoogeographical regions. Again, there were sufficient data only for passerine birds. We found no significant differences in ratio of successful to unsuccessful species ($\chi^2 = 5.59$; 0.50>P>0.25, 5df) across zoogeographical regions.

14.4.3. Are Tropical Species More or Less Likely to Succeed Than Temperate Species?

We grouped the passerine species according to their distributions depending on whether their native ranges were mostly tropical or nontropical. In grouping the species there were two species (*Paroaria capitata* and *Copsychus saularis*) we could not categorize without ambiguity, so we excluded them. Of the 47 remaining species, 24 were generally tropical and 23 were generally nontropical species. Seven of the tropical species and 10 of the nontropical species failed. These ratios were not significantly different from chance expectation ($\chi^2 = 1.05$; 0.50>P>0.25, 1 df).

14.4.4. Are There Differences Between Taxa?

Differences in the successes and failures across passerine birds, mammals and
reptiles are given in Table 14.2. We exclude amphibians because we consider
these data too unreliable. Counting each introduction to each island as a separate
datum, we find that there are significant differences in success: Birds are more
likely to fail. The appendix, however, shows that while some species succeed
on some islands and fail on others, the common pattern is to either succeed
on all islands or fail on all islands. So counting each island species combination
separately may be incorrect: the outcomes may not be independent of each
other. Also in Table 14.2 we have summarized outcomes by species, counting
successes as species that have survived on at least one of the islands to which
we know they have been introduced. Again the result holds that birds are more
likely to become extinct than the other groups.

Finally, we must anticipate a result from the next section, the number of
species already present on an island affects extinction rate. We need to correct
these results to allow for the greater number of bird introductions. So, we have
examined rates of the first 20 bird introductions. (The fate of some mammal
species was to be deliberately exterminated, so we do not count these as failures;
see the appendix.) When we make this correction, it becomes clear that birds
are more likely to fail; seven of the first 20 bird species failed, but only two of
the 20 mammals and one of the 15 reptiles failed. These numbers differ from
what we would expect by chance ($\chi^2 = 6.04$; $P<0.05$, 2df).

14.4.5. Effects of Intrinsic Growth Rates

We might expect species with higher rates of increase, r, to have a better chance
of surviving. There are at least three ways to estimate r for the passerine birds:
(1) looking at the number of eggs produced per year (O'Connor 1981); (2) looking
at the slope of log (numbers) against time; and (3) fitting the logistic growth
curve to successive censuses (Pimm 1984). Unfortunately, none of these ap-

Table 14.2. Comparison of successful and unsuccessful introductions for three
vertebrate taxa

Group	No. of Successful Introductions	No. of Unsuccessful Introductions
A. Species/island combinations		
Passeriform birds	53	31
Mammals	74	6
Reptiles	53	5
B. Species only		
Passeriform birds	30	20
Mammals	18	2
Reptiles	14	1

Appendix: Summary of Vertebrate Species Introductions to the Eight Main Hawaiian Islands

A. Mammals Species	ZR	Island							
		O	H	Ma	Mo	La	Ka	Kh	Ni
Petrogale penicillata[a]	AU	S							
Pipistrellus javanicus	PA	F							
Tadarida brasiliensis	NA/NT	F							
Rattus rattus	PA/OR	S	S	S	S	S	S	S	S
Rattus norvegicus	PA	S	S	S	S	S	S		
Rattus exulans	OR/AU	S	S	S	S	S	S		
Mus musculus	PA	S	S	S	S	S	S	S	
Canis familiaris	PA	S	S	S	S	S	S		
Herpestes auropunctatus	OR	S	S	S	S				
Felis catus	PA	S	S	S	S	S	S	S	S
Equus caballus	PA		F	F			S		
Equus asinus	PA		S		(F)[b]		(F)		
Sus scrofa	PA	S	S	S	S	F	S	F	S
Axis axis	OR	S		S	S	S			
Odocoileus hemionus	NA						S		
Antilocapra americana	NA					S			
Bulbalus bulbalis	OR				(F)				
Bos taurus	PA	(F)	S	(F)	S		S		
Capra hircus	PA	S	S	S	S	S	S	S	(F)
Ovis aries	PA		S			S			
Ovis musimon	PA		S			S			

[a] Species identification is uncertain; see text.
[b] Parentheses indicate species was exterminated by man.

B. Amphibians Species	ZR	Island							
		O	H	Ma	Mo	La	Ka	Kh	Ni
Bufo boreas	NA	F							
Bufo bufo	PA						F		
Bufo marinus	NT	S	S	S	S		S		
Dendrobates auratus	NT	S							
Rana catesbeiana	NA	S	S	S	S		S		
Rana rugosa	PA	S	S	S			S		
Rana clamitans	NA	F							
Rana nigromaculata	PA	F							

C. Reptiles Species	ZR	Island							
		O	H	Ma	Mo	La	Ka	Kh	Ni
Trionyx sinensis	PA	(S)[a]					S		
Typhlina bramina	OR	S	(S)	(S)	(S)		(S)		
Lepidodactylus lugubris	AU	S	S	S	S		S		
Gehyra mutilata	AU	S	S	S	S		S	S	
Hemiphyllodactylus typus	AU	S	S	(S)	(S)		S		
Hemidactylus garnoti	AU	S	S	S	S		S		S
Hemidactylus frenatus[a]	AU	S	(S)	(S)	(S)		(S)		
Iguana iguana[a]	NT	(S)							
Chamaeleo jacksoni[a]	ET	(S)							
Anolis carolinensis	NT	S	(S)	(S)					

Appendix *Continued*

C. Reptiles Species	ZR	Island							
		O	H	Ma	Mo	La	Ka	Kh	Ni
Cryptoblepharus boutoni	AU	S	S	S	S		S		S
Leiolopisma metallicum	AU	S	(S)	(S)	(S)		(S)		
Lipinia noctua	AU	(S)	(S)	F			F		
Emoia cyanura	AU/OR	F	F			(S)			
Phrynosoma cornutum	NT	F							

[a] Indicates new species listed in McKeown (1978) but not in Oliver and Shaw (1953).
[b] Parentheses indicate new records listed in McKeown (1978) but not in Oliver and Shaw (1953). *Phrynosoma cornutum* was listed by Hunsaker and Breese (1967), but apparently died out (McKeown 1978).

D. Passerine Birds Species	ZR	Island							
		O	H	Ma	Mo	La	Ka	Kh	Ni
Melanocorypha mongolica	PA						F		
Alauda arvensis	PA	S	S	S	S	S	F		S
Pycnonotus jocosus	OR	S							
Pycnonotus cafer	OR	S							
Mimus polyglottos	NA	S		S					
Luscinia akahige	PA	F							
Luscinia komadori	PA	F							
Copsychus saularis	OR	F					F		
Copsychus malabaricus	OR	S					S		
Cettia diphone	PA	S							
Cyanoptila cyanomelana	PA	F	F						
Rhipidura leucophrys	AU	F							
Garrulax albogularis	OR						F		
Garrulax caerulatus	?	F							
Garrulax pectoralis	OR						S		
Garrulax chinensis	OR						F		
Garrulax canorus	OR	S	S	S	S		S		
Leiothrix lutea	OR	S	S	S	S		S		
Parus varius	PA	F	F	F			F		
Zosterops japonicus	PA	S	S				S		
Grallina cyanoleuca	AU	F	F						
Acridotheres tristis	OR	S							
Gracula religiosa	OR	S							
Passer domesticus	PA	S							
Vidua macroura	ET	F							
Lagonosticta senegala	ET	F							
Uraeginthus bengalus	ET	F	S						
Estrilda caerulescens	ET	S	S						
Estrilda melpoda	ET	S							
Estrilda astrild	ET	S							
Estrilda troglodytes	ET	F							
Amandava amandava	OR	S							
Lonchura malabarica	ET		S						
Lonchura punctulata	OR	S							
Lonchura malacca	OR	S							
Lonchura oryzivora	OR	S							

Appendix *Continued*

D. Passerine Birds Species	ZR	Island							
		O	H	Ma	Mo	La	Ka	Kh	Ni
Serinus leucopygius	ET	F							
Serinus mozambicus	ET	S	S						
Carpodacus mexicanus	NA	S							
Sicalis flaveola	NT	S	S						
Tiaris olivacea	NT	S							
Paroaria coronata	NT	S					S		
Paroaria dominicana	NT	F							
Paroaria capitata	NT		S						
Cardinalis cardinalis	NA	S	S				S		
Passerina cyanea	NA	F							
Passerina ciris	NA		F						
Passerina leclancherii	NT	F							
Pezites militaris	NT					F			
Sturnella neglecta	NA	F		F			S		F

Key to abbreviations in all tables: ZR, zoogeographical region from which species came; AU, Australasian; ET, Ethiopian; NA, Nearctic; NT, Neotropical; OR, Oriental; PA, Palearctic. Islands: O, Oahu; H, Hawaii; Ma, Maui; Mo, Molokai; La, Lanai; Ka, Kauai; Kh, Kahoolawe; Ni, Niihau.
Outcomes: S, success; F, Failure; blank, species not directly introduced to that island.

proaches is satisfactory. We do not have nesting data for most of the species. And although for some species, the Hawaii Audubon Society's Christmas count data show exponential growth, for most of the species, the data are obscured by highly variable censusing efforts. The method of fitting the logistic is even more sensitive to this variability. At present we cannot overcome these statistical difficulties, so we cannot evaluate the effects of *r* on success.

14.5. What Makes a Community Easy to Invade?

We have considered the results presented in this section in some detail elsewhere (Moulton and Pimm 1983, 1986); however, we summarize them here for completeness. They involve the number of species present on each island and how it affects extinction rate, and how taxonomic and morphological similarities affect the chance that an introduced species will survive. Again, we have data only for the passerine birds.

14.5.1. Extinction Rate

By grouping extinctions on each island by decade we obtained the data of Fig. 14.2. Extinction rate *(E)*, measured in extinctions per year, rises significantly

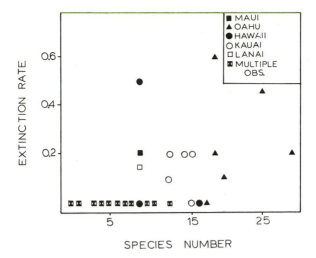

Figure 14.2. Extinction rate of passerine species (as species per year) averaged over a decade, for the six largest Hawaiian islands. Data are for the century prior to 1960.

nonlinearly with the number of species present *(S)*. Fitting a power curve to these data, $E = kS^n$, suggests a value of n = 2; thus the per species extinction rate E/S rises linearly with S (Moulton and Pimm 1983, 1986). Simply, the more species there are, the harder the system is to invade.

Why should this be so? An obvious reply is that it is because as the number of species increases so does interspecific competition. But other factors could be responsible. The first possibility is that species tend to accumulate through time. And, as man's modification of the environment may be more extensive now than in the past, this may may cause more extinctions where there are, coincidentally, more species. Clearly, we cannot estimate all possible environmental variables, but the major changes to lowland Hawaii are well-known. They involved planting sugarcane over most of the available arable land and reforesting much of the rest (data in Moulton and Pimm 1983). These changes took place rapidly from 1870 to 1920; the changes since then have been much smaller. Yet there were no extinctions until after 1920. The extinctions took place when the environment was changing the least.

A second possibility involves the species themselves. Perhaps robust species were introduced early, followed by species that were really unsuitable for Hawaii. We have rejected this explanation on a number of grounds (Moulton and Pimm 1983). First, we can divide the species into those introduced early (say prior to 1920) and those introduced late (post-1920) and see how well these species did when introduced into areas other than Hawaii. (Many species were introduced only to the Hawaiian Islands, but there were enough species introduced elsewhere to make this test.) Such a comparison shows that there is no difference in the "robustness" of the early versus late introductions (Moulton

and Pimm 1983). In addition, many unsuccessful species are widespread, common species that inhabit man-made areas (such as *Streptopelia decaocto* of Eurasia and *Grallina cyanoleuca* of Australia). And many of the common species in Hawaii were introduced late—some as recently as the 1960s.

14.5.2. The Effect of Interspecific Competition

If interspecific competition does play a role in causing extinctions, then we should expect the species that failed to be a particularly nonrandom subset of those introduced; they should be the ones that are ecologically more similar to species that have already established themselves. There are two simple ways in which to approximate ecological similarity: taxonomic relatedness and morphological similarity.

Taxonomic Relatedness

There has been considerable controversy about whether islands contain fewer species per genus than one would expect by chance. The argument is simple: On islands competition may be more severe than on the mainland, so this might weed out intensely competing congeners. But the statistics of detecting this pattern are difficult. Deciding what should happen under the null hypothesis (i.e., what is expected by chance alone) is not easy because genera differ markedly in their dispersal abilities; good dispersers may have more species per genus on islands than expected by chance.

The introduced Hawaiian birds, however, provide a particularly simple test of this idea, "Are species introduced to islands where a congener is present more likely to fail than if they are the first members of their genus to get there"? The data indicate "No." There have been 23 successful introductions out of 43 total when no congeners were present and 16 successful introductions out of 32 total when at least one congener was already present. There are no differences in the success rates (Moulton and Pimm 1986).

Morphological Relatedness

There has also been considerable controversy about whether coexisting species are morphologically overdispersed compared to what one might expect by chance. The argument parallels that for taxonomic relatedness, and the statistics are equally contentious. Again, the introduced birds provide a straightforward test. We know which species did not survive and which did, so we can test whether the morphologies of the survivors are special in any way.

For both the passerines and columbiforms (doves and pigeons), the surviving species are not a morphologically unusual subset of the total species introductions (Moulton and Pimm 1986). We have already noted the difficulties many have encountered in detecting overdispersed morphologies in natural communities. There are also many theoretical difficulties with the once popular idea of some limiting similarity between species (Roughgarden, this volume;

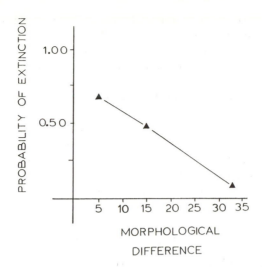

Figure 14.3. The probability that a passerine species will fail as a function of the percentage bill difference to the most similar congener already present on the island. (From Moulton MP, Pimm SL (1986) In: Diamond JM, Case TJ (eds), Community Ecology. Reproduced with permission of Harper & Row.)

Turelli 1985). We might wonder if the role of competition may not be to shape community structures in some predictable way.

Again, the experimental nature of the introduced Hawaiian birds enables us to test the idea of limiting similarity in an unusually direct way. We can ask, "How does the probability that an introduced species fails to become established vary with the morphological similarity to the species most similar to it within the established community"? In view of the often negative observations and the contentious theoretical results, we might ask if, indeed, there is any such relationship.

We have restricted our analysis to congeneric passerine species. These tend to be ecologically similar in a broad sense, a genus being largely insectivorous or gramnivorous, for example. We measured morphological dissimilarity by the percentage difference in the bill lengths. As Figure 14.3 shows, there is a dramatic and (significant) reduction in the chance that a species will fail as morphological difference increases (Moulton 1985; Moulton and Pimm 1986). Indeed, this chance would appear to approach zero at about a 40% difference—embarassingly close to the 1.3:1 ratio suggested by Hutchinson (1959).

We are in the process of testing whether this result holds for avian introductions elsewhere. But as it stands, it raises a number of points. First, morphological similarity does predict invasion success. Second, the morphological similarity required to increase the chance of a failure is considerable. Over all the passerines, we cannot detect nonrandom morphologies because most species differ by much more than 40%.

14.6. Conclusions

Our results can be summarized as follows:

1. Bird species are more likely to be successful if they have large geographical ranges.
2. Bird species from different faunal regions or species from tropical versus temperate areas do not differ in their chances of success—even though Hawaii is tropical.
3. Bird species seem more likely to be unsuccessful when compared to mammals and reptiles—even when we correct for the different numbers of species introduced.
4. The more bird species present in a system, the less likely a bird species is to invade successfully.
5. The presence of congeners does not necessarily reduce a bird species' chance of success.
6. The presence of a congener that is very similar morphologically appears to reduce the chance that a species will successfully invade.

In total, these results show that we can predict the fate of some species invasions, both from the characteristics of the species themselves and from the characteristics of the communities to which they were introduced. Moreover, the results seem to agree with, and often cleanly test, ideas of invasibility developed by theoreticians for assemblies of competing species. Thus, there appears to be some limiting similarity between coexisting species (May and MacArthur 1972); generalized species are more successful than specialized ones (May and MacArthur 1972), and the extinction rate/species number curve is nonlinear (MacArthur and Wilson 1967). This leads us to be optimistic about predicting the outcome of species invasions, but we must point to an obvious limitation.

Our detailed analyses are restricted to passerine birds; we have no way of knowing how applicable our conclusions are to other groups. The theory we have tested has generally been developed for vertebrate taxa. The comparable extinction rates for these taxa might lead us to suspect that our results might extend to other vertebrates. But it is not obvious that these results would extend to insects or plants—which comprise the vast majority of the species introduced to Hawaii.

In testing ideas on invasion it becomes critical to determine which species failed. It is often all too obvious which species succeed. We have no suggestions for documenting failed insect introductions. But for plants there seems to be an obvious solution: to compare those species that merely survive under man's direct care in gardens to those that have moved outside this protection. Orians (this volume) has done this for British trees. Hawaii's gardens are full of exotic plants from around the world, yet there are many fewer species, such as the trees *Casuarina* and *Eucalyptus,* that occupy extensive areas (even some of these species may be deliberately planted and may not be reproducing). Such

comparisons of gardens with areas away from man's care offer a unique opportunity to test ideas on species introductions.

In short, even if our results are limited to passerine birds, the potential exists to study why species succeed in other taxa—not just vertebrates, but plants too. Given the ecosystem level effects of some species introductions (such as *Andropogon;* Vitousek, this volume), it would seem to us that evaluations of why species succeed should be a high priority.

14.7. References

Atkinson IAE (1977) A reassessment of factors, particularly *Rattus rattus* L., that influenced the decline of endemic forest birds in the Hawaiian Islands. Pacif Sci 31:109–133

Berger AJ (1981) Hawaiian Birdlife, 2nd edit. University Press of Hawaii, Honolulu

Caum EL (1933) The exotic birds of Hawaii. Occas Pap Bernice P Bishop Mus 10:1–55

Diamond JM, Case TJ (1986) Overview: introductions, extinctions, exterminations, and invasions. In: Diamond JM, Case TJ (eds), Community Ecology. Harper & Row, New York

Graves GR, Gotelli NJ (1983) Neotropical landbirdge avifaunas: new approaches to null hypotheses in biogeography. Oikos 41:322–323

Hunsaker D II, Breese P (1967) Herpetofauna of the Hawaiian Islands. Pacif Sci 21:423–428

Hutchinson GE (1959) Homage to Santa Rosalia or why are there so many kinds of animals? Am Natur 93:145–159

Kirch PV (1982a) The impact of the prehistoric Polynesians on the Hawaiian ecosystem. Pacif Sci 36:1–14

Kirch PV (1982b) Transported landscapes. Nat Hist 91:32–35

Long J (1981) Introduced Birds of the World. David and Charles, London

MacArthur RH, Wilson EO (1967) The Theory of Island Biogeography. Princeton University Press, Princeton, New Jersey

May RM, MacArthur RH (1972) Niche overlap as a function of environmental variability. Proc Natl Acad Sci USA 69:1109–1113

McKeown S (1978) Hawaiian Reptiles and Amphibians. Oriental Publishing, Honolulu

Moulton MP (1985) Morphological similiarity and the coexistence of congeners: an experimental test with introduced Hawaiian birds. Oikos 44:301–305

Moulton MP, Pimm SL (1983) The introduced Hawaiian avifauna: biogeographic evidence for competition. Am Natur 121:669–690

Moulton MP, Pimm SL (1986) The extent of competition in shaping an introduced avifauna. In: Diamond JM, Case TJ (eds), Community Ecology. Harper & Row, New York

Mueller-Dombois D (1981) Some bioenvironmental conditions and the general design of IBP research in Hawaii. In: Mueller-Dombois D, Bridges KW, Carson HL (eds), Island Ecosystems: Biological Organization in Selected Hawaiian Communities. Hutchinson Ross, Stroudsburg, Pennsylvania

O'Connor RJ (1981) Comparisons between migrant and nonmigrant birds in Britain. In: Aidley DJ (ed), Animal Migration. Cambridge University Press, Cambridge

Oliver JA, Shaw CE (1953) The amphibians and reptiles of the Hawaiian Islands. Zoologica 38:65–95

Olson SL, James HF (1982) Prodromus of the Fossil Avifauna of the Hawaiian Islands. Smithsonian Cont to Zoology 365. Smithsonian Institution Press, Washington

Pimm SL (1984) Food chains and return times. In: Strong DR, Simberloff D, Abele LG, Thistle AB (eds), Ecological Communities: Conceptual Issues and the Evidence. Princeton University Press, Princeton, New Jersey

Pyle RL (1983) Checklist of the birds of Hawaii. Elepaio 44:47–58

Rock JF (1913) The indigenous trees of the Hawaiian Islands, 1st edit. (Reprinted 1974; Charles Tuttle, Rutland, Vermont)

Ruffner JA (1978) Climates of the States, Vol 1. Gale Research, Detroit

Schmitt RC (1977) Historical Statistics of Hawaii. The University Press of Hawaii, Honolulu

Stejneger L (1899) The land reptiles of the Hawaiian Islands. Proc US Nat Mus 21:783–813

St. John H (1973) List and summary of the flowering plants in the Hawaiian Islands. Pacific Tropical Botanical Garden Memoirs 1, Lawai, Hawaii

Tomich PQ (1969) Mammals in Hawaii: a synopsis and notational bibliography. Bernice P Bishop Mus Spec Pub 57:1–238

Turelli M (1985) Stochastic community theory: a partially guided tour. In: Hallam TG, Levin SA (eds), A Course in Mathematical Ecology. Springer-Verlag, Berlin

Udvardy MDF (1969) Dynamic Zoogeography. Van Nostrand-Reinhold, New York

Warner RE (1968) The role of introduced diseases in the extinction of the Hawaiian avifauna. Condor 70:101–120

15. The Invasions of Plants and Animals into California

H.A. Mooney, S.P. Hamburg and J.A. Drake

15.1. Introduction

California is a land of unusual biotic diversity. It comprises a wide range of indigenous ecosystems including a diversity of forest, woodland, scrub, and grassland types as well as numerous kinds of aquatic systems. Making up these systems are 5720 species of vascular plants (Raven and Axelrod 1978), over 200 mammal species (Ingles 1965; Williams 1979), about 28,000 insect species (Powell and Hogue 1979), 525 bird species (Small 1974), 129 amphibian and reptile species (Jennings 1983), and 132 inland fish species (Moyle 1976; Shapovalov et al. 1981).

The nature of these ecosystems is very different today than it was several hundred years ago. All have been impacted to varying degrees by anthropogenic influences. Some have been totally altered in both their structure and function, whereas others are probably not too different than they were several centuries ago. Here we focus on a particular class of ecosystem change that has been induced or greatly accelerated by the activities of humans, that of the introduction of exotic organisms. These introductions have, in certain cases, resulted in readily observable changes in ecosystem structure and hence function. In most cases, however, the ecological impact of these introductions is subtle and has not yet been determined.

What follows is an assessment of the status of exotics in California. We

catalog how many invaders have been successful, where they have came from, and how they got to the state. Where possible we note what particular characteristics have led to the success of the invaders. Unfortunately we do not have full information on many groups of organisms so we cannot give a complete assessment and thus make intergroup comparisons. However, even from the incomplete data available it is clear that there are substantial differences in the mode and success of invasions among different groups.

We then review what is known about the ecological impact of these invasions. In this regard the data are even more incomplete than those available at a taxonomic level. Yet it is evident that the impact of the invaders on the ecosystem in certain cases has been enormous. One question that we would like to answer in particular is what kinds of habitats are vulnerable to invaders? For example, can exotics become established only in disturbed habitats?

15.2. Degree of Disturbance of California's Natural Ecosystems

The state of California has a great diversity of ecosystems of which the coastal region and the Central Valley have been heavily impacted by agriculture and urbanization (Fig. 15.1.). Large areas of the state are, however, mountainous or desert and thus unsuitable for many types of agriculture though recreation is increasingly affecting even these areas. Nearly half of the state is managed by either federal or state governments. Extensive parks and wilderness areas are included in these holdings. As a result of limited access and management efforts these natural ecosystems have probably been impacted less than they might have been otherwise.

One of the difficulties in any assessment of the factors that determine the success of invading species is accounting for the degree of disturbance of the system being invaded. Certain data do give an indication of the pace and pervasiveness of anthropogenically induced change in California's ecosystems. These data indicate the level of disturbance, ranging from complete ecosystem modification due to urbanization and farming to partial disruption due to grazing. Over a third of the state is either crop or grazing lands (Fig. 15.2.). There are four to five million sheep and cattle utilizing natural ecosystems entirely or in part (Fig. 15.3.). Natural waterways have probably been the most impacted of all of California's ecosystems. The potential routes of invasions into the state are continually increasing with the construction of new roads and transmission lines. There are, for example, now over 286,000 km of roadway in California, 0.7 km of roadway for every km^2 of area (State of California 1981).

In addition to the massive and obvious ecosystem disruptions more subtle and regional disturbances have occurred such as the deterioration of air quality. It is obvious that California's ecosystems have been greatly impacted by human activities, with the greatest disturbance being concentrated in the coastal regions and in the Central Valley.

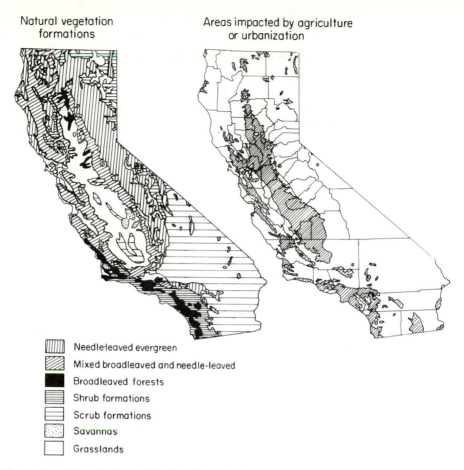

Natural vegetation
formations

Areas impacted by agriculture
or urbanization

	Needle-leaved evergreen
	Mixed broadleaved and needle-leaved
	Broadleaved forests
	Shrub formations
	Scrub formations
	Savannas
	Grasslands

Figure 15.1. Distribution of California's pristine and extant ecosystems. (From Donley MW et al. (1979) In: Atlas of California. Reproduced with permission of the Pacific Book Center, Culver City, California.)

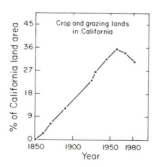

Figure 15.2. Changes in crop and grazing lands of California through time. (From U.S. Bureau of the Census, Agricultural Summaries.)

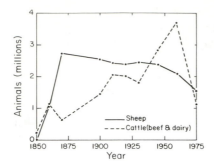

Figure 15.3. Sheep and cattle in California. (From U.S. Bureau of the Census, Agricultural Summaries.)

15.3. Taxonomic Survey of Invaders

What follows is a survey of selected taxonomic groups that occur in California, with statistics on the number of successful invaders in each. Unfortunately data on important groups, such as most of the insects, are not available.

15.3.1. Reptiles and Amphibians

Jennings (1983) has recently compiled a complete listing of the reptiles and amphibians of California. He notes that there are 129 species, of which 124 are native (Table 15.1.). There is one extinct species, the Sonoran mud turtle *(Kinosternon sonoriense)*, and five established exotics. There are 46 known unsuccessful introductions.

The successfully established species include the African clawed frog *(Xenopus laevis)* which is well established from Los Angeles County southward. The bullfrog *(Rana catesbeiana)*, which was introduced into the state from east of the Great Plains between 1914 and 1920 (Moyle 1973), is now established throughout the state except for the high mountains and deserts. It is thought that the bullfrog has been responsible for the decline of two native frog species

Table 15.1. Numbers of invading species in the California flora and fauna[a]

| | Total Species | Number of | | Percent |
		Endemics	Naturalized	
Vascular plants	6120	151	674	11
Reptiles and amphibians	129		5	4
Birds	525		9	2
Mammals	216		17	8
Inland fishes	132	25	49	19
Insects	~28,000		?	?

[a] See text for sources.

in the Central Valley (Moyle 1973). The Snapping turtle, *Chelydra serpentina,* has been found in numerous California localities. Other exotic turtles include the red-ear slider, *Chrysemys scripta,* and the texas spiny softshell *(Trionyx spiniferus).*

15.3.2. Mammals

In California there are 216 species of mammals, at least 11 of which are confirmed naturalized species with six additional questionable establishments (Williams 1979) (Table 15.1.). The naturalized species include the Virginia opossum *(Didelphis virginiana)*, Norway and black rat *(Rattus norvegicus* and *R. rattus)*, house mouse *(Mus musculus)*, gray squirrel *(Sciurus carolinensis)*, fox squirrel *(S. niger)*, wild burro *(Equus asinus)*, wild horse *(E. cabullus)*, wild pig *(Sus scrofa)*, Barbary sheep *(Ammotragus lervia)*, and Himalayan tahr *(Hemitragus jemlahicus)*.

Three species have become extinct within the last century (Williams 1979), including the grizzly bear (ca. 1925) (Ingles 1965), and the wolf (last specimen taken in 1924) (McCullough 1967).

Terrestrial mammals that are endangered include three species of kangaroo rat *(Dipodomys,)* the salt marsh havest mouse *(Reithrodontomys raviventris)*, and the San Joaquin kit fox *(Vulpes macrotis mutica)* (State of California 1980).

The ecological impact of certain of the introduced mammals has been well documented as noted below. On the other hand there is very little information on the impact of pet mammals on the native fauna. There are, for example, nearly two million registered dogs in the state (Humane Society of California, personal communication) with no doubt many more that are not registered. Dog packs frequently attack sheep as well as deer, while Cats prey on introduced as well as native birds and small mammals.

Non-native tree squirrels *(Sciurus spp.)* in California have been extremely successful, particularly in the San Francisco Bay region. Byrne (1979) suggests that the increased number of trees in urban landscapes following European settlement may be directly responsible for the success of the fox squirrel, *S. niger,* and the eastern gray squirrel, *S. carolinensis*. In urban environments where the native and non-native populations overlap the non-native species often displace the native species; however, in xeric sites, which predominate in California, the native species appear to dominate.

Feral domesticated mammals are seldom considered in the context of invasions, but once removed from the controlling inluence of humans many species are easily naturalized. Within California several species of feral livestock, all originating from the Old World, are well established within the wild lands of the state.

Wild burros *(Equus asinus)* represent the most abundant of these species (approximately 8000 in 1960; McKnight 1961) and may have contributed to the decline of bighorn sheep *(Ovis canadensis)* (McMichael 1964 as cited by Woodward 1976). Wild burros are so well established that state laws now protect

them, a curious twist, given the breadth of laws designed to prevent the introduction of new non-native species into the state.

The horse has also escaped from control by humans, resulting in small wild herds that over-graze range lands and disrupt watering holes, thus impacting some desert systems. The wild horse is protected from selected types of hunting by federal legislation.

Feral cattle, sheep, and goats are all found in California, but are few in number. For the most part these species have not survived when predators are present, and have been largely restricted to the Channel Islands.

Feral swine have fared well in California, particularly in northern California, where McKnight (1961) estimated the population to be about 1000.

Feral, as well as domesticated dogs running in packs, can cause serious depredation of domestic livestock and wildlife. Estimating the extent or importance of these interactions is difficult, though certainly interactions such as those between the deer and dogs may influence deer populations in some parts of the state.

15.3.3. Birds

California has 525 species of birds of which nine are well established introductions including the ring-necked pheasant, chukar, gray partridge, rock dove, spotted dove, ring-turtle dove, starling, house sparrow, and turkey (Small 1974). Most of these have been brought in purposefully for sport hunting. There are in fact continuing attempts to introduce birds for sport (bobwhite, European partridge, woodcock, etc.) as well as continuing escapes of exotic species, some of which have successfully nested but have not yet become established (e.g., parrots, parakeets, cardinals, etc.) (Hardy 1973). The most abundant bird species in the state are the introduced house sparrow and starling. There are 14 endangered bird species in California (State of California 1980) including, among others, sparrows, rails, terns, owls, and the California condor of which there are less than 30 individuals.

15.3.4. Fishes

Shapovalov et al. (1981) lists 66 species of native freshwater and anadromous fishes in California (Table 15.1.). The native fish fauna has been greatly impacted by introductions, with over of a third of the fauna now being non-native. There are 13 endangered fish species including trout, chubs, pupfish, and suckers (State of California 1980).

An example of the rate of change in the fish fauna in California is shown for Clear Lake (Fig. 15.4.) In 1973 only a third of the species in the lake were native.

Moyle (1976) notes that there have been fish introductions to improve sport and commercial fishing, provide forage for game fishes and bait for fisherman, control insects and weeds, use for aquaculture, and as pets. There have also

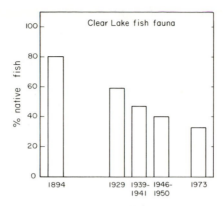

Figure 15.4. Increase in the percentage of exotic fish species in Clear Lake, California. (From Moyle PB (1976) In: Inland Fishes of California. Reproduced with permission of University of California Press.)

been accidental introductions. These introduced species have eliminated native fishes through competition, predation, habitat modification, and hybridization. In addition man has directly influenced and eliminated the native fish fauna through massive habitat change including altering waterways and by polluting streams and lakes. Fishing also has contributed to the loss of native fishes.

15.3.5. Molluscs

The molluscs of California are few in number, relative to the insects, but provide an interesting view of invasions and introductions of invertebrates. Of the introduced molluscs of the western United States described by Hanna (1966) 24 species of terrestrial and freshwater molluscs appear to have naturalized in California. Of these 24 species it appears that only two or three are established within native undisturbed ecosystems, most are restricted to highly disturbed ecosystems or modified systems such as irrigation canals.

Several species of land snails have been deliberately introduced in an effort to control the non-native brown garden snail, (*Helix aspera* Muller), the first non-native mollusc introduced into western North America (Hanna 1966). These deliberate introductions have met with little success in either establishing naturalized colonies or decreasing the population of the brown garden snail.

Most, but not all, of the non-native species of molluscs are of limited ecological or economic importance. Of particular concern has been the Asiatic clam *(Crobicula fluminea* Muller) (Ingram 1959). Since its first observation in 1938 in the western United States it has spread widely in California with large populations in the state's many irrigation canals (Hanna 1966). In two decades up to a meter of clam shells and associated silt have been reported deposited on the bottom of irrigation canals as a result of the presence of the Asiatic clam (Fitch 1953 cited by Hanna 1966). This tremendous productivity in an anthropogenically created ecosystem poses interesing questions concerning the com-

parative ability of native and non-native species to exploit energetically rich, but new, habitats. The Asiatic clam does not displace a native species in most areas where it is found, but rather exploits an unused resource.

Race (1982) has recently described a case where a native mud snail, *Cerithidea californica,* has been competitively displaced, but not excluded, by an ecologically equivalent invading species, *Ilyanassa obseleta.*

15.3.6. Insects

The number of insect species that have invaded California has not yet been tabulated; however, California has been a major entry point for insects that have invaded North America. Sailer (1978) has documented the insect and mite invasions into the United States (excluding Alaska and Hawaii) and found that from about 1640 to 1977 over 1300 non-native species have become established. This number represents a little over 1.3% of the total native insect fauna.

Since 1980 there have been 48 documented cases of insect species that have invaded California (USDA Pest Management Unit, Sacramento, CA). Over the last 5 years this represents a rate of about 9.6 species invasions per year. The rate of insect species invasion of the contiguous United States is only about 1.4 times greater at 13.3 species invasions per year from 1900 to 1977 (Sailer 1978). Although it is currently not possible to partition this rate of invasions into invaders of California and invaders of the remaining 47 states, it is nevertheless important to note that California appears to have been particularly vulnerable to insect invaders. Sailer (personal communication) has data that show that the number of newly arriving insect species is currently accelerating. To what extent the relatively high rate of species invasions into California reflects this increase is unknown.

The influence of insect invaders on Californian agroecosystems has been substantial, causing serious economic losses (DeBach 1964; Huffaker and Messenger 1976). Many insect species have also invaded natural ecosystems in California (e.g., Tilden 1968; Gibo and Metcalf 1978; Stone 1980; Arnaud 1983). However, the effects of these invaders are largely unknown. Arnaud (1983), for example, described the invasion of a thrips species (*Cartomothrips* spp.) from Australia or New Zealand to the San Bruno Mountains in California around 1976. Essentially nothing is known about the effect this species has had on this ecosystem, partially because little is known about the structure and function of the ecosystem itself. Arnaud suggests that the population size of this species will increase rapidly during the next few years. Because this species feeds on fungi of decaying vegetation it has the potential to influence ecosystem processes such as decomposition and nutrient cycles.

The social wasp *(Polistes apachus),* a recent invader of California, has established a number of viable colonies (Gibo and Metcalf 1978). The potential effect of this species on the systems it invades is not clear; however, it appears to have escaped a heavy parasite load which influences the dynamics of the source population. Predation pressure from birds appears to have kept this species from increasing substantially (Gibo and Metcalf 1978).

15.4. Plants: A Detailed Survey

There have been a number of accounts detailing the changing structure of the Californian flora during the past decade. Frankel (1977) gives an excellent summary of the data on which these accounts are based and the difficulties of interpretation of the available sources. Frankel notes the sharp increase in the number of established alien plants through time (Fig. 15.5). He estimates that during the Spanish colonization from 1769 to 1824 16 introduced species became established, 63 species during the Mexican occupation from 1825 to 1848, and an additional 55 species during the American pioneer period ending in 1860, giving a total of 134 established alien species by 1860. Hendry (1934) documented from Spanish language source literature the introduction of 147 taxa into Spanish America. He does not speculate on how many of these became naturalized. By 1968, 975 alien plant species were recorded in the state (Howell 1972). Raven and Axelrod (1978) take a somewhat more conservative view and conclude that there are only 674 truly naturalized species, contrasting with the 5046 native species. By their calculations, of the naturalized dicotyledonous plants, only two are trees *(Ailanthus* and *Robinia)*, five are vines, 29 are shrubs, 172 are herbaceous perennials, and 296 are annuals species. For the naturalized flora as a whole, 57% of the species are annuals, 20% are grasses, and 17% are composites. There are 478 grass species represented in the flora (322 perennial and 156 annual) of which over a third (175) have been introduced (Crampton 1974). Of the 674 naturalized species in the total flora, 559, or nearly three-fourths, are of Old World origin (Raven and Axelrod 1978).

15.4.1. Habitats Invaded

Frankel (1977) proposed that the increase in habitats invaded has been due to both a greater source of exotic species as well as a greater extent of disturbed habitats. Frankel made a detailed study of the vegetation of roadsides in central

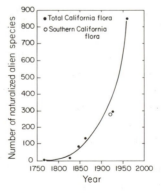

Figure 15.5. Change in the numbers of established vascular plants during the period of 1769–1959. (From Frankel RE (1977) In: University of California Publications in Geography 20:1–163. Reproduced with permission.).

and northern California that encompassed a wide range of climatic and vegetation types. He found, as might be expected, that introduced species sharply decreased in proportion to distance from the roadway and that annuals were the predominate life form found in the most disturbed microsites. There were, however, important differences in the percentages of introduced species found in the various plant communities. Communities that lack a closed tree canopy, such as grasslands, woodlands, and marshes, had a very high percentage of introduced species in the roadside habitats.

In the analysis of his roadside data Frankel also noted that the percentage of aliens dropped markedly with elevation (Table 15.2.). These trends can also be seen in an analysis of flora of various regions of the state (Table 15.3.). Urbanized coastal areas such as San Francisco and Santa Barbara have about 40% adventives whereas interior mountainous regions have only about 5% aliens.

15.4.2. Characteristics of Invaders

There have been a number of attempts to characterize traits associated with successful invaders, the most recent of these being by Baker (1974). These include germination, reproductive, and dispersal characteristics. Unfortunately there are insufficient data to analyze the hundreds of Californian invaders in relation to Baker's outline. We look at a few cases of Californian invaders in order to make very general points.

15.4.3. Modes of Invasion

The California Department of Food and Agriculture has collected distribution information about noxious weeds, those non-native species that pose an economic threat to agricultural production, using field reports and herbarium specimens. The patterns and rates of spread of four of the approximately 120 noxious weeds currently listed by the state are shown in Fig. 15.6. The isograms drawn represent the best approximation of the area of infestation for the years indicated, however, the definition of these population fronts is limited by the data available and uneven reporting intensities. These species, all of Eurasian origin,

Table 15.2. Percentages of aliens and annuals in the roadside flora in California in relation to elevation

Community Type	Elevational Range (ft)	Percent	
		Aliens	Annuals
Foothill woodland	1500–3000	60	85
Yellow pine	3000–6000	32	49
Red fir	6000–8000	22	40
Lodgepole	8000–9500	15	50

Source: Frankel RE (1977) In: University of California Publications in Geography 20:1–163. Reprinted with permission.

Table 15.3. Vascular plant exotics in California local floras

Location	Area (km²)	Elevation, Low (m)	Elevation High (m)	No. Total Species	Percent Introduced	Source
Kern County	21,171	62	2668	1713	14.6	1
Santa Cruz Mountains	3591	0	1160	1799	30.7	2
White Mountains	2280	1220	4343	811	5.9	3
Lassen Park	~518	1463	3187	715	5.2	4
Monterey County	8611	0	—	1713	17.0	5
San Francisco County	236	0	284	1126	41.3	6
Tiburon Peninsula	15	0	228	495		7
Mt. Hamilton Range	3890	61?	1283	761	6.0	8
Santa Barbara Region	7111	0	1982	1390	36.0	9
Mt. Diablo	142	~ 100	1174	644	14.1	10
Marin County	1370	0	796	1313	23.5	11
San Luis Obispo County	8615	0	1556	1287	23.0	12
Vaca Mountains	250	60	860	586	26.8	13
San Bruno Mountains	12.1	0	401	542	29.2	14
Trinity Alps	~400	1525	2332	571		15
Santa Barbara Island	3	0	193	96	29.2	16
Santa Monica Mountains		0	861	805	25.6	17
Angel Island	3	0	238	406	30.5	18
Santa Catalina		0	631	559	29.6	19

Sources:
1. Twisselmann (1967)
2. Thomas (1961)
3. Lloyd and Mitchell (1973)
4. Gillett et al. (1961)
5. Howith and Howell (1964)
6. Howell et al. (1958)
7. Penalosa (1963)
8. Sharsmith (1945)
9. Smith (1976)
10. Bowerman (1944)
11. Howell (1970)
12. Hoover (1970)
13. Willoughby (1981)
14. McClintock and Knight (1968)
15. Ferlatte (1974)
16. Philbrick (1972)
17. Raven and Thompson (1966)
18. Ripley (1980)
19. Thorne (1967)

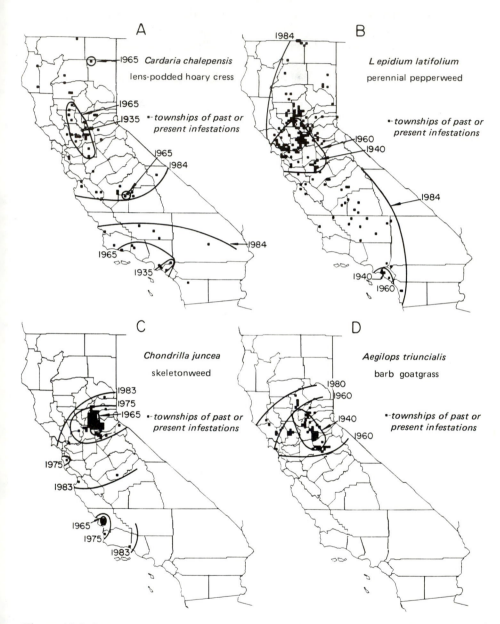

Figure 15.6. Patterns of spread of four invading plants in California. (From State of California Department of Food and Agriculture, Division of Plant Industry files.)

illustrate the two major patterns of invasions described by Baker in this volume, steady advance and satellite. *Aegelops triuncialis* shows a steady advance. Though the exact time and location of the introduction is not known, the continuous spread seems evident. From documentary records it is not always easy to discern a steady advance from satellite distributions, however the spread of *Cardaria chalepensis* appears to show distinct satellite populations that were later out flanked by the broader less dramatic advance of an increasing population.

Also seen among the four invasions maps presented in Fig. 15.6. is the close connection between anthropogenic disturbance and invasions. For three of the four species examined the points of initial establishment were both the Los Angeles and San Francisco and/or San Joquin regions. Though the species selected may represent a biased view of plant invasions due to their economic importance, the role of anthropogenic factors seems clear. Whether recent increases in the rates of spread of the four species presented is an artifact of increased observations, or rather of an infection pressure, as discussed by Baker in this volume, is unclear.

15.4.4 Genetics

There has been considerable work on the genetics of invading plants in California. From these studies a number of conclusions can be made. One, apparently there can be rapid genetic change of the colonizers subsequent to establishent. For example, there are numerous examples of annuals as well as perennials changing their characteristics subsequent to invasion. Martins and Jain (1980) found that the annual *Trifolium hirtum*, which was introduced as a pasture species during the 1940s, has diverged from ancestral gene pools. They note that, "colonization has produced rapid microevolutionary shifts, not due isolation or drift, but most likely due to local selective forces." Similarly Pritchard (1960) found that the perennial *Hypericum perforatum*, which was a serious weed in northern California before being controlled biologically, has formed races that are taller than populations of this species found in eastern Canada, Great Britain, and New Zealand, where it is not a weed.

In addition to the rapid genetic change that can occur through selection there can be genetic change through hybridization of the invaders with native species. As an example, there is circumstantial evidence that *Helianthus annuus,* which is a widespread weed in the U.S., was introduced into California in fairly recent times by the American Indian (Heiser 1949). It appears that since introduction it has hybridized with the native *H. bolanderi,* which is found on serpentine outcrops, resulting in introgression between these species and the formation of races of both species which are successful weeds in California. There are similar examples of hybridization of cultivated and weedy species of radish resulting in new colonizing types (Panetsos and Baker 1968).

In a general survey of the most widespread Californian weeds Heiser and Whitaker (1948) found that they were evenly divided between diploids and polyploids and with a somewhat greater percentage of annuals over perennials.

Over two-fifths of the weeds were members of either the Compositae or Gramineae. In comparisons of these families as a whole with those members which comprise the non-native community Heiser and Whitaker conclude that "annuals generally, and annual polyploids in particular, are superior to other species in these two families in their capacity to become weeds."

15.4.5. Ecological Characteristics

There have been a few studies that have examined ecological characteristics of invading species. Baker (1972), for example, noted that the seeds of introduced herbs were generally heavier than those of comparable native species even when the habitat types were carefully matched (Table 15.4.). He attributed this to the fact that disturbance generally makes the site somewhat more xeric and that there is a trend for increased seed size with increasing aridity in the Californian flora as a whole.

A valuable approach to the study of the ecological characteristics of successful invaders is a comparison of their traits with those of closely related noninvasive species. Two species of *Cortaderia* (pampas grass) have been brought into cultivation into California (Lippmann 1977). One, *C. selloana*, was grown commercially in Santa Barbara county near the end of the last century for "plumes." It was later planted by the Soil Conservation Service for erosion control. This species shows little tendency to escape from planted areas. On other hand, *C. jubata*, which in 1958 was reported to be colonizing areas from Ventura to Monterey counties, now is found throughout coastal California with extenive populations developing on eroding banks and cliffs and logged areas in the Redwood Forest.

The two species of Cortaderia differ in a number of properties but the principal one is their breeding systems. *C. jubata*, the weedy species, is apomictic and hence all seeds produced by a plant are genetically identical. The non-weedy species is, in contrast, dioecious.

The weedy pampas grass *(C. jubata)* produces vast amounts of very small

Table 15.4. Mean seed weights of native versus established introduced herbs for roughly matched community types in California

	Mean Seed Wt[a]		Mean Seed Wt
Freshwater marsh	5.15	Ponds and marshes	5.70
Alkali sink	5.62	Saline waste places	5.72
Coastal prairie	6.26	Coastal grassland	6.42
Valley grassland	6.26	Valley grassland	7.12
Northern Oak	6.45	Arable land	6.65
woodland		Orchards	6.73
Mixed Evergreen forest	6.44	Woodlots	6.53

Source: Baker HG (1972) Ecology 53:997–1010. Reprinted with permission.
[a] Seed weights represent weight classes: 5, 0.10–0.315 mg; 6, 0.316–0.999 mg; 7, 1.00–3.161 mg.

seeds that can germinate on a variety of soil types. Survival is often high in the absence of competition. This lack of competitive capacity in the seedling stage along with a lack of frost tolerance restricts the ultimate distribution of this species.

15.5. Ecosystem Impact of Plant Invaders

Unfortunately there have been few detailed studies of ecosystem function in California. Thus, it is difficult to make generalizations about the ecosystem impact of various invasive organisms. However, there is some information about specific systems which is instructive.

15.5.1. The Grasslands

One of the most dramatic examples of the modification of a native ecosystem by invaders is that of the grasslands of California. Although there is some question as to the original nature of this grassland, there is no doubt that its composition today is unlike that of several centuries ago. Today the grassland is dominated almost entirely by introduced annuals. The conversion has been so complete that Heady (1977) has stated that the "alien species should be considered as new and permanent members of the grassland rather than aliens. Their elimination from the California prairie is inconceivable."

It is thought by some that in the past this grassland was dominated by native perennial bunchgrasses. Heady (1977), for example, has stated that the bunchgrass "*Stipa pulchra,* beyond all doubt, dominated the valley grassland." Others have shared this view of a pristine valley grassland dominated by perennial bunchgrasses (Shantz and Zon 1924; Clements 1934; Burcham 1957; Baker 1978). More recently, however, there has been some doubt cast on this interpretation. Wester (1981) claims that bunchgrasses were dominant only in the Coastal Ranges, Sierran foothills, and in especially favorable sites in the Central Valley. Wester felt that in most of the San Joaquin Valley the original grassland was composed mostly of annuals. In the driest sites xerophytic shrubs and annuals co-occurred. Recently, Bartolome and Gemmill (1981) have presented evidence that *Stipa pulchra* most likely played a role as a disturbance species in the original vegetation rather than as a dominant.

The source of invading species has been reasonably well documented. The early plant colonizers were brought in accidently as well as purposely. Crampton (1974) has proposed that wild oats (*Avena* spp.) and wild rygrass *(Lolium multiflorum)* and maybe even species of bromegrass (*Bromus* spp.) were imported for animal forage. He also proposed that the annual legume, *Medicago polymorpha,* and the filarees (*Erodium* spp.) may have been imported as sheep forage. There are purposeful introductions being made even today into the grassland with rangeland seedings of annual clovers (*Trifolium* spp.) and introductions of such mediterranean-region perennial grasses as *Phalaris turberosa.*

However, a great number of the species came in inadvertently in packing material, ship ballast, seed impurities, etc.

As noted above most of the invaders of the grassland have been annuals of Mediterranean origin. Jackson (1985) has recently compared the ecology of the invaders in California versus that in their native habitats. She found that in the Mediterranean basin these species are ruderal or successional (e.g., cultivated fields and vineyards, degraded steppe, and scrub vegetation). They do not form stable annual grasslands as in California. Evidently strong grazing pressures and continuous human disturbance in the Mediterranean Basin selected for species with a flexible life history pattern (high reproductive effort, low root/shoot ratios, large seed banks) which preadapted them to exploit the California grassland habitats once they were disrupted by overgrazing and severe drought in the 1860s. The annuals can outcompete both native or introduced perennial grasses in California.

Jackson (1985) notes that the Mediterranean basin has a climate that is dissimilar to that of California and one that apparently does not lead to stable annual communities. California has either longer droughts or wetter winters than Mediterranean counterparts.

It is noteworthy that comparable annual grassland associations composed principally of invaders from the Mediterranean basin are found in both Chile (Gulmon 1977) and Australia (Rossiter 1966), which according to Jackson (1985) also have the requisite wet winters and long dry summers. Gulmon (1977) noted, for example, that of 19 species that were shared between the annual grasslands near Santiago, Chile and San Diego, California, 17 were of European origin, one was of North American origin, and the remaining species was bihemispheric.

A clue to the basis of invasion of the California annual grasslands can be seen in comparisons of the grassland vegetations on serpentine-derived soils and on more fertile soil types (Table 15.5.). Serpentine-derived soils have a high percentage of dominant annual species, most of which are endemic. Observations indicate that the serpentine species are poor competitors and do not survive well off of serpentine, particularly on more fertile soils. The invaders appear to be species that can tolerate the low fertility and high heavy metal content of the serpentine but they are poor competitors under such conditions (H.A. Mooney, unpublished observations).

15.5.2. Dunes and Beaches

In a sample of California dune vegetation Barbour et al. (1981) found that 43% of the flora was composed of California endemics. Fourteen percent were introduced species, mostly annuals. They proposed that the "annual niche" is underexploited by native species, citing the lower annual-to-perennial ratio of the Californian dune flora in contrast to similar dunes in Mediterranean-climate Israel.

Marram grass *(Ammophila arenaria)* was introduced in 1869 to stabilize dunes in the San Francisco area. It has since escaped and has become fully naturalized

Table 15.5. Comparative Biomass of Grasslands Occupying Different Soil Types

Species	Sandstone				Serpentine			
	North-east	North-west	South-east	South-west	North-east	North-west	South-east	South-west
Avena fatua[a]	44.4		8.6					
Medicago hispida[a]	0.7		3.7					
Avena barbata[a]			1.7					
Torilis nodosa[a]		1.8	0.9					
Centaurea melitensis[a]			2.0					
Festuca megalura[a]			0.2					
Lolium multiflorum[a]	7.0	7.4		3.5				
Erodium botrys[a]	4.4	0.8	5.9	8.3				
Bromus rigidus[a]	35.4	67.8	42.6	22.6				
Bromus mollis[a]	7.0	21.1	21.3	38.1	7.9	19.6	27.3	53.2
Clarkia purpurea	1.1	1.1	4.7	1.2	3.3	1.2	2.7	
Hemizonia luzulaefolia			1.2	8.3	11.8			5.9
Eriastrum abramsii				3.0				0.8
Lotus subpinnatus			0.2	3.7	4.0	6.5	3.6	14.4
Stipa pulchra				16.4	40.7	41.0	26.4	10.5
Eschscholzia californica					11.8	11.6	13.4	1.3
Festuca grayii					1.1	1.2	10.2	1.4
Plantago erecta					2.0	4.1	5.6	1.8
Medica californica					8.0	11.2		
Linanthus androsaceus					1.7	0.45		
Brodiaea laxa					0.35	3.0		
Calochortus venustus					3.3			1.0
Agroseris heterophylla					0.31			2.4
Festuca dertonensis[a]					0.30		2.1	
Achillea millefoliuma[a]				1.4				
Polypogon monospeliensis[a]					0.20			
Trifolium tridentatum						0.10		
Brodiaea pulchella							4.0	
Sitanion jubatum							3.6	
Lomatium utriculatum							6.5	
Madia gracilis						1.0	1.6	
Poa scabrella							0.6	

Source: McNaughton, SJ 1968. Structure and function in California grasslands. Ecology 49:962–972.
[a] Introduced.

north of San Francisco (Barbour and Johnson 1977). It is a very aggressive species that, because of its growth habit, alters the light environment as well as changes the original dune orientation from that induced by the native dominant, *Elymus mollis*. The dense cover of marram grass excludes many native taxa, reducing species richness by half. In addition to the marram grass there are numerous other naturalized dune and beach species that play a major role in the dynamics of these systems. *Cakile* and *Mesembryanthemum* are particularly important introduced components of the species-poor beach ecosystem.

15.5.3. Chaparral

The chaparral, a dense scrub community, is widespread throughout California. Since this community is of little economic use, as well as a fire hazard, there have been repeated attempts to convert it to either a less flammable vegetation and/or one of use to livestock. These conversions have generally utilized such exotics as perennial grasses or shrubs of low flammability. Another management practice has involved the use of fast-growing exotic annuals, such as annual ryegrass *(Lolium multiflorum)* for erosion control on burned chaparral slopes. Zedler et al. (1983) have recently demonstrated that this practice can have profound effects on ecosystem structure. They describe a chaparral site in San Diego that was seeded with ryegrass after fire. A subsequent wet year supported a lush growth of this invader, much denser than normally produced by the native herbaceous vegetation. The ryegrass supported a fire the second year, which is a highly unusual event in the native chaparral. Seed stocks of the natives were depleted by post-fire germination and had not yet been restocked prior to the second fire which killed shrubs before they reproduced. This series of events, promoted by the invaders, resulted in a complete shift in community structure.

There are other examples of the use of exotics in the management of chaparral that have not had such a profound effect on the dynamics of adjacent non-manipulated communities. For example, species of the genus *Cistus* originating from the Mediterranean Basin have been utilized for erosion control and fire hazard reduction. Montgomery and Strid (1976) found that these plants are not aggressive enough to naturalize within the chaparral.

15.5.4. Forest Lands

The forest lands of California, which are composed of a diversity of forest types, account for 39% (US Department of Commerce 1984) of the state's land area. These forest ecosystems do not appear to be impacted to any large degree by invasions of non-native species (US Forest Service, Pacific Southwest Region and California State Foresty Department, personal communication). Annually >1% of the state's forest lands (state, federal, and private) are cut to varying degrees. Between 1977 and 1983 the state issued permits to cut an average of 160,000 ha per year. In 1983 55,000 ha of Forest Service lands were cut. Even following clear-cutting there are few invasions of non-native plant species (US Forest Service, Pacific Southwest Region, personal communication). Although broom *(Cytisus)* and pampas grass *(Cortaderia)* may be present they do not appear to be important in hindering forest regeneration.

15.6. Legislation

The State of California is concerned with biological invasions into the state and has strict laws controlling biological introductions. The focus of this legislation

is the state's vast agricultural industry and the potential economic impacts of biological invasions. The Department of Food and Agriculture has a legislative mandate to "promote and protect the agricultural industry of the State...prevent the introduction and spread of injurious insect or animal pests, plant diseases, and noxious weeds" (State of California, Food and Agriculture Code). Actual control efforts are influenced by financial and political considerations using a rating system. Species to be eradicated whenever encountered are deemed Class A, and those of adverse economic impact, but so widely dispersed to make control unrealistic, are rated Class C. A rating of Q is applied to taxa not yet present, but deemed to be potential invaders. Introductions are controlled by inspections of all vehicles and shipments of biotic material, including seeds, entering the state, as well as through the maintenance of a staff of taxonomists to identify and monitor the presence of noxious weeds. The Department of Food and Agriculture maintains a Plant Quarantine manual that lists all known non-native pests, and includes rules governing importation of plants and animals, including all foodstuffs, from other parts of the United States and foreign nations.

The fauna of California as well as the native flora come under the protection of the California Department of Fish and Game, though recent legislation defines "plants in native stands...as part of the agricultural industry for the purpose of any law that provides for the protection of the agricultural industry from pests," thus involving the Department of Food and Agriculture in maintaining native ecosystems (Barbe 1985). Economic realities, however, make the follow-through on laws protecting native ecosystems very difficult. The California Department of Parks and Recreation has a program of ecosystem restoration that involves the removal of non-native species (Barry 1985); however, non-natives can be removed from only a few areas as the financial commitment is relatively small.

15.7. Conclusions

The flora and fauna of California have undergone substantial changes during the last 200 years. Many non-native species have become firmly established, often at the expense of native species. Interestingly, some habitats such as grasslands and coastal strand and dune communities have been particularly susceptible to invasion and subsequent modification of community structure. Other habitats, such as chaparral and some forest ecosystems, have been invaded, but with minimal impact on community structure. Such differences in susceptibility appear to be linked to the extent of anthropogenic disturbance. In California (and most ecosystems worldwide) human activity has greatly increased invasion success by creating ecologically "open" habitats. At the same time global travel has provided a means of rapid dispersal for many species. Generally human travel through the most disturbed habitats has increased the potential for successful invasions. Unfortunately the type and extent of disturbance that increases the vulnerability of an ecosystem to invasion is still unknown.

Many species of both plant and animals, have successfully invaded California, yet some groups of organisms have been more successful invaders than others. For example, relatively few reptiles, amphibians, and birds have been successful colonists, whereas many more plant and insect species have colonized. The reason for these differences is not known, but it is likely related to rates of introduction by humans. There also appears to be differences in colonization success between these taxonomic groups; however, there does not appear to be any clear pattern.

The effects of various types of invaders on the ecosystems they invade is highly variable. Although a quantitative assessment of differential impacts has not been attempted, some general trends are clear. Both insects and plants have been particularly successful invaders, and in many cases have altered community structure drastically. Plants in particular have been successful in displacing native species. On the other hand, non-native birds appear to have had little impact on the systems they have invaded. Clearly, many plants and insects produce more locally abundant populations than birds or mammals. Propensity for producing large relatively stationary populations may result in stronger ecosystem effects.

15.8. Acknowledgments

We are grateful to the following persons for providing information: B. Dow, D.M. Graber, W.L. Halvorson, C. Hunter, L.L. Norris, and R.I. Sailer. We thank D. Barbe of the California Department of Food and Agriculture who provided encouragement and access to Department records. B. Weil helped locate much of the material incorporated into this paper.

15.9. References

Arnaud PH (1983) The collection of an adventive exotic thrips—*Cartomothrips* sp (Thysanoptera: Phlaeothripidae)—in California. Proc Entomol Soc Wash 85:622–624

Baker HG (1972) Seed weight in relation to environmental conditions in California. Ecology 53:997–1010

Baker HG (1974) The evolution of weeds. Annu Rev Ecol Syst 5:1–24

Baker HG (1978) Invasion and replacement in Californian and neotropical grasslands. In: Wilson JR (ed), Plant Relations in Pastures. CSIRO, pp 368–384

Barbe, D (1985) The role of the Calfironia Department of Food and Agriculture. Fremontia 13:13–14

Barbour MG, Johnson AF (1977) Beach and dune. In: Barbour MG, Major J (eds), Terrestrial Vegetation of Calfironia. John Wiley, New York, pp 223–261

Barbour MG, Shmida A, Johnson AF, Holton B Jr (1981) Comparison of coastal dune scrub in Israel and California: physiognomy, association patterns, species richness, phytogeography. Isr J Bot 30:181–198

Barry WJ (1985) Ecosystem restoration in the California State Park system. Department of Parks and Recreation, unpublished manuscript

Bartolome JW, Gemmill B (1981) The ecological status of *Stipa pulchra* (Poaceae) in California. Madrono 28:172–184

Bowerman ML (1944) The flowering plants and ferns of Mount Diablo, California. The Gillick Press, Berkeley, California, 290 p

Burcham LT (1957) California Rangeland. California Division of Forestry, Sacramento

Burcham LT (1970) Ecological significance of alien plants in California grasslands. In: Proceedings, Association of American Geographers 2:36–39

Burcham LT (1975) Climate, structure, and history of California's annual grassland ecosystem. In: Love RM (ed), The California Annual Grassland Ecosystem. Institute of Ecology Publication 7, University of California, pp 7–14

Byrne S (1979) The distribution and ecology of the non-native tree squirrels *Sciurus carolinensis* and *Sciurus niger* in northern California. University of California, Berkeley, PhD Dissertation, 190 p

Clements FE (1934) The relict method in dynamic ecology. J Ecol 22:39–68

DeBach P (1964) Biological Control of Insect Pests and Weeds. Reinhold, New York

Donley MW, Allen S, Caro P, Patton CP (1979) Atlas of California. Pacific Book Center, Culver City, California

Ferlatte WJ (1974) A Flora of the Trinity Alps of Northern California. University of California Press, Berkeley, California, 206 p

Frankel RE (1977) Ruderal Vegetation Along Some California Roadsides. University of California Publications in Geography 20:1–163

Gibo DL, Metcalf RA (1978) Early survival of *Polistes apachus* (Hymenoptera: Vespidae) colonies in California: a field study of an introduced species. Can Entomol 110:1339–1343

Gillett GW, Howell JT, Leschke H (1961) A flora of Lassen Volcanic National Park, California. Wasmann J Biol 19:1–185

Gulmon SL (1977) A comparative study of the grassland of California and Chile. Flora 166:261–278

Hanna GD (1966) Introduced molluscs of western North America. Occ Papers Calif Acad Sci 48:1–108

Hardy JW (1973) Feral exotic birds in southern California. Wilson Bull 85:506–512

Heady HF (1977) Valley grassland. In: Barbour MG, Major J (eds), Terrestrial Vegetation of Calfironia. John Wiley, New York, pp 491–514

Heiser CB Jr (1949) Study in the Evolution of the Sunflower Species *Helianthus annuus* and *H. Bolanderi*. University of California Publications in Botany. 23:157–208

Heiser CB Jr, Whitaker TW (1948) Chromosome number, polyploidy, and growth habit in California weeds. Am J Bot 35:179–186

Hendry GW (1934) The source literature of early plant introductions into Spanish America. Agric Hist 8:64–71

Hoover RF (1970) The Vascular Plants of San Luis Obispo County, California. University of California Press, Berkeley, California, 350 p

Howell JT (1970) Marin flora: Manual of the Flowering Plant and Ferns of Marin County, California. University of California Press, Berkeley, California, 366 p

Howell JT (1972) A statistical estimate of Munz' supplement of a California flora. Wasmann J Biol 30:93–96

Howell JT, Raven PH, Rubyzoff P (1958) A Flora of San Francisco, California. University of San Francisco, San Francisco, California, 157 p

Howith BF, Howell JT (1964) The vascular plants of Monterey County, California. Wasmann J Biol 22:1–184

Huffaker CV, Messenger PS (1976) Theory and Practice of Biological Control. Academic Press, New York

Ingles LG (1965) Mammals of the Pacific States. Stanford University Press, Stanford California, 506 p

Ingram WM (1959) Asiatic clams as potential pests in California water supplies. Am Water Works Assoc J 51:363–370

Jackson LE (1985) Ecological origins of California's mediterranean grasses. J Biogeogr 12:349–361

Jennings MR (1983) An annotated check list of the amphibians and reptiles of California. Calif Fish Game 69:151–171

Lippmann MC (1977) More on the weedy "pampas grass" in California. Fremontia 4:25–27

Lloyd RM, Mitchell RS (1973) A Flora of the White Mountains, California, Nevada. University of California Press, Berkeley, California, 202 p

Martins PS, Jain SK (1980) Interpopulation variation in rose clover. J Hered 71:29–32

McClintock E, Knight W (1968) A Flora of the San Bruno Mountains, San Mateo, California. Proceedings of the California Academy of Sciences

McCullough DR (1967) The probable affinities of a wolf capture near Woodlake, California. Calif Fish Game 53:239–244

McKnight T (1961) A survey of feral livestock in California. Assoc Pacif Coast Geo J 23:28–42

McNaughton, SJ (1968) Structure and function in California grasslands. Ecology 49:962–972

Montgomery KR, Strid TW (1976) Regeneration of introduced species of *Cistus* (Cistaceae) after fire in southern California. Madrono 23:417–427

Moyle PB (1973) Effects of introduced bullfrogs, *Rana cartesbeiana,* on the native frogs of the San Joaquin Valley, California. Copeia 1:18–22

Moyle PB (1976) Inland Fishes of California. University of California Press, Berkeley/Los Angles/London, 405 p

Panetsos C, Baker HB (1968) The origin of variation in "wild" *Raphanus sativus* (Cruciferae) in California. Genetica 38:243–274

Penalosa J (1963) A flora of the Tiburon Peninsula, Marin County, California. Wasmann J Biol 21:1–74

Philbrick RN (1972) The plants of Santa Barbara Island, California. Madrono 21:329–393

Powell J, Hogue C (1979) California Insects. University of California Press, Berkeley and Los Angles, 388 p

Pritchard T (1960) Race formation in weedy species with special reference to *Euphorbia cyparissias* L. and *Hypericum perforatum* L. In: Harper JL (ed), The Biology of Weeds. Oxford, pp 61–66

Race MS (1982) Competitive displacement and predation between introduced and native mud snails. Oecologia (Berl) 54:337–347

Raven PH, Axelrod DI (1978) Origin and relationships of the California flora. University of California Publications in Botany 72:134

Raven PH, Thompson HJ (1966) The Flora of the Santa Monica Mountains, California. University of California, Los Angeles (photocopy), 185 p

Ripley JD (1980) Plants of Angel Island, Marin County, California. Great Basin Natur 40:385–407

Rossiter RC (1966) Ecology of the mediterranean annual-type pasture. Adv Agron 18:1–56

Sailer RI (1978) Our immigrant insect fauna. Bull Entomol Soc Am 24:3–11

Shantz HL, Zon R (1924) Atlas of American Agriculture. Section E: Natural Vegetation. USDA Bureau of Agricultural Economy

Shapovalov L, Cordone AJ, Dill WA (1981) A list of the freshwater and anadromous fishes of California. Calif Fish Game 67:4–38

Sharsmith HR (1945) Flora of the Mount Hamilton Range of California. Am Midl Natur 34:289–382

Small A (1974) The Birds of California. Collier Books, New York, 310 p

Smith C (1976) A Flora of the Santa Barbara Region, California. Santa Barbara Museum of Natural History, Santa Barbara, California, 331 p

State of California (1980) Endangered, Rare, and Threatened Animals of California. Department of Fish and Game, Sacramento

State of California (1981) California Statistical Abstract. Economic Development Agency, Sacramento

Statistical Abstract of the United States (1984) Superintendent of Documents, US Government Printing Office, Washington, DC

Stone MW (1980) Notes on the life history of 3 *conoderus* species of wireworms in California, USA (Coleoptera: Elateridae) Pan-Pac. Entomology 56:157–160

Thomas JH (1961) Flora of the Santa Cruz Mountains of California. Stanford University Press, Stanford, California, 434 p

Thorne RF (1967) A flora of Santa Catalina Island, California. Aliso 6:1–77

Tilden JW (1968) Records of 2 species of exotic Lepidoptera captured in California. J Lepidopt Soc 22:187

Twisselmann EC (1967) Flora of Kern County of California. Wasmann J Biol 25:1–395

US Department of Commerce, Bureau of the Census (1984)

Wester L (1981) Composition of native grasslands in the San Joaquin Valley, California. Madrono 28:231–241

Williams DF (1979) Checklist of California mammals. Ann Carnegie Museum 48:425–433

Willoughby JW (1981) A Flora of the Vaca Mountains, California. Ma Thesis, California State University, Sacramento, 238 p

Woodward SL (1976) Feral Burros of the Chemehuevi Mountains, California: the Biogeography of a Feral Exotic. University of California, Los Angeles, PhD Dissertation, 178 p

Zedler PH, Gautier CR, McMaster GS (1983) Vegetation change in response to extreme events: the effect of a short fire interval between fires in a California chaparral and coastal scrub. Ecology 64:809–818

6. Control of Invaders

16. Control of Invaders

D.L. Dahlsten

16.1. Introduction

Invasions of organisms have been of great concern to those involved in the various aspects of food and fiber production. In many cases, introduced species have been deliberately released, such as most of the plants used in commercial agriculture and the parasitoids and predaceous insects that are used in classical biological control attempts. However, many invaders are accidentally introduced, such as insects and pathogens, and many of these are a threat to commercially important commodities. Of the 444 species of insects and mites listed by Metcalf et al. (1951) as crop pests in the United States, it is estimated that approximately 36% of them are introductions (van den Bosch 1971). Many applied scientists, pest managers, and growers or producers spend much of their time concerning themselves with controlling pests, many of which are invading species. Usually the well-established invaders are treated as native pests and conventional control techniques are used. On the other hand, some pest management strategies have been developed to deal specifically with invading species and these are eradication and classical biological control. In many cases, the biology and ecology of the invaders is poorly understood and as a result the control programs are poorly conceived. This is particularly true with eradication programs.

Federal and state institutions have developed around these strategies to deal with invaders. The Division of Biological Control at the University of California

at Berkeley and Riverside deal primarily with insect problems as do most of the agencies. There are international agencies as well, such as the Commonwealth Institute for Biological Control. In the United States, the U.S. Department of Agriculture, Agricultural Research Service has a biological control program and operates a number of national and international laboratories. A listing of all biological control programs and laboratories (state and federal) is compiled annually (Coulson and Hagan 1984).

Eradication programs are planned and funded in part by the Animal and Plant Health Inspection Service (APHIS) of the U.S. Department of Agriculture. This agency is responsible for detection, quarantine, and the development of eradication techniques for invading species. The individual states oftentimes have a comparable agency for the same activities and usually fund part of the programs as well. The vigor with which eradication is pursued varies from state to state. The California Department of Food and Agriculture (CDFA) is an unusually vigorous and efficient agency in its response to invading species. For example, in 1984, the CDFA was involved in seven eradication programs (apple maggot, gypsy moth, Japanese beetle, cotton boll weevil, Mexican fruit fly, oriental fruit fly, and the Caribbean fruit fly). The approach taken by the agencies responsible is similar with each insect and this will be discussed below. Many of these programs have stirred considerable public controversy because they are largely based on the use of chemical insecticides.

The federal government, through APHIS, maintains a quarantine program to keep out a number of potential plant pests. The CDFA also maintains an inspection service and border stations on major highways entering California as part of their program. Most of the plant pests intercepted are insects. For example, from 1 October 1978 through 30 September 1979, APHIS intercepted 18,644 plant pests at the various ports of entry in the United States (Anonymous 1981). Of these, 14,002 (75%) were insects. In California, numerous interceptions are made at the border stations each year. The number of interceptions has risen in recent years but it is not certain if this is due to greater vigilance on the part of CDFA staff or to more insects and other pests entering the state because of increased movement. The mobility of modern society and the amount of travel by air, sea, and land unquestionably provides more opportunity for insects and pests to move from place to place. Modern jets are capable of moving pests around the world within hours, so the potential for invaders is even greater. Introductions could also be a function of pest population density. For example, the increased movement of gypsy moth into California was no doubt partly related to the recent (1980–1982) outbreaks in the eastern United States. The total numbers of live and dead gypsy moths (of various stages) intercepted at border stations and discovered as a result of county quarantine inspections since 1980 are shown in Table 16.1.

An interesting, though largely unanswerable, question is, What proportion of the invaders are actually intercepted? Suffice it to say that if the federal or California quarantine and inspection programs get 10% of the organisms actually coming in, they are doing well. The quarantine probably discourages people from bringing in potentially infested items, so, on balance, such programs are

Table 16.1. Total numbers of gypsy moth individuals intercepted by California state border stations and discovered by county quarantine inspections in California, 1980–1985

Year	1980	1981	1982	1983	1984	1985
Border Stations	21	47	146	267	348	375
County Inspection	112	210	198	200	109	84
Total	133	257	344	467	457	459

Source: Alan Clark and Dick Brown, California Department of Food and Agriculture, Sacremento, California, personal communication.

worthwhile. The next question is, what becomes of all the pests that are introduced? Obviously most of them die out, or perhaps remain at undetectable levels for several years before dying out. Those that are detected will often become the object of an eradication program.

There are great philosophical differences between the proponents of eradication and those proposing biological control. These have been discussed at length by Perkins (1982) and will not be discussed here. However, this controversy can be put into proper perspective if such differences are understood at the outset. There are some other approaches to dealing with invading species, such as habitat management: either make the habitat undesirable for the invader or make the habitat desirable for the natural enemies of the invaders. These will be discussed below.

16.2. Biological Control

As with any definition of a field as broad as biological control, there is considerable controversy as to what biological control really encompasses (DeBach 1964a). To some, natural control of any population by biotic agents is considered naturally occurring biological control. However, the classical definition defines the field of endeavor as practiced by many agencies and this involves the activities of humans. Biological control is the importation of natural enemies (parasitoids or protelean parasites, predators and pathogens) to control introduced insect or plant pests. A contemporary definition of the field includes not only the importation activities but also basic studies of population dynamics, insect behavior, etc. In addition, there are also programs to control native pests by introduction of natural enemies or by augmentation and conservation of native natural enemies. The modern or contemporary definition of biological control is the study, importation, augmentation, and conservation of natural enemies. Biological control does not include any of the autocidal approaches such as the sterile male technique and the use of pheromones or the development of pest-resistant plants. There are many publications in the field, ranging from basic publications on theories of population regulation to reports of biological control attempts. There are also several textbooks that describe the many phases of biological control, from theory to quarantine and importation procedures to

case histories of many biological control programs (DeBach 1964a; Huffaker 1971; Huffaker and Messenger 1976; van den Bosch et al. 1982).

An important component of biological control efforts is the control of invaders. The rationale is that a non-native species introduced into an area can be brought under control if its "ecology" is understood. This usually involves determining the country of origin of the invading species and then going to that country and searching for the pests' natural enemies. The natural enemies when found are then imported, put into quarantine to eliminate secondary parasitoids, and then released or put into an insectary for mass rearing. In agriculture the invading pests are usually attacking a deliberately released invading plant. The natural enemies that are released then can also be considered invaders. In forestry, the invading pest species are usually attacking native trees.

Almost all of the work in biological control has been done with insects and mites that are parasitoids or predators, and with phytophagous insects. Little has been done with pathogens as biological control agents. Pathogens such as *Bacillus thuringiensis* are being used primarily as microbial insecticides, and this is a rapidly developing area of pest control. There are examples of biological control work with exotic pathogens such as the inadvertent introduction of a nucleopolyhedrosis virus for control of the European spruce sawfly, *Diprion hercyniae* (Hartig), in Canada (McGugan and Coppel 1962).

It is difficult to get an estimate of the percentage of success versus the number of attempts at releasing parasitoids and predators. In part this is due to poor record keeping, but it is also due to inadequate sampling and evaluation procedures. Many times this is a result of the length of time necessary for the establishment and buildup of the natural enemies. By 1971, biological control had been attempted against 223 species of insects and some degree of success has been recorded for over half of these species. In the Coccoidea (scale, mealybugs, etc.), there has been some degree of success with 50 of 64 species attempted (78% successful). With all the other insect groups, some degree of success has been attained with 70 of 159 species (44% successful) (DeBach et al. 1971). There have been some startling successes and several of these will be discussed below. The biological control strategy gained considerable impetus in the late 1800s with the successful control program against the cottony cushion scale in California (see below). Since that time there have been many introductions, primarily of beneficial insects to control arthropod and plant pests worldwide. A review of 80 years of literature on these introductions showed introductions against arthropod pests in 12 orders and 74 families and against plants in 16 families (Clausen 1978). This, of course, does not include any of the unpublished introduction attempts.

There have been many successes and to 1976 Laing and Hamai (1976) listed 327 examples of partial to complete successes of biological control of insects and 57 examples of plants for every location into which importation has occurred. DeBach (1964a) drew some general conclusions regarding the various successes to 1964; of 107 species controlled, 41 were coccidae (Homoptera), 21 Lepidoptera, 18 Coleoptera, and 16 were Homoptera other than Coccidae. He found with these species that control was usually due to one dominant natural

enemy and that parasitoids were four times more successful than predators as control agents. DeBach concludes that the successes were as common in temperate regions as in the climatically less variable tropical and subtropical areas. Biological control has proven to be a viable control strategy in North America as illustrated by the case histories described below.

16.2.1. Cottony Cushion Scale: *Icerya purchasi Maskell*

The biological control project against the cottony scale in California established biological control as a valid pest control technique. The history of biological control antedates this project but the project is considered by most as *the* milestone for biological control. The scale was believed to have been introduced near Menlo Park, California, around 1868 and initially fed upon Acacia. Its native home was thought to be Australia. By 1887, the scale was threatening the growing California citrus industry. At the recommendation of C.V. Riley, who was Chief of the Division of Entomology in the U.S. government at the time, a search for the natural enemies of the cottony cushion scale in its native home was conducted. Albert Koebele went to Australia in August of 1888 to begin the search. It is interesting that W.G. Klee, the California State Inspector of Fruit Pests, had arranged independently with Frazer Crawford in Australia to send natural enemies of the scale to California. Crawford sent some parasitic flies, *Cryptochetum iceryae* (Williston) (Cryptochetidae), to Klee in 1888 and these flies were liberated in San Mateo County in 1888 before Koebele sailed for Australia (Doutt 1954). This fly is still an effective parasite of the scale in several counties of the San Francisco Bay Area in addition to some areas in southern California.

Koebele was successful in his search and sent close to 12,000 individuals of *Cryptochetum* to California. Later, in 1888, he discovered the coccinellid, *Rodolia (Vedalia) cardinalis* (Mulsant) and from November of 1888 to January of 1889 he sent in a total of 129 of these ladybird beetles. D.W. Coquillet received the beetles and placed them under a tent on an infested orange tree in Los Angeles. In April the tent was removed and the beetles began moving to other scale-infested trees. By the middle of June, Coquillet and the owner of the orange grove, J. W. Wolfskill and his foreman, Alexander Craw, had distributed 10,555 beetles to 228 different orchardists (Doutt 1964). The distribution of the beetles continued from one grower to another and by the end of 1889 many of the growers were reporting that the beetles had literally cleaned their trees of the scale. The rail shipments of oranges from Los Angeles County alone jumped from 700 to 2000 car loads in one year (Doutt 1954). The cost of this fantastic success was about $1500.

The two natural enemies are still effective in controlling the cottony cushion scale throughout California. The ladybird beetle, because of the spectacular success in southern California, is referred to more often than the parasitic fly. However, it has been found that in the desert areas the beetle is dominant and displaces *Cryptochetum*. On the coast, the reverse is true and the fly displaces

Rodolia. The two coexist in interior areas (DeBach et al. 1971). Since the California experience, the ladybird beetle has been used successfully in 25 other countries where complete control has been achieved and in four countries where there has been a substantial degree of control. The underlying biological and ecological explanations for the success of these programs are not really known. It remains the goal of biological control workers to find these explanations and then to use this information to develop further the biological control technique.

16.2.2. Olive Scale: *Parlatoria oleae* (Colvee)

The olive scale became a major pest of olives in California around 1934. It also became a pest of a number of deciduous fruit crops as well as ornamental trees and shrubs. By 1961, the scale had spread throughout the San Joaquin and Sacramento Valleys and into scattered sites in southern California (DeBach et al. 1971).

On olive, the first instars, or crawlers, of the second generation tend to settle and develop on the fruit. This means that relatively low scale densities in the first generation can result in considerable fruit infestation in the early fall. This is important to the olive growers since scale-discolored fruits are subject to cullage. A biological control program to be successful would have to reduce the scale to extremely low densities.

An intensive biological control program was initiated in 1949 and a small parasitic wasp, *Aphytis maculicornis* (Masi) (Aphelinidae) was introduced from Egypt. From 1951 to 1953 an effort was made to collect parasites from olive scale throughout its range of distribution. Several species of parasites were obtained and released including four "strains" or sibling species of *Aphytis* that were morphologically indistinguishable from *A. maculicornis*. The "strains" were distinct biologically, however, and were therefore reared and released separately (Doutt 1954; Hafez and Doutt 1954).

The "Persian strain" of *A. maculicornis,* which was obtained from Iran and Iraq and colonized in 1952, was found to be the only parasite giving some degree of control. Efforts were concentrated on this parasite and over 27 million individuals were released between 1952 and 1960 at numerous sites in 24 counties in California (Huffaker et al. 1962). Rosen and DeBach (1976) eventually classified the "Persian strain" as a sibling species, *A. paramaculicornis*.

This parasite was very effective and reached parasitization rates of at least 90% on the olive scale. However, since only one scale on a fruit may cause it to be culled, these drastic reductions in scale were not enough. The solution was therefore not an economical one for the growers. Higher rates of parasitization were not possible as the parasites could not tolerate the hot, dry summers of the Central Valley. It was also found that the unusually low fall or spring temperatures reduced the overall effectiveness of this parasite (Huffaker et al. 1962).

The decision was made to look for additional natural enemies of the olive scale. In 1957, P. DeBach rediscovered *Coccophagoides utilis* Doutt (Aphelinidae) in Pakistan on olive scale on apple (van den Bosch et al. 1982). This parasite had been discovered in the early 1950s but since it has a complex life

cycle and is difficult to rear, the colony was lost. *C. utilis* became established and showed promise to improve the control of olive scale. It was colonized and mass-reared and over four million parasites were released at 170 sites in 24 of California's counties from 1962 to 1964 (Kennett et al. 1966). This parasite was also collected from heavily infested trees in the field and distributed to other sites. This is similar to the type of distribution that took place with the *Vedalia* ladybird beetle and cottony cushion scale (see above). The use of field "insectary trees" is an efficient and economical way to raise and subsequently distribute natural enemies.

The introduction of *C. utilis* proved to be an effective addition and in combination with *A. paramaculicornis* resulted in complete control of the olive scale. The development of *C. utilis* is well synchronized with olive scale development and averages about 40% parasitization on the two generations of scale each year. However, it probably could not have controlled the scale alone (DeBach et al. 1971). Since C. utilis was able to survive the summers without undue losses and contribute to scale mortality in the fall, it was an ideal complement to *A. paramaculicornis*.

The olive scale project took much longer and was not nearly as spectacular as the cottony cushion scale project but it was every bit as successful. Also, a number of important contributions were made to biological control methodology. It was shown that multiple introductions of parasites could be successful and that natural enemies could attain an acceptable level of economic control in a commercial crop. The importance of a broad search for natural enemies throughout the range of the host was demonstrated. The existence of "biological strains" or sibling species and their comparative efficacy in controlling the host was found, showing, in part, the importance of taxonomy to the field of biological control.

16.2.3. St. Johnswort or Klamath Weed: *Hypericum perforatum L.*

Invading plants that are determined to be pests have also been the subject of biological control attempts. Worldwide to 1976 there have been 57 partial to complete successful attempts at biological control of plants (Laing and Hamai 1976). Most attempts have been made with phytophagous insects but there is increasing interest in the use of pathogens (Andres et al. 1976). There have been two startling successes: the control of prickly pear in Australia with a cactus-feeding moth, *Cactoblastis cactorum* (Berg), which was imported from Argentina, and the control of St. Johnswort in California. The latter of these major successes is the topic of this section.

Biological control of weeds involves rigorous feeding tests with candidate insects to ensure that any introduced species will not feed on economically important crops or ornamental plants. The control of St. Johnswort is an example of a properly and carefully conducted biological control project.

St. Johnswort (Hypericaceae) is a perennial native of Europe but it has spread throughout many of the temperate areas of the world. It is considered a major plant pest in several areas including northwestern North America. The plant has two other common names in the western United States. In Washington and

Oregon it is called goat weed. In California it is called Klamath weed because it was first reported in the vicinity of the Klamath River in the northern part of the state around 1900.

This plant not only crowds out useful range plants but also has undesirable effects on certain range animals. Those animals with unpigmented skin become very sensitive to sunlight. The white skin areas of the animals become irritated; eventually sore, scrubby patches form on the hide. The plant also acts as an irritant in the mouth, and animals appear to suffer discomfort while drinking water (Holloway 1964). The end result is that the animals do not do as well as other animals because they cannot maintain a normal appetite and do not gain weight.

Biological control against *H. perforatum* was initiated by the Australians in 1920. The initial search efforts were concentrated in England. On completion of starvation tests three chrysomelid beetle species in the genus *Chrysolina* were released. Only *C. hyperici* (Forster), a leaf feeder, became established but the rate of increase and dispersion was slow. The search was extended to southern France and after further testing two beetles from France became established, *C. quadrigemina* (Suffrian) and *Agrilus hyperici* (Creutzer) (Buprestidae), a root feeder. In all, eight insect species were introduced into Australia (Andres et al. 1976). Although good control has been achieved in some localized areas, overall the programs can be rated only partially successful. However, the program in Australia turned out to be very important to the California biological control attempts that were started in 1944.

Klamath weed had spread to 30 counties in California and was occupying approximately 2 million acres of useful rangeland by 1944 (Holloway 1964). The war in Europe in 1944 made importation from France impossible, so arrangements were made with the Australian government to bring in the two *Chrysolina* species and *Agrilus*. The first problem was to get the insects in phase with the seasons in the Northern Hemisphere. This was solved with the aestivating *Chrysolina* adults by applying fine sprays of water which resulted in the beetles becoming active within 3 weeks after their arrival and thus synchronized with the seasons in California. Problems were encountered with the *Agrilus* root borers and it was decided to import them from the Northern Hemisphere after the war. Further starvation tests were done with the two species of *Chrysolina* and the releases started in California in 1945 and 1946.

The two *Chrysolina* species became established and importations were no longer necessary after 2 years. It was found that *C. quadrigemina* was increasing much faster than *C. hyperici*. After three generations in the field it was possible to collect thousands of beetles for redistribution from the original location where 5000 beetles were released (Holloway 1964). In 1950, three million adult beetles were collected from this site and redistributed. The success of *C. quadrigemina* is due to its better synchronization with the growth phases of the plant. Both *Chrysolina* species are brought out of aestivation by autumn rainfall, but *C. quadrigemina* requires less rain than *C. hyperici* and therefore starts feeding and egg laying sooner when the plants are more suitable as food (Holloway 1964).

The root borer, *Agrilus,* was sent to California from Europe in 1947 for further feeding studies. By 1950 releases were made and the root borer was established. The borers were overwhelmed by *C. quadrigemina* in the colonization areas but borers are still present in certain locations.

Within 10 years after the release of *Chrysolina* the Klamath weed was reduced to less than 1% of its former abundance and was removed from California's list of noxious weeds (Huffaker and Kennett 1959). The *Chrysolina* beetles have been released in British Columbia in Canada, New Zealand, South Africa, Chile, Hawaii, and Washington, Oregon, Idaho, and Montana in the northwestern United States with varying degrees of success (Holloway 1964; Andres et al. 1976; Laing and Hamai 1976).

16.2.4. Winter Moth: *Operophthera brumata (L.)*

The winter moth is a well known pest of fruit and deciduous trees in Europe and is an example of an invading species that poses a threat to indigenous forest hardwoods as well as to introduced fruit trees in North America. The winter moth, a geometrid, was introduced, presumably from Europe, and was first reported in 1949 on the south shore of Nova Scotia after being confused for some years with the spring cankerworm, *Paleacrita vernata* Peck (McGugan and Coppel 1962). In a span of 10 years the damage in the oak forest due to the winter moth was calculated to be 26,000 cords per year in two counties. If the moth would have continued unchecked the losses would have seriously affected the forest economy in Nova Scotia (Embree 1971a).

The winter moth control project was initiated in 1954. Because at that time the moth was considered only a minor pest of shade trees and apple orchards and since it dispersed slowly, it was possible to institute a biological control program. If the moth had been recognized 10 to 15 years later when paper and hardwood mills were in existence, a chemical insecticide would most certainly have been used (Embree 1971a).

From 1955 to 1960 six species of parasites (three tachnids and three ichneumonids) were introduced from Europe. Four of the parasites did not become established but a tachinid fly, *Cyzenis albicans* (Fallen), and a ichneumonid wasp, *Agrypon flaveolatum* (Gravely), became established within 3 years. To assist in the establishment of the parasites, large numbers of host larvae were collected and reared in areas where release was desired and the parasites were obtained from this material (Embree 1971a).

There are 63 known parasites of the winter moth (Wylie 1960) but only six were selected for release. They were apparently selected based on abundance and when it appeared that two were going to be successful, further releases of other species were discontinued to avoid potential competition. By 1961 the winter moth had become a minor pest due to the action of the two introduced parasites. A nuclear polyhedrosis virus was found in 1961 and is now generally present over the range of the winter moth in eastern Canada. It has caused the decline of at least two infestations (Embree 1971b).

The winter moth project is an example of how a control project should be

conducted. The population dynamics of the moth had been studied in England for several years (Varley and Gradwell 1968), and by Embree (1971a) in Canada just prior to the release of the parasites. Embree made his initial parasite releases at one point on the edge of his main study area. This enabled him to study the efficacy of the parasite as well as its dispersal ability. After 3 years, in 1958, the parasites were released in other locations in Nova Scotia (Embree 1971a). Detailed life table analysis showed that prior to the parasite introductions, the lack of early instar larval hatch synchronization with bud burst and subsequent starvation was the key factor in the dynamics of the winter moth. In 1958 parasitism was 8% but by 1960 it was 72% and in 1961 the winter moth population collapsed. Embree (1971a) was therefore able to demonstrate that parasitism alone was responsible for the collapse of the winter moth populations.

The two parasites that were established are compatible and supplementary in that *C. albicans* is the most efficient at high densities and can react rapidly to host increases. On the other hand, *A. flaveolatum* is more efficient at low host densities as it oviposits directly on the host. Both species can survive at low host densities. *C. albicans,* which oviposits near feeding damage and not on the host directly, can better survive multiparasitism because it develops more rapidly within the host (Embree 1971a).

The winter moth still occurs throughout Nova Scotia and in isolated pockets of New Brunswick and Prince Edward Island (Embree 1971b) . It is not now considered to be a major pest of hardwood forests in the area. It is occasionally a pest in apple orchards and populations are controlled with chemical insecticides. The winter moth is well synchronized with bud burst on apple. On occasion outbreaks occur in abandoned apple orchards but these are of short duration. The two parasites reach high densities in these apple orchards but generation survival is perhaps higher due to synchronization with apple bud burst.

The winter moth is poorly controlled in Europe even though there are many parasites. This may be due to hyperparasites, which reduce the effectiveness of parasites, or because of competition between the parasites (Pschorn-Walcher et al. 1969). Another explanation for the ineffectiveness of *C. albicans* in England has been posed by Hassell (1978). Beetles and shrews can cause high mortality of winter moth pupae in the soil, because *C. albicans* pupates in the soil. It too may be subject to such predation. Winter mortality of the winter moth, and presumably *C. albicans,* is much lower in Novia Scotia. As a result winter moth population levels at equilibrium are 10 times higher in England than in Nova Scotia because of the greater parasite winter mortality in England.

Although this project demonstrated the importance of careful documentation in biological control, we are still not certain why the two compatible parasitoids were so successful in Nova Scotia but not in Europe. The synchronization of the larvae with bud burst may be one of the factors in the hardwood forests, but the lack of hyperparasitism, predation, and competition may actually be the most important factor. As with the olive scale (see above), the importance of two supplementary parasites is demonstrated. Also, a case can be made for careful selection and introduction of a few parasites rather than multiple species

releases and selection of the fittest. However, since *C. albicans* and *A. flaveolatum* do not effect good control in Europe, would they have even been selected for use elsewhere? In all successful biological control cases to date, serendipity has certainly played a role. The future should see increased evaluation on all biological control introductions.

16.2.5. European Spruce Sawfly: *Diprion hercyniae (Hartig)*

The European spruce sawfly is another example of an invading species attacking a native forest tree. This defoliator was considered a major threat to spruce forests in eastern Canada in the 1930s. In 1931 over 2000 square miles of forest was defoliated in the interior of the Gaspe Peninsula of Quebec. In the following years, *D. hercyniae* was found in increasing numbers in New Brunswick and further west in Quebec (McGugan and Coppel 1962). Concern for the forest economy in the region led to one of the largest and subsequently most successful biological control programs in Canada.

Arrangements were made to collect cocoons of *D. hercyniae* in Europe to rear for parasites and subsequent release in Canada. However, *D. hercyniae* and a closely related species, *D. polytomum* (Hartig), were so uncommon in Europe that collectors turned to collecting cocoons of other sawflies *(D. frutetorum, D. pallidus, D. virens, D. pini, D. similis, D. abieticola, Neodiprion sertifer, Pachynematus scutellatus,* and *Lygaeonematus abietinus)* that were known to be parasitized by the parasites of *D. hercyniae* and *D. polytomum* (McGugan and Coppel 1962). Between 1933 and 1951, 27 parasite species numbering 890 million individuals were released in the field. Of these 882 million were the eulophid wasp *Dahlbominus fuscipennis* (Zett.). Initially there was great encouragement about the recovery of *D. fuscipennis* and the Belleville, Ontario, Biological Control Institute developed a highly successful propagation program to facilitate the distribution of the parasites (McGugan and Coppel 1962).

The other two most commonly released parasites were a tachinid fly *Drino bohemica* (Mesn.) and an ichneumonid *Exenterus amictorius* (Panz).

The success of this effort shows that a multiple species release program can be successful. Three species of parasites gained considerable importance at high sawfly densities, *Dahlbominus fuscipennis, Exenterus amictorius,* and *E. confusus* (McGugan and Coppel 1962). By 1942 sawfly populations were beginning to decline due to parasite activity but also due to a nucleopolyhedrosis virus that was discovered in 1938. It is speculated that the virus was inadvertently released with the parasites. The virus disease spread and it was one of the first times that virus disease was recognized as being important in the control of insect populations.

The virus, according to Bird and Elgee (1957), caused the sawfly populations to crash before the full potential of the parasites was realized. They felt, however, that the parasites might have been able to bring the populations under control without the disease. By 1945 the sawfly was not considered to be a threat to forests of eastern Canada and by 1958 it was no longer considered to

be an economically important forest insect (Neilson et al. 1971). The release
program was terminated after 1951. It is interesting that two other species of
introduced parasites (*Drino bohemica* and *Exenterus vellicatus* (Cash.) are important in maintaining the sawfly at low population levels (Bird and Elgee 1957).
As the European spruce sawfly became less and less important, attention was
focused on native sawflies and many of the parasites were also tested against
seven species of *Neodiprion* and *Pikonema alaskensis* (Rob.). Approximately
900,000 parasites were released between 1937 and 1951 to control these sawflies.

The effectiveness of the virus and parasites was demonstrated once again
in the 1960s in New Brunswick. For 3 consecutive years (1960–1962) one of
the study areas was sprayed with DDT for the spruce budworm. As a result
sawfly densities were reduced to the lowest level ever recorded and the parasites
and virus were for all purposes eliminated from the area. Sawfly populations
began to recover and within five generations after spraying stopped, reached
densities of the outbreak years. Parasitism was evident three sawfly generations
after spraying ceased and parasitism increased from 12 to 37% in the six following sawfly generations. The virus appeared in the seventh post-spray generation and by the second generation of 1966 the sawfly populations were reduced to abundances found during pre-spray years (Neilsen et al. 1971).

The European spruce sawfly program is only one of 20 agricultural and forest
insect and eight plant biological control programs promoted by Agriculture
Canada (Commonwealth Agriculture Bureau 1962 and 1971). In spite of the
successes, interest in biological control lessened and in 1955 there was a definite
change in policy. This may have been due, in part, to the increasing importance
of chemical insecticides in pest control at that time. The responsibilities of the
staff at the Institute for Biological Control at Belleville, Ontario were reduced
in 1955 and in about 10 years this highly successful unit was closed. During
the 1960s biological control efforts reached a low point in Canada, but since
then emphasis on this approach has increased markedly (Hall 1981).

16.3. Habitat Management

Another approach to pest control is through habitat management. Prior to the
advent of the modern chemical approach, cultural techniques for controlling
pests were commonly practiced in agriculture and with many crops this is still
true (Pimentel et al. 1978). Interest in this approach has increased since the
1970s, when it was realized that chemicals were not a panacea. As with biological
control, habitat management would not be a technique chosen to combat invading species as a first choice. The approach does offer some possibilities for
making a particular habitat less vulnerable to invaders. However, not nearly
as much work has been done with this compared to biological control techniques.

There are two ways that habitat modification can be approached in both
agriculture and forest systems. The first is habitat modification aimed at encouraging, enhancing, or conserving natural enemies of invading species and
this is closely related to activities in biological control. The second is habitat

modification to make an agricultural or forest community less susceptible to invading pest species and is a preventative approach.

16.3.1. Habitat Modification to Encourage Natural Enemies

Many of the modern agricultural practices such as the removal of competing plants and trees, use of pesticides, large monocultures, use of fertilizers, irrigation, and planting of uniform genotypes have created simplified environments that are now susceptible to invading species but are unfavorable to natural enemy activity. The trend in modern forestry with respect to fiber production is also toward habitat simplification and is likely to result in problems similar to those seen in agriculture (Dahlsten 1976).

There has been relatively little research done on habitat management to provide alternate food sources, shelter, breeding sites and other ecological requisites for natural enemies. This area of study, and the ecological theory underlying the approach, have been reviewed by Altieri (1983) and van den Bosch and Telford (1964). Adding diversity to agricultural environments by growing crop plants together in polycultures may provide the necessary ecological requisites for some natural enemies. Diversity in the form of undesirable plants may also be important. Although many plants compete with the crop plants and may harbor insect pests or diseases, there also are positive attributes of non-crop plants (Altieri 1983). Non-crop plants can be the source of food for natural enemies either by providing alternate prey or pollen or nectar. Insects, such as aphids, living on non-crop plants may produce honeydew and this, too, can be used as food by various parasites. Some non-crop plants may provide refuges for natural enemies while others may attract natural enemies to an area. It has been shown that parasitization of corn earworm eggs by small wasps (*Trichogramma* sp.) can be increased by spraying fields with extracts of a plant (*Amaranthus* sp.) (Altieri 1983). Other examples of such non-crop plants enhancing the biological control of specific crop pests are given by Altieri and Letourneau (1982). In forestry, it has been shown that native parasites of an invading species, the European pine shoot moth, *Rhyacionia buoliana* (Denis and Schiffermuller), had increased longevity and fecundity due to herbaceous plants, some of which are considered noxious weeds (Syme 1975). The two parasites, *Exeristes comstocki* Cressan and *Hyssopus thymus* Girault, have been effective at times in depressing shoot moth populations and their success depends, in part, on the availability of an energy source. Nectar from the wild flowers provides this energy source and Syme (1975) feels that the status of certain plants as noxious weeds should be reviewed, particularly when they occur in the confines of pine plantations.

In forest communities ground cover and dead trees (snags) may provide protection, roosting sites, feeding sites, and nesting sites for generalist vertebrate predators, mostly small mammals and birds. For the most part, the importance of vertebrate predators in controlling pest arthropods has been little studied (van den Bosch and Telford 1964; Dickson et al. 1979). Larvae of the gypsy moth, *Lymantria dispar* (L.), an invading species, feed primarily at night but

during the day retire to cool, moist secluded places to rest and eventually pupate. Vegetation cover on the forest floor is important to the gypsy moth as well as to small mammals and arthropods that feed on the gypsy moth. Vegetative cover on the forest floor has been correlated with predator efficacy and as a result gypsy moth defoliation is seldom severe where there is good undergrowth and abundant litter (Bess et al. 1947).

Artificial structures have been used to encourage and increase natural enemies of pests. Attempts have been made to create nesting shelters for arthropods such as *Polistes* wasps which are predators of some agricultural pests, and artificial nesting sites for ants, *Formica rufa* L., in forest environments in Europe (van den Bosch and Telford 1964). Nesting boxes designed to encourage cavity-nesting insectivorous birds have been used extensively in the forests of Europe (Bruns 1960) and more recently in the western United States (Dahlsten and Copper 1979). It has been demonstrated that bird populations can be increased rather readily with nest boxes; however, the documentation of the impact of the birds on selected forest insect populations is sparse. Some ecologists consider birds to be inverse density-dependent factors and that birds are important in maintaining some forest pests at low densities.

16.3.2. Habitat Modification to Reduce Vulnerability of Habitats to Invaders

Pest control by modifying agricultural or forest habitats to make them less susceptible to pests, some of which may be invading species, has its roots in ecology and in recent years has received increased attention. The idea that diversity leads to stability in ecological systems is appealing to many applied and basic ecologists. In agriculture the fields can be made more diverse by leaving non-crop plants or by using polycultures as compared to the clean monocultures so commonly used in agriculture today in developed countries. Polycultures are common and have been used for many years in less-developed countries (Altieri 1983). There are a number of examples of the benefits of multicropping systems in preventing insect outbreaks (Altieri and Letourneau 1982). Diversity in agriculture does not mean there will be no pest problems but a review of 150 experiments by a number of researchers is encouraging (Risch et al. 1983). Approximately 53% of the insect pests in these experiments were reduced in their incidence in diversified systems whereas only 18% of the pest insects increased in the diverse systems.

The forest, by comparison to agricultural ecosystems, is an already diverse habitat. However, recent efforts in forestry are to simplify forests for efficiency in production of timber (Dahlsten 1976). The forester and forest pest manager can, through various silvicultural procedures, effect changes in species composition, age composition, and stocking (density) of forest stands. Graham (1956) states that, in theory, mixed species, mixed-age, non-overstocked stands are least subject to damage by insect and disease pests. This is the basis of silvicultural control and a way that forests could be protected from invading insects and diseases.

In the past, the role of phytophagous species has always been viewed as

negative in forest ecosystems. However, Mattson and Addy (1975) have given a fresh perspective to the ecological role of insects in forest ecosystems. They argue that, in fact, some insects play a positive role in maximizing primary production in forest ecosystems. In addition, there is increasing evidence that some forest insects occur on poor sites, where trees are overstocked (over-crowded) or where trees are overmature and declining from old age (Dahlsten and Rowney 1983; Mattson and Addy 1975). Finally, it appears that many tree species are adapted to disturbances for their existence and insects, along with fire, windstorms, and other disturbances, play important roles in succession and the dynamics of forest communities (Mattson and Addy 1975; Smith 1976; McLoed 1980).

The gypsy moth, *Lymantia dispar,* is a well-publicized invader from Europe whose larvae feed on a wide variety of tree species but prefer oaks, particularly white and chestnut oaks. The moth was initially introduced in North America in Massachusetts and has been moving north, south, and west since 1869. Currently there are active eradication and control programs for the gypsy moth in a number of southern states and in the west—British Columbia, Washington, Oregon, and California. It appears that the gypsy moth will eventually spread over much of North America. A closer look at the effects of the gypsy moth on trees and forests is necessary. In the eastern hardwood forests the moth alters the species composition on some sites by selective mortality (Campbell and Sloan 1977). Damage was most severe on weaker trees in the lower crown classes. Subsequent outbreaks of the gypsy moth have been less damaging on these sites. Others have also associated the gypsy moth with disturbances and poor sites (Bess et al. 1947; Houston 1979). Disturbances due to fire and logging have hampered forest development on some sites and have favored the preferred food of the gypsy moth. Urbanization has also created additional susceptible sites (Houston 1979). Recent outbreaks in New England have occurred in stands with low moisture availability. Trees on dry ridges or drained sandy soils are often the preferred hosts, whereas stands on moist sites are faster growing and contain mixtures of preferred and non-preferred hosts. Slower growing and/or exposed trees on ridges, because of bark texture, for example, provide more refuges for pupation than faster growing, straight trees on moist sites. (Houston 1979).

16.4. Eradication

The idea of totally eliminating a species, a pest, from an area has had great appeal to the pest-control profession. The basic simplicity of removing the problem completely and not having to worry about management is a panacea to some but unrealistic to others. The debate is not new and is renewed every-time an eradication program is proposed. As early as the beginning of this cen-tury Congress appropriated funds to eradicate the European corn borer, *Ostrinia nubialis* (Hubner) (Cox 1978).

Technological innovations in recent years have given renewed hope to pest

managers that eradication of certain pests can become a reality. The continuing development of chemical pesticides and application strategies is no doubt the greatest impetus since chemical pesticides are the most commonly used technique in eradication programs. Rapid development in the field of insect attractants, pheromones, etc. has facilitated the trapping of adults of invading species. Effective trapping is the necessary component in the early discovery of invading potential insect pests. In the state of California attempts are made to trap 14 insect species on an annual basis. (Table 16.2.)

One of the important developments among the eradication arguments was the sterile insects approach. Knipling had great success eradicating the screwworm fly *Cochliomyia hominivorax* (Coquerel), a pest of cattle from the small island of Curacao. Knipling subsequently became the most prominent spokesperson for eradication or total pest management (Perkins 1982). The screwworm fly project probably generated too much optimism among applied entomologists for the application of the sterility approach to eradicating pests (Newsom 1978). According to Knipling (1978) the possibliity of eradication should be continually re-evaluated based on new technological developments as they became available.

Eradication as a concept is defined in a number of different ways by applied entomologists. The most realistic definition is one given by Newsom (1978):

> Eradication is the destruction of every individual of a species from an area surrounded by naturally occuring or man-made barriers sufficiently effective to prevent reinvasion of the area except through the intervention of man.

Because of the differences in definition of the term, the evaluation of the success of many projects is not clear. By the dictionary definition of eradication there is not a single example of an insect pest being eradicated. There have been some successful programs if modifications are made as to a prescribed area and a specific period of time (Newsom 1978). The successful programs are those where the invading species has been detected relatively soon after the introduction and the suspected area of infestation is relatively small. In California, for example, there have been 27 successful eradication programs against nine insect species. Three of the insects have been eradicated twice (Khapra beetle, Japanese beetle, and the Mediterranean fruit fly) and one (Oriental fruit fly) has been eradicated 15 times (Frankie et al. 1982).

There have been several large-scale eradication attempts against well-established invading species. It is questionable whether the eradication approach should really be attempted against this category of pests. There has been an extremely high failure-to-success ratio and in addition these projects are very costly in terms of money and manpower (Newsom 1978). Since chemical pesticides are almost always used, other problems such as human health and environmental side effects are also magnified by comparison to the smaller projects. Examples of controversial large-scale programs are the boll weevil *Anthonomus grandis* (Boheman) and the imported fire ant *Solenopsis saevissima richteri* (Forel) eradication projects, both in the southern United States.

Table 16.2. 1984 California statewide trap counts for 14 insects on the dates with highest trap density and lowest trap density

	Statewide Trap Counts	
Species Trapped	High August 25, 1984	Low Jan. 21, 1984
Mediterranean fruit fly	38,843	32,697
Oriental fruit fly	12,693	10,559
Melon fly		
Gypsy moth	32,455	361
Apple maggot	14,426	94
Japanese beetle	15,396	338
Cotton boll weevil	7348	6316
Europeanpine shoot moth	157	30
European corn borer	206	22
Mexican fruit fly	8815	5210
Carribbean fruit fly		
Peach fruit fly		
Khapra beetle	6688	6573
Comstock mealy bug	76	0
Total	137,103	62,200

Source: Del Clark, California Department of Food and Agriculture, Sacramento, California, personal communication). 9

16.4.1. Characteristics of Eradication Programs

The approach to eradication has become highly institutionalized, no doubt due to the extremely high costs in terms of money and human resources. The federal agency in the Department of Agriculture is APHIS, and in the states usually a department of agriculture, as in California. Most eradication programs, whether large or small, have much in common and they can be characterized generally. The generalizations that follow, however, are intended to apply to recent relatively localized invasions or attempts to prevent invasions.

The first line of defense against invading species in detection and exclusion. The federal government maintains inspection services at ports of entry throughout the United States. In California there are border stations on all major highways entering the state. Workers have a list of various undesirable pests that they look for or they confiscate certain plant material, fruit, or anything that might be infested with various pests. For the gypsy moth in California, moving vans coming from potentially infested areas must have a certificate that the contents of the van were inspected at the time of loading, otherwise the entire van must be unloaded and inspected. In addition, the addresses of those moving into California from areas infested with gypsy moth are put on a computer and usually new arrivals will have a gypsy moth trap placed on their property. Certainly these procedures discourage the public from trying to bring

potentially infested material into the state, but the question is, What percentage of the material is caught at the borders or the ports of entry and is this costly exercise economically justified?

Detection of invaders by using pheromones or chemical attractants is a rapidly developing field. Trapping is becoming more and more common throughout the United States as traps become more efficient and new traps become available. Early detection, although important, perhaps could also lead to many more eradication programs for some insects that could never become established. There is an increasing tendency to initiate an eradication program based on the trapping of adults alone, however, it is important in many cases that a second life-state (eggs, larvae, or pupe) be found before initiating a program that may not be necessary. In California, this has been done with the gypsy moth but there has been pressure from the Gypsy Moth Science Advisory Committee to base the decision to treat on the trapping of adult males alone. No doubt many pests have invaded parts of the United States on many occasions in the past and certainly prior to the use of the traps. Some pests such as the cereal leaf beetle *Oulema melanopus* (L.) and the citrus blackfly *Aleurocanthus waglumi* (Ashby) are examples of pests that became firmly established over large areas prior to discovery (Newsom 1978). At some point serious thought must be given to the whole approach. The seven concurrent eradication programs in California in 1984 may be only the beginning of a continuing and potentially escalating approach by state and federal agencies to attempt the elimination of invading potential plant pests. The decision to eradicate should be considered very seriously as will be discussed below.

Quarantine is another form of exclusion. The threat of quarantine can force other states or countries into programs of inspection, fumigation, and eradication that otherwise might not be undertaken. The threat of quarantine on California fruit by several states and Japan no doubt expedited and expanded the aerial malathion-bait application for Mediterranean fruit fly in 1981 and 1982.

Once the decision has been made to eradicate, the next decision is the selection of the means of eradication and how the material is to be applied. Almost always a chemical pesticide is selected because they are fast acting and easily applied by aerial means. The use of aircraft in urban environments also avoids contact with uncooperative homeowners. However, ground application may be a better choice even though much more material is applied because there is more control over where the chemical is placed. Logistically, ground application is not easy in many cases but if small localized areas are to be treated it is feasible. This has been done in California with the gypsy moth for 4 years (1980–1984) with carbaryl. Ground application of malathion bait spray was also used for the Mediterranean fruit fly along with fruit stripping and the release of sterile flies in California in 1980. However, some non-sterile flies were accidentally released and the infestation exploded in Santa Clara County and then aerial application was used in 1981 and 1982. It appeared that the initial program was successful but the release of the non-sterile flies necessitated a change in strategy because of the extent of the infestation.

The Mediterranean fruit fly is an interesting example of an insect that has

been successfully eradicated from several locations in the United States since 1929. Although both ground and air applications of several different chemicals and Malathion bait sprays have been used, the projects have all been successful. The bait traps used for the Med-fly are not very efficient and it could be with this insect as with others, that populations are present but at such low levels as to be undetectable. Since experience with most insecticides and baits has shown that they are rarely, if ever, 100% effective, how can every eradication program be successful? It suggests that perhaps with some insects there are temporary extensions of range and these invading populations die out from other causes. Sparse insect populations may die out from natural causes or from failure to find mates, and with plant diseases the amount of inoculum may become so reduced that host resistance and environmental influences are sufficient to prevent infection from persisting (Smith 1933). Undoubtedly, many undetected introductions of insects and diseases do not result in establishment because of these factors. Further, the overkill that is common in so many eradication programs may be unnecessary. Once population abundance is reduced the population may not persist because of the natural causes mentioned above. Surely more research on invading insects would help answer some of these questions.

Scientific evaluation of eradication projects is impossible as check plots cannot be used since the goal is total eradication. Also, sparse populations are extremely difficult to work with. All eradication projects involve a great amount of guess work and since information cannot be obtained on the insect in the habitat being invaded, the programs start at the beginning each time. There is little, if any, accumulation of knowledge on these insects. Most often the control techniques are selected based on performance against dense outbreak populations in areas that can be substantially different from the habitat being invaded. The choice of carbaryl for gypsy moth eradication is a good example of this. Carbaryl is used on dense gypsy moth populations in the eastern mixed hardwood forests. The urban areas in California where the gypsy moth has been trapped are very different and are much warmer. The temperature causes the larvae to hatch earlier and over an extended period of time in California (California Department of Food and Agriculture, unpublished data). Larval hatch was determined from very few egg masses but the question of spray timing must be considered as a result. In California a few of the indigenous egg masses were caged in the field to determine timing. With so little information it is difficult to tell if the spray has been timed correctly or not. Further, there is no way to determine why the insects are eliminated. Factors such as weather, suitable hosts, and larval hatch synchronization with bud burst must be considered as possibilities for the ability of a population of the gypsy moth to sustain itself in an area. Again, the assumption is that a chemical, carbaryl is this case, which is rarely 100% effective at high-level populations, is as effective at low-level populations.

The selection of the chemical is also arbitrary. Using the same evaluation procedures it appears that a microbial insecticide, *Bacillus thuringiensis*, is equivalent to carbaryl for gypsy moth control, yet in California the Department

of Food and Agriculture will not use *Bacillus* except in areas where label restrictions prevent the application of carbaryl. Using the same non-scientific criteria for eradication success, it appears that *Bacillus* was very effective in Washington and Oregon programs in 1984.

It can be argued that there is a disproportionate amount of money spent on eradication compared to what is spent on monitoring, research, or, for that matter, control projects. Although some chemical monitoring is done by the Calfiornia Department of Food and Agriculture in California, there is little, if any, biological monitoring. The chemical monitoring consists of soil, air, water, and vegetation analysis for residues of the chemical being used. Biological monitoring should be a search for any potential side effects. Aquatic organisms should be sampled before and after treatments as should soil organisms. Certain species that are known to be disrupted by chemical sprays should also be sampled. Many of the homopterans such as diaspine scales, white flies, mealybug, and aphids could be excellent "index" species as well as some phytophagous mite species. Pollinators should also be monitored on these projects. In addition to the monitoring, the eradication projects offer unique opportunities for research.

If only a small percentage of the money devoted to the eradication projects were used for short-term research associated with the application, much could be learned. A larger percentage of the money should be used for long-term research considering the apparent importance of the invading species. The lack of both short- and long-term research is one of the major shortcomings of these programs. This is undoubtedly part of the reason that eradication projects are generally biologically and ecologically unsound in addition to being costly and possibly uneconomical in the long run.

The few short-term studies that have been done indicate that at the very least investigations of potential side effects should be a regular part of all eradication programs. Following the start of the aerial application of malathion-bait spray for the Mediterranean fruit fly (Med-fly) in Santa Clara Valley in 1981 there were numerous calls from homeowners to the California Department of Food and Agriculture (unpublished) on increases of aphids, white flies, mealybugs, etc. As a result, in 1982, the California Department of Food and Agriculture funded a series of studies to look for side effects. The Department is to be commended for undertaking this first-of-its-kind program. The results showed there were both positive and negative effects of the bait spray in the areas studied.

One study focussed on the ice plant scale, *Pulvinariella mesembryanthemi* (Vallot) and *P. delotti* (Gill), an introduced pest along highways which is the object of a biological control program in part of the treated areas (Washburn et al. 1983). The balance of this system shifted to favor the scale or the natural enemies depending on the frequency and seasonal timing of the application. The effect was therefore mixed, favoring the pest sometimes and sometimes not. In another study, it was found that a diverse group of arthropods were killed by sprays and that aphid and white flies were higher in the spray zone

as compared with unsprayed areas, indicating an effect on the natural enemies of these homopterens (Troetschler 1983).

In a third study in Stanislaus County, increases in several scale species on olive and citrus were documented in the treated areas as this was attributed to the destruction of natural enemies (Ehler and Endicott 1984). It was also found that a mealybug and two scale species were not affected by the treatment while one scale on olive was actually supressed by the bait spray.

A fourth study showed the bait spray killed a variety of insects, particularly aphids, on live oak, walnut, and Monterey pine (Dahlsten, unpublished). Populations of scale and aphids on California bay and a scale on Monterey pine were lower in the treated areas, while two species of white flies increased on California bay in the treated areas. From all of these studies there can be little question that the Malathion bait sprays kill a wide variety of non-target insects, and that in some cases the potential exists for secondary outbreaks of insects that are also not desirable to homeowners.

Environmental disruptions due to one eradication program were inadvertently documented in southern California as the eradication zone included an area that had been under study as part of a biological control program (DeBach and Rose 1977). In 1973 an eradication program for the Japanese beetle was attempted in San Diego County with carbaryl, chlordane, and dicofol. The woolly white fly had been the subject of a successful biological control program on citrus and several other pests were also under good biological control at the start of the Japanese beetle eradication program. Within a year after eradication attempts were started, woolly whitefly, citrus red mite and purple scale were causing serious damage to citrus in the eradication zone. This is the best documented example of the potential for disruption by chemical eradication programs, clearly the decisions to use chemical pesticides in eradication efforts should not be taken lightly. Many of the eradication programs occur in urban areas since these are the primary sites of introduction. Of 30 successful eradication programs in California listed by Frankie et al. (1982), all but five were in urban environments. Since the general public has developed very strong environmental concerns over the past 15 years, most communities are not eager to be sprayed with chemicals or to have helicopters flying overhead. Citizens have also learned to use the legal system to try to stall or halt eradication programs. The response by federal and state agencies has been to develop smooth public relations operations and to use their legal arm very effectively to avoid any delays in treatment applications. In California, for example, the Department of Food and Agriculture tries to convince the general public that even though there is relatively little scientific information available they are pursuing the correct course. There is little local or citizen involvement and science advisory committees normally support the course of action desired by state officials. These advisory committees supposedly give the projects scientific credibility and they are used precisely in this way.

Since very little money is put into research or evaluation each eradication program becomes a highly institutionalized goal-oriented program. The biology

and ecology of the target pest and the efficacy of the treatment become secondary to public relations, and organization of the project becomes an end in itself. The information to the public is geared to portraying the invader as an important pest and simplistic economic analyses are used to show how great the losses will be. The agencies insist on a single narrow viewpoint rather than an open, honest appraisal of the problem.

16.4.2. Criteria for Eradication Decision

The decision to attempt eradication of a pest should not be made lightly and there are several factors that should be seriously considered before embarking on a program. Since most of the programs are in densely populated areas and usually involve the use of chemical pesticides, the potential human health hazards of such campaigns must be considered first and foremost. Although barely considered in past programs, potential environmental disruptions should be evaluated thoroughly. DeBach (1964b) suggests several additional aspects of eradication of a pest that should be considered before initiating a program.

The first is a careful economic analysis of the cost as compared to the returns and savings that would result from the attempted eradication. The cost of the 1981–1982 Med-fly eradication program in California was approximately 150 million dollars. The gypsy moth programs in California have cost between one-half and one and one-half million dollars annually. These are both very different examples since one is potentially a pest of agriculture and the other of urban shade trees. The cost of eradication should also be weighed against the ease of reintroduction. The Med-fly, although an important pest of fruit in other countries, is controlled or managed in the same manner as other fruit pests in these countries. The same would no doubt be true in Calfiornia and other parts of the United States. In fact, many of the fruit pest control programs that are already in place may be sufficient for the Med-fly too. It may also be true that the Med-fly is incapable of permanent establishment in most parts of the U.S. due to the pest control activities for other insects as well as climate and other environmental factors. The economic analyses that were done assumed that the Med-fly would be a pest of many fruits in California and did not take into consideration any of the external costs such as the creation of new pests or the elimination of competing pests, human health effects, etc. The same type of economic analyses are done for each pest. The gypsy moth, for example, was assumed to be a pest of the mixed conifer forests in California as well as urban shade trees, chapparal, and agriculture (fruit trees).

The costs of these programs are largely borne by the taxpayers but the beneficiaries of the program are agriculture, forestry, etc. The taxpayers would benefit in the case of the gypsy moth if it were to be kept out of the urban areas. As pointed out above, the urban areas are usually the points of introduction and in many cases bear the brunt of the cost as well as exposure to the pesticides for potential agricultural pests. From an economic perspective it may be far cheaper in the long run to try eradication on an extremely limited basis for some insects but to incorporate others into the pest management pro-

grams if and when they become pests. Good biological and ecological research would help to resolve the economic threats that are posed by many of the potential pests.

A second issue is to look for any conflicts of interest in the eradication program. The example used by DeBach (1964b) is yellow star thistle. This weed is detrimental to the cattle industry but a benefit to the bee industry. There are other examples that could be drawn from phytophagous insects such as the gypsy moth. As has been shown, some phytophagous insects may have a positive influence on net primary productivity in the forest (Mattson and Addy 1975). Other positive aspects may be the displacement of known pests by the invading species or changes in the plant community that may be desirable. The gypsy moth could conceivably displace the California oak worm, *Phryganidia californica* (Pack), in urban areas and although the gypsy moth has a broader host range than the oak worm the end result may be swapping one pest for another at no additional cost. Foresters may find the gypsy moth a positive influence since the black oak in California and red alder and willow in Oregon and Washington are desired hosts. Although the gypsy moth feeds on conifers at high population levels, it may result in reducing the number of hardwoods in mixed conifer stands and save the foresters the money that they would normally spend on herbicides to control the hardwoods.

The relative potential danger of the new pest must be evaluated. As can be seen from the discussion above, the gypsy moth cannot be considered nearly as dangerous as the Med-fly, and every invading plant pest poses a somewhat different hazard. Well-planned research programs would help to resolve the potential threats of invaders.

The ease with which an invader can be eradicated should be considered carefully. Those species that escape detection and become well established and widely distributed should be taken off the eradication list. Localized invasion of pests could be the subject of eradication efforts but then effective traps must be available as well as control techniques that are near 100% effective on low-level populations. In addition, the control chosen should be the least disruptive to the environment and not a danger to human health. There are very few invading species for which all these criteria can be met. Since it is impossible to evaluate eradication efforts scientifically, the ease of eradication may never be known anyway. There are so many factors involved in the successful invasion of an area that it is only circumstantial evidence if an invader does not appear in the detection traps a year or two after a treatment of an area. With the gypsy moth in Michigan, it was not certain if low-level survival from eradication attempts or reinfestation from the eastern United States was the reason for the eventual establishment of this invader (Morse and Simmons, 1978, 1979). It was concluded that a policy of spraying when defoliation was visible was the least costly economically and environmentally.

The availability of alternative control strategies should be evaluated for each invader. It appears that certain types of invading species, for example, the Homoptera, are more amenable to biological control than other species. Eradication of the California red scale in the Fillmore area of southern California

was attempted for many years and then considered a failure (DeBach, 1964b). A successful biological control program is now in existence in this area. The woolly whitefly is another example of a chemical eradication program that failed for which biological control was a successful alternative (DeBach and Rose 1976, 1977). The decision to attempt biological control for the Japanese bayberry whitefly, *Parabemisia myricae* (Kuwana), in California was based on the success of this approach with the woolly whitefly, in addition to the high cost of chemical eradication due to the widespread distribution of this whitefly in California (Frankie et al. 1982). However, this program has not been successful as yet. This project demonstrates that it is possible to select alternative strategies for dealing with invading species. The gypsy moth program in Michigan is an example of another alternative to eradication (Morse and Simmons 1978).

The last important attribute to be evaluated is the case of recolonization. Species that have already successfully invaded parts of the United States should not be the subjects of eradication programs in other areas of the country. Four of the programs in California were conducted against insects in this category: the Japanese beetle from the eastern United States; the gypsy moth from the east, Michigan, and possibly Oregon as of 1984, the apple maggot from Oregon; and the boll weevil from Arizona. These programs seem destined to failure even if the eradication attempts are successful. Recolonization from bordering states seems almost inevitable, consequently the high costs of eradication cannot be justified. Species that cannot be detected easily are also good candidates for recolonization. Other species such as the fruit fly come in infested fruit, often because people have fruit in their luggage when traveling.

Although there are no simple answers as to when to attempt eradication, the criteria discussed above should be an important part of the decision-making process. Alternatives to eradication are available and in the long run these may be economically and environmentally less costly. In the future there must be a better evaluation of the need for eradication, more consideration of environmental and health effects, ongoing efforts to evaluate the worth of the effort, and the institutionalization of the eradication response should be de-emphasized.

16.5. Conclusions

Introduced organisms have in many cases created problems for commercial agriculture and forestry. Those organisms that have become well established and/or widespread in their distribution are usually treated with conventional pest control techniques. However, two approaches have been designed by pest control specialists for dealing specifically with introduced organisms and these are biological control and eradication. There are other possibilities such as habitat modification to either favor the natural enemies of introduced organisms or to make the environment unfavorable for the proliferation of the introduced species.

Philosophically and ecologically the two dominant approaches are completely different. Biological control is a more environmentally compatible strategy that

will keep introduced species at population densities that do not cause economic damage when it is successful. The goal is not to eliminate the invading species. Eradication, on the other hand, is designed to eliminate the invading species totally.

Eradication programs for invading species are highly institutionalized, very costly, and usually involve the use of toxic chemicals. Most of these programs are in urban areas and there has been increasing concern about the use of chemicals in densely populated regions. The eradication programs are usually surrounded by considerable controversy.

The approach to be taken with respect to invading species, be it biological control, eradication, or habitat management, should be evaluated carefully. Much more emphasis needs to be placed on biological and ecological research of invading species so that the decision on a course of action can be made with an increasing wealth of information. More emphasis must be placed on using techniques that have long-term economic benefit and those that are not environmentally disruptive.

16.6. Acknowledgments

I thank my colleagues L.E. Caltagirone, D.L. Rowney, J.W. Lownsbery, M.C. Whitmore, and S.H. Dreistadt for thoughtful, constructive reviews and help in gathering information. I also thank Ms. Karen Branson and Ms. Nettie Mackey for typing and editing.

16.7. References

Altieri MA (1983) Agroecology. Division of Biological Control, University of California, Berkeley
Altieri MA, Letourneau DK (1982) Vegetation management and biological control in agroecosystems. Crop Protect 1:405–430
Andres LA, Davis CJ, Harris P, Wapshere AJ (1976) Biological control of weeds. In: Huffaker CB, Messenger PS (eds), Theory and Practice of Biological Control. Academic Press, New York, pp 481–499
Anonymous (1981) List of intercepted plant pests from October 1, 1978 through September 30, 1979. USDA, APHIS publication 82–7
Bess HA, Spurr SH, Littlefield EW (1947) Forest site conditions and the gypsy moth. Harvard Forest Bull 22
Bird FT, Elgee DE (1957) A virus disease and introduced parasites as factors controlling the European spruce sawfly, Diprion hercyniae (Htg.) in central New Brunswick. Can Entomol 89:371–378
Bruns H (1960) The economic importance of birds in forests. Bird Study 7:193–208
Campbell RW, Sloan RJ (1977) Forest stand responses to defoliation by the gypsy moth. For Sci Monogr 19
Clausen CP (1978) Introduced parasites and predators of arthropod pests and weeds: a world review. USDA Agricultural Research Service, Agriculture Handbook No 480
Commonwealth Agricultural Bureau (1962) A review of the biological control attempts against insects and weeds in Canada. Technical Communications Bulletin No 2

Commonwealth Agricultural Bureau (1971) Biological control programs against insects and weeds in Canada, 1959–1968. Technical Communications Bulletin No 4

Coulson JR, Hagan JH (eds) (1984) U.S. biological control programs using natural enemies against insects, weeds and other pests. USDA Agricultural Research Service, Agricultural Research Service Biological Control Documentation Center, Beltsville, Maryland. Document No 00061

Cox HC (1978) Eradication of plant pests—pros and cons. Bull Entomol Soc Am 24:35.

Dahlsten DL (1976) The third forest. Environment 18:35–42

Dahlsten DL, Copper WA (1979) The use of nesting boxes to study the biology of the mountain chickadee *(Parus gambeli)* and its impact on selected forest insects. In: Dickson JG, Conner RN, Fleet RR, Kroll JC, Jackson JA (eds), The Role of Insectivorous Birds in Forest Ecosystems. Academic Press, New York, pp 217–260

Dahlsten DL, Rowney DL (1983) Insect pest management in forest ecosystems. Environ Manage 7:65–72

DeBach P (ed) (1964a) Biological Control of Insect Pests and Weeds. Reinhold, New York

DeBach P (1964b) Some ecological aspects of insect eradication. Bull Entomol Soc Am 10:221–224

DeBach P, Rose M (1976) Biological control of woolly whitefly. Calif Agric 30(5): 4–7

DeBach P, Rose M (1977) Environmental upsets caused by chemical eradication. Calif Agric 31(7):8–l0

DeBach P, Rosen D, Kennett CE (1971) Biological control of coccids by introduced natural enemies. In: Huffaker CB (ed), Biological Control. Plenum Press, New York, pp 165–194

Dickson JG, Connor RN, Fleet RR, Jackson JA, Kroll JG (eds) (1979) The Role of Insectivorous Birds in Forest Ecosystems. Academic Press, New York

Doutt RL (1954) An evaluation of some natural enemies of the olive scale. J Econ Entomol 47:39–43

Doutt RL (1964) The historical development of biological control. In: DeBach P (ed), Biological Control of Insect Pests and Weeds. Reinhold, New York, pp 2l–42

Ehler LE, Endicott PC (l984) Effect of malathion-bait sprays on biological control of insect pests of olive, citrus and walnut. Hilgardia 52:l–47

Embree DG (1971a) The biological control of the winter moth in eastern Canada by introduced parasites. In: Huffaker CB (ed), Biological Control. Plenum Press, New York, pp 217–226

Embree DG (1971b) *Operophtera brumata* (L.), winter moth (Lepidoptera: Geometridae). Biological Control Programmes Against Insects and Weeds in Canada, 1959-1968. Commonwealth Institute of Biological Control, Trinidad, Technical Communication No 4, pp 167–175

Frankie GW, Gill R, Koehler CS, Dilly D, Washburn JO, Hamman P (1982) Some considerations for the eradication and management of introduced pests in urban environments. In: Battenfield SL (ed), Proceedings of Symposium on the Imported Fire Ant, June 7–10, 1982, Atlanta, Georgia. US Environmental Protection Agency and USDA, Animal, Plant Health Inspection Service, pp 237–255

Graham SA (1956) Forest insects and the law of natural compensation. Can Entomol 88: 44–55

Hafez M, Doutt RL (1954) Biological evidence of sibling species in *Aphytis maculicornis* (Masi) (Hymenoptera, Aphelinidae). Can Entomol 86:90–96

Hall RH (1981) A new approach to pest control in Canada. Canadian Environmental Advisory Council, Ottawa, Report No 10

Hassell MP (1978) The Dynamics of Arthropod Predator-Prey Systems. Princeton University Press, Princeton, New Jersey

Holloway JK (1964) Projects in biological control of weeds. In: DeBach P (ed), Biological Control of Insect Pests and Weeds. Reinhold, New York, pp 650–670

Houston DR (1979) Classifying forest susceptibility to gypsy moth defoliators. U S Department of Agriculture Handbook 542

Huffaker CB (ed) (1971) Biological Control. Plenum Press, New York

Huffaker CB, Kennett CE (1959) A ten-year study of vegetation changes associated with biological control of Klamath weed. J Range Manage 12:69–82

Huffaker CB, Kennett CE, Finney GL (1962) Biological control of the olive scale, *Parlatoria oleae* (Colvee), in California by imported *Aphytis maculicornis* (Masi) (Hymenoptera: Aphelinidae). Hilgardia 32:541–636

Huffaker CB, Messenger PS (eds) (1976) Theory and Practice of Biological Control. Academic Press, New York

Kennett CE, Huffaker CB, Finney GL (1966) The role of an autoparasitic aphelinid, *Coccophagoides utilis* (Doutt), in the control of *Parlatoria oleae* (Colvee). Hilgardia 37:255–282

Knipling EF (1978) Eradication of plant pests—pro. Bull Entomol Soc Am 24:44–52

Laing JE, Hamai J (1976) Biological control of insect pests and weeds by imported parasites, predators, and pathogens. In: Huffaker CB, Messenger PS (eds), Theory and Practice of Biological Control. Academic Press, New York, pp 685–743

Mattson WJ, Addy ND (1975) Phytophagous insects as regulators of forest primary production. Science 190(4214):515–522

McGugan BM, Coppel HC (1962) A review of the biological control attempts against insects and weeds in Canada. Part II. Biological control of forest insects 1910-1958, pp 35–216. Commonwealth Agricultural Bureau, Farnham Royal, England

McLoed JM (1980) Forests, disturbances and insects. Can Entomol 112:1185–1192

Metcalf CL, Flint WD, Metcalf RL (1951) Destructive and Useful Insects, 3RD edit. McGraw-Hill, New York

Morse JG, Simmons GA (1978) Alternatives to the gypsy moth eradication program in Michigan. Great Lakes Entomol 11:243–248

Morse JG, Simmons GA (1979) Simulation model of gypsy moth introduced into Michigan forests. Environ Entomol 8:293–299

Neilson MM, Martineau R, Rose AH (1971) *Diprion hercyniae* (Hartig), European spruce sawfly (Hymenoptera: Diprionidae). Biological Control Programmes Against Insects and Weeds in Canada, 1959–1968. Commonwealth Institute for Biological Control, Technical Communication No 4, pp 136–143

Newsom LD (1978) Eradication of plant pests—con. Bull Entomol Soc Am 24:35–40

Perkins JH (1982) Insects, Experts and the Insecticide Crisis. Plenum Press, New York

Pimentel D, Krummel JK, Gallahan D, Hough J, Merrill A, Schreiner I, Vittum P, Koziol F, Back E, Yen D, Fiance S (1978) Benefits and costs of pesticide use in US food production. Bioscience 28:772, 778–783

Pschorn-Walcher H, Schroder D, Eichhorn O (1969) Recent attempts at biological control of some Canadian forest insect pests. Commonwealth Institute of Biological Control Tech Bull 11:1–100.

Risch SJ, Andow D, Altieri MA (1983) Agroecosystem diversity and pest control: data, tentative conclusions and new research directions. Environ Entomol 12:625–629

Smith DM (1976) Changes in eastern forests since 1600 and possible effects. In: Anderson JF, Kaya HK (eds), Perspectives in Forest Entomology. Plenum Press, New York, pp 3–20

Smith HS (1933) The efficacy and economic effects of plant quarantine in California. University of California Agricultural Experiment Station, Berkeley, Bulletin 553, 276 p

Syme PD (1975) The effects of flowers on the longevity and fecundity of two native parasites of the European pine shoot moth in Ontario. Env Entomol 4:337–346

Troetschler RG (1983) Effects of malathion bait sprays used in California to eradicate the Mediterranean fruit fly, *Ceratitis capitata* (Weidemann) (Diptera: Tephritidae). Environ Entomol 12:1816–1822

van den Bosch R (1971) Biological control of insects. Annu Rev Ecol System 2:45–66
van den Bosch R, Messenger PS, Gutierrez AP (1982) An Introduction to Biological Control. Plenum Press, New York
van den Bosch R, Telford AD (1964) Environmental modification and biological control. In: DeBach P (ed), Biological Control of Insect Pests and Weeds. Reinhold, New York, pp 459–488
Varley GC, Gradwell GR (1968) Population models for the winter moth. Symp R Entomol Soc Lond 4:132–142
Washburn JA, Tassan RL, Grace K, Bellis, E, Hagen KS, Frankie GW (1983) Effects of malathion sprays on the ice plant insect system. Calif Agric 37(1-2):30–32
Wylie HG (1960) Insect parasites of the winter moth, *Operophtera brumata* (L.) (Lepidoptera: Geometridae) in western Europe. Entomophaga 5:111–129

Species Index

Topical Index